WAVES AND OSCILLATIONS

(SECOND EDITION)

R N CHAUDHURI

Ph.D.

Retired Professor and Head
Department of Physics, Visva-Bharati University
Santiniketan, West Bengal

Publishing Globally

NEW AGE INTERNATIONAL (P) LIMITED, PUBLISHERS

New Delhi • Bangalore • Chennai • Cochin • Guwahati
Hyderabad • Kolkata • Lucknow • Mumbai
Visit us at **www.newagepublishers.com**

Published by New Age International (P) Ltd., Publishers
First Edition: 2001
Second Edition: 2010
Reprint: 2018

GLOBAL OFFICES

- **New Delhi** — **NEW AGE INTERNATIONAL (P) LIMITED, PUBLISHERS**
7/30 A, Daryaganj, New Delhi-110002, (INDIA)
Tel.: (011) 23253771, 23253472, **Telefax:** 23267437, 43551305
E-mail: contactus@newagepublishers.com • Visit us at www.newagepublishers.com
- **London** — **NEW AGE INTERNATIONAL (UK) LTD.**
27 Old Gloucester Street, London, WC1N 3AX, UK
E-mail: info@newacademicscience.co.uk • Visit us at www.newacademicscience.co.uk
- **Nairobi** — **NEW AGE GOLDEN (EAST AFRICA) LTD.**
Ground Floor, Westlands Arcade, Chiromo Road (Next to Naivas Supermarket), Westlands, Nairobi, KENYA
Tel.: 00-254-713848772, 00-254-725700286
E-mail: kenya@newagepublishers.com

BRANCHES

- **Bangalore** — 37/10, 8th Cross (Near Hanuman Temple), Azad Nagar, Chamarajpet, Bangalore- 560 018
Tel.: (080) 26756823, **Telefax:** 26756820, **E-mail: bangalore@newagepublishers.com**
- **Chennai** — 26, Damodaran Street, T. Nagar, Chennai-600 017, **Tel.:** (044) 24353401, **Telefax:** 24351463
E-mail: chennai@newagepublishers.com
- **Cochin** — CC-39/1016, Carrier Station Road, Ernakulam South, Cochin-682 016, **Tel.:** (0484) 2377303, **Telefax:** 4051304
E-mail: cochin@newagepublishers.com
- **Guwahati** — Hemsen Complex, Mohd. Shah Road, Paltan Bazar, Near Starline Hotel, Guwahati-781 008,
Tel.: (0361) 2513881, **Telefax:** 2543669, **E-mail: guwahati@newagepublishers.com**
- **Hyderabad** — 105, 1st Floor, Madhiray Kaveri Tower, 3-2-19, Azam Jahi Road, Near Kumar Theater, Nimboliadda Kachiguda, Hyderabad-500 027, **Tel.:** (040) 24652456, **Telefax:** 24652457
E-mail: hyderabad@newagepublishers.com
- **Kolkata** — RDB Chambers (Formerly Lotus Cinema) 106A, 1st Floor, S N Banerjee Road, Kolkata-700 014
Tel.: (033) 22273773, **Telefax:** 22275247, **E-mail: kolkata@newagepublishers.com**
- **Lucknow** — 16-A, Jopling Road, Lucknow-226 001, **Tel.:** (0522) 2209578, 4045297, **Telefax:** 2204098
E-mail: lucknow@newagepublishers.com
- **Mumbai** — 142C, Victor House, Ground Floor, N.M. Joshi Marg, Lower Parel, Mumbai-400 013
Tel.: (022) 24927869, **Telefax:** 24915415, **E-mail: mumbai@newagepublishers.com**
- **New Delhi** — 22, Golden House, Daryaganj, New Delhi-110 002, **Tel.:** (011) 23262368, 23262370, **Telefax:** 43551305
E-mail: sales@newagepublishers.com

ISBN: 978-81-224-2679-3
C-17-07-10609

Printed in India at Ajit Printing Press, Delhi.
Typeset at Abro Enterprises, Delhi.

NEW AGE INTERNATIONAL (P) LIMITED, PUBLISHERS
7/30 A, Daryaganj, New Delhi-110002
Visit us at **www.newagepublishers.com**
(CIN: U74899DL1966PTC004618)

WAVES AND OSCILLATIONS

Foreword

In order to understand the physical world around us, it is absolutely necessary to know the basic features of Physics. One or the other principle of Physics is at work in objects of daily use, *e.g.*, a ceiling fan, a television set, a bicycle, a computer, and so on. In order to understand Physics, it is necessary to solve problems. The exercise of solving problems is of immense help in mastering the fundamentals of the subject. Keeping this in mind, we have undertaken a project to publish a series of books under the broad title *Basic Physics Through Problems.* The series is designed to meet the requirements of the undergraduate students of colleges and universities, not only in India but also in the rest of the third world countries.

Each volume in the series deals with a particular branch of Physics, and contains about 300 problems with step-by-step solutions. In each book, a chapter begins, with basic definitions, principles, theorems and results. It is hoped that the books in this series will serve two main purposes: (*i*) to explain and derive in a precise and concise manner the basic laws and formulae, and (*ii*) to stimulate the reader in solving both analytical and numerical problems. Further, each volume in the series is so designed that it can be used either as a supplement to the current standard textbooks or as a complete text for examination purposes.

Professor R. N. Chaudhuri, the author of the present volume in the series, is a teacher of long standing. He has done an excellent job in his selection of the problems and in deriving the solutions to these problems.

Kiran C. Gupta
Professor of Physics
Visva-Bharati
Santiniketan

Preface to the Second Edition

It is a great pleasure for me to present the second edition of the book after the warm response of the first edition. There are always important new applications and examples on Waves and Oscillations. I have included many new problems and topics in the present edition. It is hoped that the present edition will be more useful and enjoyable to the students.

I am very thankful to New Age International (P) Ltd., Publishers for their untiring effort to bringing out the book within a short period with a nice get up.

R.N. Chaudhuri

Preface to the First Edition

The purpose of this book is to present a comprehensive study of waves and oscillations in different fields of Physics. The book explains the basic concepts of waves and oscillations through the method of solving problems and it is designed to be used as a textbook for a formal course on the subject. Each chapter begins with the short but clear description of the basic concepts and principles. This is followed by a large number of solved problems of different types. The proofs of relevant theorems and derivations of basic equations are included among the solved problems. A large number of supplementary problems at the end of each chapter serves as a complete review of the theory. Hints are also provided in the case of relatively complex problems.

The topics discussed include simple harmonic motion, superposition principle and coupled oscillations, damped harmonic oscillations, forced vibrations and resonance, waves, superposition of waves, Fourier analysis, vibrations of strings and membranes, Doppler effect, acoustics of buildings, electromagnetic waves, interference and diffraction. In all, 323 solved and 350 supplementary problems with answers are given in the book.

This book will be of great help not only to B.Sc. (Honours and Pass) students of Physics, but also to those preparing for various competitive examinations.

I thank Professor K.C. Gupta for going through the manuscripts carefully and for suggesting some new problems for making the book more interesting and stimulating.

R.N. Chaudhuri

Contents

1 Simple Harmonic Motion

1.1 PERIODIC MOTION

When a body repeats its path of motion back and forth about the equilibrium or mean position, the motion is said to be periodic. All periodic motions need not be back and forth like the motion of the earth about the sun, which is periodic but not vibratory in nature.

1.2 THE TIME PERIOD (T)

The time period of a vibrating or oscillatory system is the time required to complete *one full cycle* of vibration of oscillation.

1.3 THE FREQUENCY (ν)

The frequency is the number of complete oscillations or cycles per unit time. If T is the time for one complete oscillation.

$$\nu = \frac{1}{T} \qquad \text{...(1.1)}$$

1.4 THE DISPLACEMENT (*X* OR *Y*)

The displacement of a vibrating body is the distance from its equilibrium or mean position. The maximum displacement is called the *amplitude.*

1.5 RESTORING FORCE OR RETURN FORCE

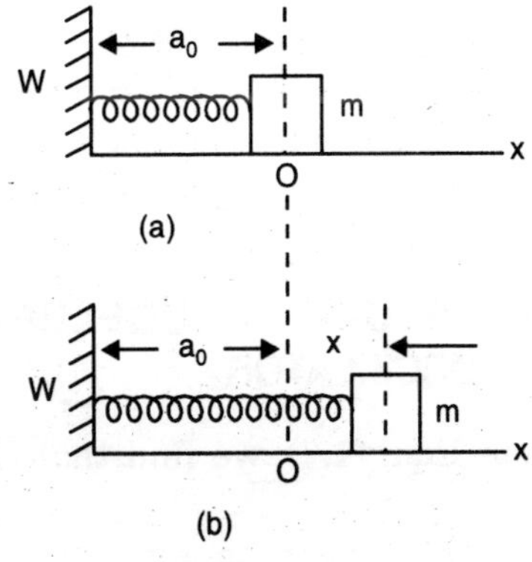

Fig 1.1

The mass m lies on a frictionless horizontal surface. It is connected to one end of a spring of negligible mass and relaxed length a_0, whose other end is fixed to a rigid wall W [Fig. 1.1 (*a*)].

If the mass m is given a displacement along the x-axis and released [Fig. 1.1 (*b*)], it will oscillate back and forth in a straight line along x-axis about the equilibrium position O. Suppose at any instant of time the displacement of the mass is x from the equilibrium position. There is a force tending to restore m to its equilibrium position. This force, called the *restoring force* or return force, is proportional to the displacement x when x is not large:

$$\vec{F} = -k\,x\,\hat{i} \qquad ...(1.2)$$

where k, the constant of proportionality, is called the *spring constant* or *stiffness factor,* and $\hat{i}$ is the unit vector in the positive x-direction. The minus sign indicates that the restoring force is always opposite in direction to the displacement.

By Newton's second law Eqn. (1.2) can be written as

$$m\ddot{x} = -kx \text{ or, } \ddot{x} + \omega^2 x = 0 \qquad ...(1.3)$$

where $\omega^2 = k/m$ = return force per unit displacement per unit mass. ω is called the angular frequency of oscillation.

1.6 SIMPLE HARMONIC MOTION (SHM)

If the restoring force of a vibrating or oscillatory system is proportional to the displacement of the body from its equilibrium position and is directed opposite to the direction of displacement, the motion of the system is simple harmonic and it is given by Eqn. (1.3). Let the initial conditions be $x = A$ and $\dot{x} = 0$ at $t = 0$, then integrating Eqn. (1.3), we get

$$x(t) = A\cos\omega t \qquad ...(1.4)$$

where A, the maximum value of the displacement, is called the amplitude of the motion. If T is the time for one complete oscillation, then

$$x(t+T) = x(t)$$

or

$$A\cos\omega(t+T) = A\cos\omega t$$

or

$$\omega T = 2\pi$$

or

$$T = \frac{2\pi}{\omega} = 2\pi\sqrt{\frac{m}{k}} \qquad ...(1.5)$$

and

$$\nu = \frac{1}{T} = \frac{\omega}{2\pi} \quad \text{or,} \quad \omega = 2\pi\nu.$$

The general solution of Eqn. (1.3) is

$$x(t) = C\cos\omega t + D\sin\omega t \qquad ...(1.6)$$

where C and D are determined from the initial conditions. Euqation (1.6) can be written as

$$x(t) = A\cos(\omega t - \phi) \qquad ...(1.7)$$

where $C = A\cos\phi$ and $D = A\sin\phi$. The amplitude for the motion described by Eqn. (1.7) is now $A = (C^2 + D^2)^{1/2}$ and the angular frequency is ω which is uneffected by the initial conditions. The angle ϕ called the *phase angle or phase constant or epoch* is given by $\phi = \tan^{-1}(D/C)$, where ϕ is chosen in the interval $0 \le \phi \le 2\pi$.

1.7 VELOCITY, ACCELERATION AND ENERGY OF A SIMPLE HARMONIC OSCILLATOR

From Eqn. (1.7), we find that the magnitude of the velocity v is

$$v = |-A\,\omega\sin(\omega t - \phi)| = A\omega(1 - x^2/A^2)^{1/2}$$

or

$$v = \omega(A^2 - x^2)^{1/2} \qquad ...(1.8)$$

and the acceleration of the particle is

$$a = \ddot{x} = -A\omega^2 \cos(\omega t - \phi) = -\omega^2 x \qquad ...(1.9)$$

We see that, in simple harmonic motion, the acceleration is proportional to the displacement but opposite in sign.

If T is the kinetic energy, V the potential energy, then from the law of conservation of energy, in the absence of any friction-type losses, we have

$$E = T + V = \text{constant}$$

where E is the total energy of the oscillator.

Also, Force $\quad \vec{F} = -\vec{\nabla} V$

or $\quad -\dfrac{dV}{dx} = -kx$

or $\quad V = \dfrac{1}{2}kx^2 + c$

or $$V = \frac{1}{2}m\omega^2 A^2 \cos^2(\omega t - \phi) + c \qquad ...(1.10)$$

where c is an arbitrary constant.

The kinetic energy of the oscillator is

$$T = \frac{1}{2}m\dot{x}^2 = \frac{1}{2}m\omega^2 A^2 \sin^2(\omega t - \phi) \qquad ...(1.11)$$

If $V = 0$ when $x = 0$, then $c = 0$ and

$$E = \frac{1}{2}m\omega^2 A^2 \qquad ...(1.12)$$

(*i*) At the end points $x = \pm A$,
The velocity of the particle $v = 0$,
Acceleration $\quad a = \omega^2 A$ directed towards the mean position,
kinetic energy $T = 0$

potential energy $V = \dfrac{1}{2}m\omega^2 A^2 = E$

(*ii*) At the mid-point ($x = 0$),

$v = \omega A,\ a = 0,\ T = \dfrac{1}{2}m\omega^2 A^2 = E,\ V = 0$

(*iii*) At $x = \pm A/\sqrt{2}$, $T = V = E/2$.

1.8 REFERENCE CIRCLE

Suppose that the point Q is moving anticlockwise with uniform angular velocity ω along a circular path with O as the centre (Fig. 1.2). This circle is called the reference circle for simple harmonic motion. BOB' is any diameter of the circle. $B'OB$ is chosen to be along the x-axis. From Q, a perpendicular QP is dropped on the diameter $B'B$. When Q moves with uniform angular velocity along the circular path, the point P executes simple harmonic motion along the diameter BB'. The amplitude of the back and forth motion of the point P

about the centre O is OB = the radius of the circle = A. Suppose Q is at B at time $t = 0$ and it takes a time t for going from B to Q and by this time the point P moves form B to P. If $\angle QOB = \theta$, $t = \theta/\omega$ or, $\theta = \omega t$, and $x = OP = OQ \cos \theta = A \cos \omega t$.

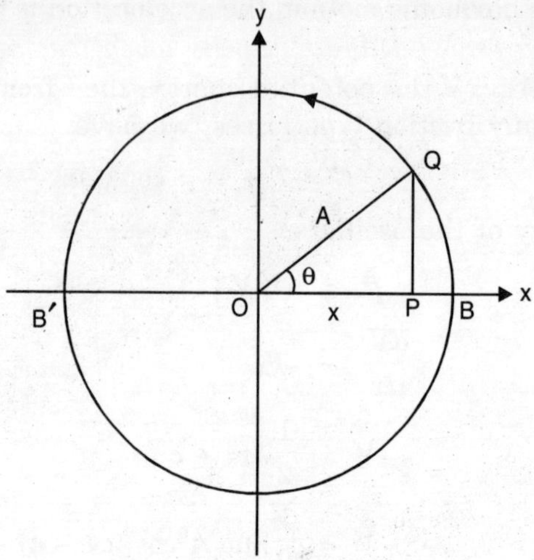

Fig. 1.2

When Q completes one revolution along the circular path, the point P executes one complete oscillation. The time period of oscillation $T = 2\pi/\omega$. If we choose the circle in the xy plane, the position of Q at any time t is given by

$$\vec{r} = A \cos \omega t \ \hat{i} + A \sin \omega t \ \hat{j}.$$

1.9 THE SIMPLE PENDULUM

The bob of the simple pendulum undergoes nearly SHM if its angle of swing is not large. The time period of oscillation of a simple pendulum of length l is given by

$$T = 2\pi\sqrt{l/g} \quad \text{...(1.13)}$$

where g is the acceleration due to gravity.

1.10 ANGULAR SIMPLE HARMONIC MOTION (TORSIONAL PENDULUM)

A disc is suspended by a wire. If we twist the disc from its rest position and release it, it will oscillate about that position in angular simple harmonic motion. Twisting the disc through an angle θ in either direction, introduces a restoring torque

$$\Gamma = -C\theta \quad \text{...(1.14)}$$

and the period of angular simple harmonic oscillator or torsional pendulum is given by

$$T = 2\pi\sqrt{I/C} \quad \text{...(1.15)}$$

where I is the rotational inertia of the oscillating disc about the axis of rotation and C is the restoring torque per unit angle of twist.

SOLVED PROBLEMS

1. *A point is executing SHM with a period πs. When it is passing through the centre of its path, its velocity is 0.1 m/s. What is its velocity when it is at a distance of 0.03 m from the mean position?*

Solution

When the point is at a distance x from the mean position its velocity is given by Eqn. (1.8):

$$v = \omega(A^2 - x^2)^{1/2}.$$

Its time period, $T = 2\pi/\omega = \pi$; thus $\omega = 2\ s^{-1}$. At $x = 0$, $v = A\omega = 0.1$; thus $A = 0.05$ m. When $x = 0.03$ m, $v = 2\ [(0.05)^2 - (0.03)^2]^{1/2} = 0.08$ m/s.

2. *A point moves with simple harmonic motion whose period is 4 s. If it starts from rest at a distance 4.0 cm from the centre of its path, find the time that elapses before it has described 2 cm and the velocity it has then acquired. How long will the point take to reach the centre of its path?*

Solution

Amplitude A = 4 cm and time period $T = 2\pi/\omega = 4$ s. The distance from the centre of the path $x = 4–2 = 2$ cm. Since $x = A \cos \omega t$, we have $2 = 4 \cos \omega t$. Hence $t = 2/3$ s and the velocity $v = \omega\sqrt{A^2 - x^2} = \pi/2\sqrt{4^2 - 2^2} = \pi\sqrt{3}$ cm/s. At the centre of the path $x = 0$ and $\omega t = \pi/2$ or, $t = 1$ s.

3. *A mass of 1 g vibrates through 1 mm on each side of the middle point of its path and makes 500 complete vibrations per second. Assuming its motion to be simple harmonic, show that the maximum force acting on the particle is π^2 N.*

Solution

$A = 1$ mm $= 10^{-3}$ m, $\nu = 500$ Hz and $\omega = 2\pi\nu$.

Maximum acceleration $= \omega^2 A$.

Maximum force $= m\omega^2 A = 10^{-3} \times 4\pi^2\ (500)^2 \times 10^{-3} = 2\ \pi^2 N$.

4. *At t = 0, the displacement of a point x (0) in a linear oscillator is –8.6 cm, its velocity v (0) = – 0.93 m/s and its acceleration a (0) is + 48 m/s^2. (a) What are the angular frequency ω and the frequency ν? (b) What is the phase constant? (c) What is the amplitude of the motion?*

Solution

(*a*) The displacement of the particle is given by

$x(t) = A \cos(\omega t + \phi)$

Hence,
$$x(0) = A \cos \phi = -8.6 \text{ cm} = -0.086 \text{ m}$$
$$v(0) = -\omega A \sin \phi = -0.93 \text{ m/s}$$
$$a(0) = -\omega^2 A \cos \phi = 48 \text{ m/s}^2$$

Thus,
$$\omega = \sqrt{-\frac{a(0)}{x(0)}} = \sqrt{\frac{48}{0.086}} = 23.62 \text{ rad/s}$$

$$\nu = \omega/2\pi = \frac{23.62}{2\pi} = 3.76 \text{ Hz}$$

(b) $\dfrac{v(0)}{x(0)} = -\,\omega \tan \phi$

or
$$\tan \phi = -\frac{v(0)}{\omega x(0)} = -\frac{0.93}{23.62 \times 0.086} = -0.458$$

Hence ϕ = 155.4°, 335.4° in the range $0 \le \phi < 2\pi$. We shall see below how to choose between the two values.

(c) $A = \dfrac{x(0)}{\cos \phi} = \dfrac{-0.086}{\cos \phi}.$

The amplitude of the motion is a *positive* constant. So, ϕ = 335.4° cannot be the correct phase. We must therefore have

$$\phi = 155.4°$$

$$A = \frac{-0.086}{-0.909} = 0.0946 \text{ m.}$$

5. *A point performs harmonic oscillations along a straight line with a period T = 0.8 s and an amplitude A = 8 cm. Find the mean velocity of the point averaged over the time interval during which it travels a distance A/2, starting from (i) the extreme position, (ii) the equilibrium position.*

Solution

We have

$$x(t) = A \cos(\omega t - \phi)$$

(*i*) The particle moves from $x = A$ to $x = A/2$,

or
$$\omega t - \phi = 0 \text{ to } \omega t - \phi = \frac{\pi}{3},$$

or
$$t = \frac{\phi}{\omega} \text{ to } t = \frac{\phi}{\omega} + \frac{\pi}{3\omega}.$$

The average value of velocity over this interval is

$$< v > = \frac{1}{\pi/3\omega} \int_{\phi/\omega}^{\phi/\omega + \pi/3\omega} \dot{x}\, dt$$

$$= \frac{3A\omega^2}{\pi} \left[\frac{\cos(\omega t - \phi)}{\omega} \right]_{t = \frac{\phi}{\omega}}^{t = \frac{\phi}{\omega} + \frac{\pi}{3\omega}}$$

$$= \frac{3A\omega}{\pi} \left(\frac{1}{2} - 1 \right) = -\frac{3A}{T}.$$

(*ii*) The particle moves from $x = 0$ to $x = A/2$

or,
$$t = \frac{\phi}{\omega} + \frac{\pi}{2\omega} \text{ to } t = \frac{\phi}{\omega} + \frac{\pi}{3\omega}$$

$$< v > = \frac{6A}{T}$$

The magnitude of the average velocity is

(i) $$\frac{3A}{T} = \frac{3 \times 8}{0.8} \text{ cm/s} = 30 \text{ cm/s}$$

(ii) $$\frac{6A}{T} = 60 \text{ cm/s}$$

6. *A particle performs harmonic oscillations along the x-axis according to the law*

$$x = A \cos \omega t.$$

Assuming the probability P of the particle to fall within an interval from –A to A to be equal to unity, find how the probability density dP/dx depends on x. Here dP denotes the probability of the particle within the interval from x to x + dx.

Solution

The velocity of the particle at any time t is

$$\dot{x} = -A\omega \sin \omega t.$$

Time taken by the particle in traversing a distance from x to $x + dx$ is

$$\frac{dx}{|\dot{x}|} = \frac{dx}{A\omega\sqrt{1 - x^2/A^2}} = \frac{dx}{\omega\sqrt{A^2 - x^2}}.$$

Time taken by the particle in traversing the distance $-A$ to A is $T/2$.

Thus, $$dP = \frac{1}{T/2}\frac{dx}{\omega\sqrt{A^2 - x^2}} = \frac{dx}{\pi\sqrt{A^2 - x^2}}.$$

Hence $$\frac{dP}{dx} = \frac{1}{\pi\sqrt{A^2 - x^2}}.$$

7. *In a certain engine a piston executes vertical SHM with amplitude 2 cm. A washer rests on the top of the piston. If the frequency of the piston is slowly increased, at what frequency will the washer no longer stay in contact with the piston?*

Solution

The maximum downward acceleration of the washer = g. If the piston accelerates downward greater than this, this washer will lose contact.

The largest downward acceleration of the piston

$$= \omega^2 A = \omega^2 \times 0.02 \text{ m/s}^2.$$

The washer will just separate from the piston

when $$\omega^2 \times 0.02 = g = 9.8 \text{ m/s}^2.$$

Thus, $$\nu = \frac{\omega}{2\pi} = \frac{1}{2\pi}\sqrt{\frac{9.8}{0.02}} = 3.52 \text{ Hz}.$$

8. *A light spring of relaxed length a_0 is suspended from a point. It carries a mass m at its lower free end which stretches it through a distance l. Show that the vertical oscillations of the system are simple harmonic in nature and have time period, $T = 2\pi\sqrt{l/g}$.*

Solution

The spring is elongated through a distance l due to the weight mg. Thus we have

$$kl = mg$$

where k is the spring constant. Now the mass is further pulled through a small distance from its equilibrium position and released. When it is at a distance x from the mean position (Fig. 1.3), the net upward force on the mass m is

$$k(l + x) - mg = kx = mgx/l.$$

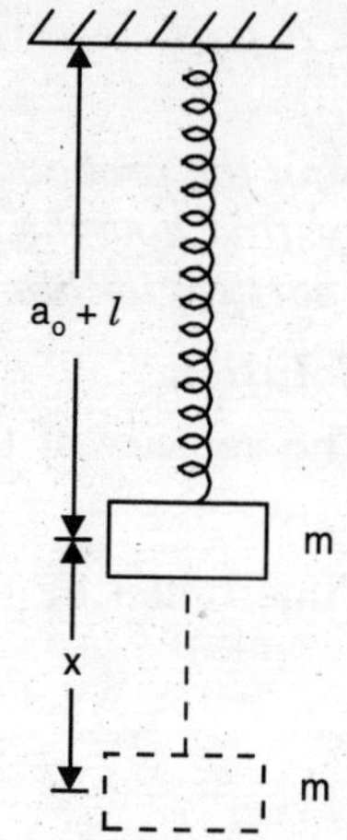

Fig. 1.3

Upward acceleration = $gx/l = \omega^2 x$, which is proportional to x and directed opposite to the direction of increasing x. Hence the motion is simple harmonic and its time period of oscillation is

$$T = \frac{2\pi}{\omega} = 2\pi\sqrt{l/g}.$$

Note: Young's modulus of the material of the wire is given by

$$Y = \frac{mg}{A}/(l/L) = \frac{mgL}{Al},$$

where L is the length of the wire and A is the cross-sectional area of the wire.

Thus, $\dfrac{mg}{l} = \dfrac{AY}{L} = k$ = spring constant of the wire.

9. *A 100 g mass vibrates horizontally without friction at the end of an horizontal spring for which the spring constant is 10 N/m. The mass is displaced 0.5 cm from its equilibrium and released. Find: (a) Its maximum speed, (b) Its speed when it is 0.3 cm from equilibrium. (c) What is its acceleration in each of these cases?*

Solution

(*a*) $\omega = \sqrt{k/m} = \sqrt{10/0.1} = 10\ s^{-1}$ and $A = 0.005$ m.

The maximum speed = $A\omega$ = 0.05 m/s

(*b*) $|v| = \omega\sqrt{A^2 - x^2} = 0.04$ m/s

(*c*) Acceleration $a = -\omega x$

(*i*) At $x = 0$, $a = 0$

(*ii*) At $x = 0.3$ cm, $a = -0.03$ m/s^2.

10. *A mass M attached to a spring oscillates with a period of 2 s. If the mass is increased by 2 kg, the period increases by one second. Find the initial mass M assuming that Hooke's law is obeyed.*

(I.I.T. 1979)

Solution

Since $T = 2\pi\sqrt{m/k}$, we have in the first case $2 = 2\pi\sqrt{M/k}$ and in the second case $3 = 2\pi\sqrt{(M+2)/k}$. Solving for M from these two equations we get $M = 1.6$ kg.

11. *Two masses m_1 and m_2 are suspended together by a massless spring of spring constant k as shown in Fig. 1.4. When the masses are in equilibrium, m_1 is removed without disturbing the system. Find the angular frequency and amplitude of oscillation.* (I.I.T. 1981)

Solution

When only the mass m_2 is suspended let the elongation of the spring be x_1. When both the masses $(m_2 + m_1)$ together are suspended, the elongation of the spring is $(x_1 + x_2)$.

Thus, we have

$$m_2 g = kx_1$$

$$(m_1 + m_2)g = k(x_1 + x_2)$$

where k is the spring constant.

Hence $$m_1 g = kx_2.$$

Thus, x_2 is the elongation of the spring due to the mass m_1 only. When the mass m_1 is removed the mass m_2 executes SHM with the amplitude x_2.

Amplitude of vibration = $$x_2 = m_1 g/k$$

Angular frequency $$\omega = \sqrt{k/m_2}.$$

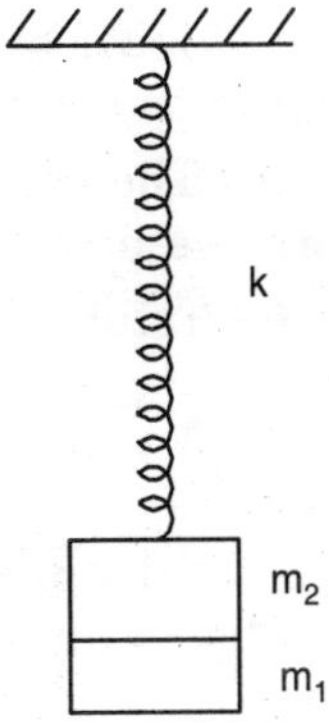

Fig. 1.4

12. *The 100 g mass shown in Fig. 1.5 is pushed to the left against a light spring of spring constant k = 500 N/m and compresses the spring 10 cm from its relaxed position. The system is then released and the mass shoots to the right. If the friction is ignored how fast will the mass be moving as it shoots away?*

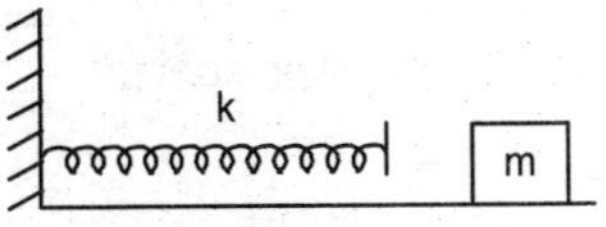

Fig. 1.5

Solution

When the spring is compressed the potential energy stored in the spring is

$$\frac{1}{2}kx^2 = \frac{1}{2} \times 500 \times (0.1)^2 = 2.5 \text{ J}.$$

After release this energy will be given to the mass as kinetic energy. Thus

$$\frac{1}{2} \times 0.1 \times v^2 = 2.5$$

from which $v = \sqrt{50} = 7.07$ m/s.

13. *In Fig. 1.6 the 1 kg mass is released when the spring is unstretched (the spring constant k = 400 N/m). Neglecting the inertia and friction of the pulley, find (a) the amplitude of the resulting oscillation, (b) its centre point of oscillation, and (c) the expressions for the potential energy and the kinetic energy of the system at a distance y downward from the centre point of oscillation.*

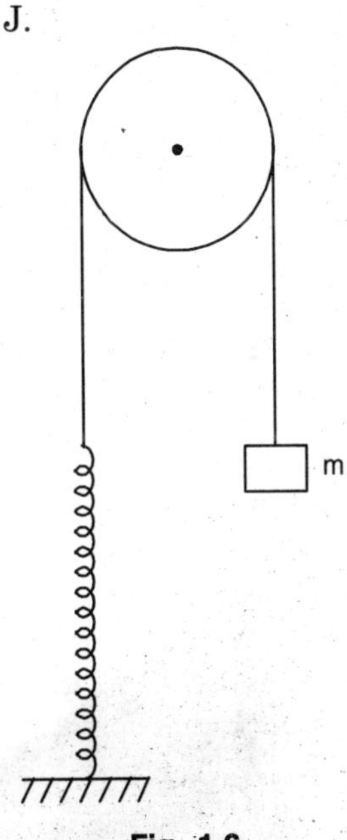

Fig. 1.6

Solution

(*a*) Suppose the mass falls a distance h before stopping. The spring is elongated by h. At this moment the gravitational potential energy (mgh) the mass lost is stored in the spring.

Thus, $$mgh = \frac{1}{2} kh^2$$

or $$h = \frac{2mg}{k} = \frac{2 \times 1 \times 9.8}{400} = 0.049 \text{ m}.$$

After falling a distance h the mass stops momentarily, its kinetic energy $T = 0$ at that moment and the PE of the system $V = 1/2\ kh^2$, and then it starts moving up. The mass will stop in its upward motion when the energy of the system is recovered as the gravitational PE (mgh). Therefore, it will rise 0.049 m above its lowest position. The amplitude of oscillation is thus 0.049/2 = 0.0245 m.

(*b*) The centre point of motion is at a distance $h/2 = 0.0245$ m below the point from where the mass was released.

(*c*) Total energy of the system

$$E = mgh = \frac{1}{2} kh^2.$$

At a distance y downward from the centre point of oscillation, the spring is elongated by $(h/2 + y)$ and the total potential energy of the system is

$$V = \frac{1}{2} k \left(\frac{h}{2} + y\right)^2 + mg\left(\frac{h}{2} - y\right) = \frac{1}{2} k\left(y^2 + \frac{3}{4} h^2\right)$$

and the kinetic energy

$$T = E - V = \frac{1}{2} k\left(\frac{1}{4} h^2 - y^2\right), \quad -\frac{h}{2} \le y \le \frac{h}{2}.$$

14. *A linear harmonic oscillator of force constant 2×10^6 N/m and amplitude 0.01 m has a total mechanical energy of 160 J. Show that its (a) maximum potential energy is 160 J (b) maximum kinetic energy is 100 J.* (I.I.T. 1989)

Solution

From Eqns. (1.10) to (1.12), we have total mechanical energy $= 1/2\ kA^2 + c$

$$= \frac{1}{2} \times 2 \times 10^6 \times (0.01)^2 + c = 100 \text{ J} + c = 160 \text{ J}$$

(*a*) Maximum P.E. $= \frac{1}{2} kA^2 + c = 160$ J

(*b*) Maximum K.E. $= \frac{1}{2} kA^2 = 100$ J.

15. *A long light piece of spring steel is clamped at its lower end and a 1 kg ball is fastened to its top end (Fig. 1.7). A force of 5 N is required to displace the ball 10 cm to one side as shown in the figure. Assume that the system executes SHM when released. (a) Find the force constant of the spring for this type of motion. (b) Find the time period with which the ball vibrates back and forth.*

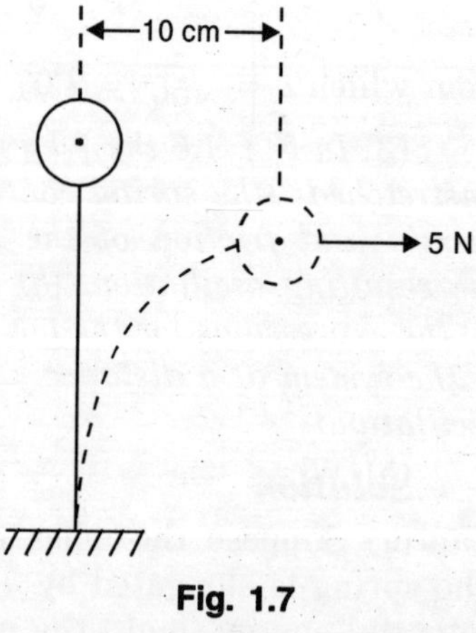

Fig. 1.7

Solution

(*a*) $k = \dfrac{\text{External Force}}{\text{Displacement}} = \dfrac{5\,\text{N}}{0.1\,\text{m}} = 50$ N/m

(*b*) $T = 2\pi\sqrt{m/k} = 2\pi\sqrt{1/50} = 0.89$ s.

16. *Two blocks (m = 1.0 kg and M = 11 kg) and a spring (k = 300 N/m) are arranged on a horizontal, frictionless surface as shown in Fig. 1.8. The coefficient of static friction between the two blocks is 0.40. What is the maximum possible amplitude of the simple harmonic motion if no slippage is to occur between the blocks?*

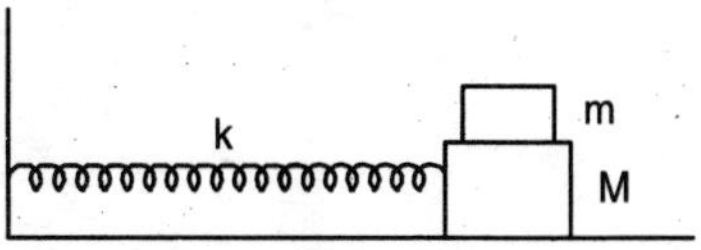

Fig. 1.8

Solution

Angular frequency of SHM = $\omega = \sqrt{300/12}$

Maximum force on the smaller body without any slippage is $m\omega^2 A = \mu m g$

Thus, $$A = \frac{\mu g}{\omega^2} = \frac{0.4\times 9.8\times 12}{300}\,\text{m} = 15.68 \text{ cm}.$$

17. *Two identical springs have spring constant k = 15 N/m. A 300 g mass is connected to them as shown in Figs. 1.9(a) and (b).*

Find the period of motion for each system. Ignore frictional forces.

Solution

(*a*) When the mass m is given a displacement x, one spring will be elongated by x, and the other will be compressed by x. They will each exert a force of magnitude kx on the mass in the direction opposite to the displacement. Hence, the total restoring force $F = -2\,kx = m\ddot{x}$. So,

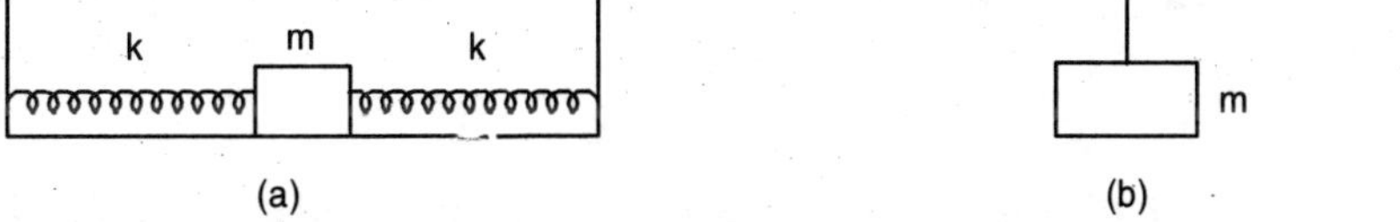

Fig. 1.9

$$\omega = \sqrt{2k/m} = \sqrt{2\times 15/0.3} = 10 \text{ s}^{-1}$$

$$T = 2\pi/\omega = 0.63 \text{ s}.$$

(*b*) When the mass is pulled a distance y downward, each spring is stretched a distance y. The net restoring force on the mass = $-2\,ky$, $\omega = \sqrt{2k/m}$ and the period is also 0.63 s.

18. *Two massless springs A and B each of length a_0 have spring constants k_1 and k_2. Find the equivalent spring constant when they are connected in (a) series and (b) parallel as shown in Fig. 1.10 and a mass m is suspended from them.*

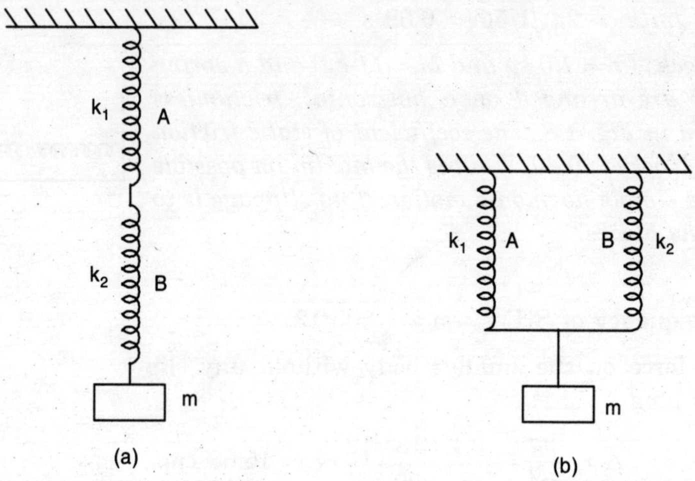

Fig. 1.10

Solution

(*a*) Let x_1 and x_2 be the elongations in springs A and B respectively. Total elongation $= x_1 + x_2$.

$$mg = k_1x_1 \text{ and } mg = k_2x_2$$

Thus,
$$x_1 + x_2 = mg\left(\frac{1}{k_1}+\frac{1}{k_2}\right).$$

If k is the equivalent spring constant of the combination (*a*), we have

$$x_1 + x_2 = mg/k$$

or
$$\frac{1}{k} = \frac{1}{k_1} + \frac{1}{k_2} \quad or,\ k = \frac{k_1k_2}{k_1+k_2}.$$

(*b*) Let x be the elongation in each spring.

$$mg = (k_1 + k_2)x$$

If k is the equivalent spring constant of the combination (*b*), we have

$$mg = kx$$

Thus,
$$k = k_1 + k_2.$$

19. *Two light springs of force constants k_1 and k_2 and a block of mass m are in one line AB on a smooth horizontal table such that one end of each spring is on rigid supports and the other end is free as shown in Fig. 1.11. The distance CD between the free ends of the springs is 60 cm. If the block moves along AB with a velocity 120 cm/s in between the springs, calculate the period of oscillation of the block.*

(k_1 = 1.8 N/m, k_2 = 3.2 N/m, m = 200 g) (I.I.T. 1985)

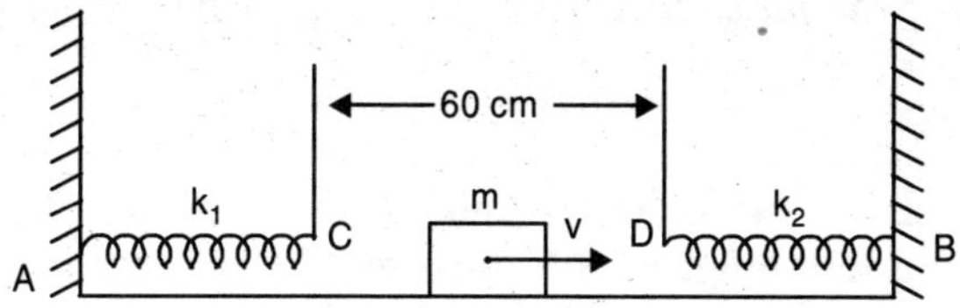

Fig. 1.11

Solution

The time period of oscillation of the block = time to travel 30 cm to the right from mid-point of *CD* + time in contact with the spring k_2 + time to travel *DC* (60 cm) to the left + time in contact with spring k_1 + time to travel 30 cm to the right from *C*

$$= \frac{30}{120} + \frac{1}{2}\left[2\pi\sqrt{\frac{m}{k_2}}\right] + \frac{60}{120} + \frac{1}{2}\left[2\pi\sqrt{\frac{m}{k_1}}\right] + \frac{30}{120}$$

$$= 1 + \pi\left[\sqrt{0.2/3.2} + \sqrt{0.2/1.8}\right] = 1 + \pi\left[\frac{1}{4} + \frac{1}{3}\right]$$

$$= 2.83 \text{ s.}$$

20. *The mass m is connected to two identical springs that are fixed to two rigid supports (Fig. 1.12). Each of the springs has zero mass, spring constant k, and relaxed length a_0. They each have length a at the equilibrium position of the mass. The mass can move in the x-direction (along the axis of the springs) to give longitudinal oscillations. Find the period of motion. Ignore frictional forces.*

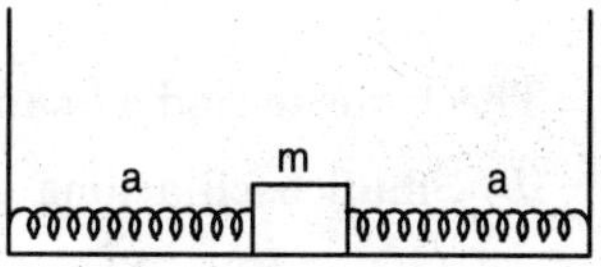

Fig. 1.12

Solution

At the equilibrium position each spring has tension $T_0 = k(a - a_0)$. Let at any instant of time x be the displacement of the mass from the equilibrium position. At that time the net force on the mass due to two springs in the *+ve x*-direction is

$$F_x = -k(a + x - a_0) + k(a - x - a_0) = -2kx.$$

Thus, $\quad m\ddot{x} = -2kx$ and $\omega^2 = 2k/m$

and $\quad T = 2\pi\sqrt{m/(2k)}.$

21. *A mass m is suspended between rigid supports by means of two identical springs. The springs each have zero mass, spring constant k, and relaxed length a_0. They each have length a at the equilibrium position of mass m [Fig. 1.13(a)]. Consider the motion of the mass along the y-direction (perpendicular to the axis of the springs) only. Find the frequency of*

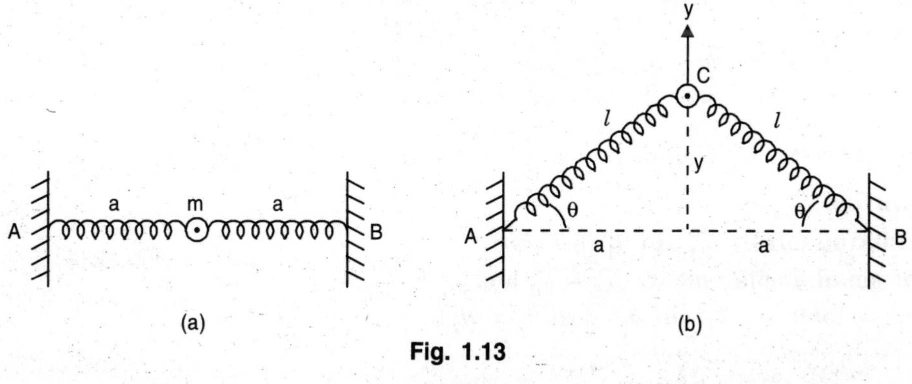

Fig. 1.13

transverse oscillations of the mass under (a) slinky approximation ($a_0 << a$), (b) small oscillations approximation ($y << a$).

Solution

At equilibrium each spring exerts tension $T_0 = k\ (a - a_0)$. In the general configuration (Fig. 1.13(b)) each spring has length l and tension $T = (l - a_0)$ which is exerted along CA or CB. The y-component of this force is $-T \sin \theta$. Each spring contributes a return force $T \sin \theta$ in the –ve y–direction. Using Newton's second law, we have

$$m\ddot{y} = -2T \sin\theta = -2k(l - a_0)y/l. \quad \text{...(1.16)}$$

The x-components of the two forces due to two springs balance each other so that there is no motion along the x-direction. Thus, we have

$$m\ddot{y} = -2ky\left(1 - \frac{a_0}{\sqrt{a^2 + y^2}}\right). \quad \text{...(1.17)}$$

The above equation is not exactly in the form that gives rise to SHM.

(a) Slinky approximation ($a_0/a << 1$): Since $l > a$, $a_0/l << 1$ and we get from Eqn. (1.16)

$$\ddot{y} = -\frac{2k}{m}y = -\omega^2 y$$

The time period is same as longitudinal oscillation. Time period = $2\pi\sqrt{m/2k}$.

(b) Small oscillations approximation ($y << a$): Under this approximation, we have

$$\frac{a_0}{\sqrt{a^2 + y^2}} \approx \frac{a_0}{a}\left(1 - \frac{y^2}{2a^2}\right).$$

Thus,
$$m\ddot{y} = -2ky\left(1 - \frac{a_0}{a} + \frac{a_0 y^2}{2a^3}\right).$$

we neglect $(y/a)^3$ term in this equation, we get

$$\ddot{y} = -\frac{2ky}{ma}(a - a_0) = -\frac{2T_0}{ma}y.$$

Hence
$$\omega^2_{tr} = \frac{2T_0}{ma} = \frac{2k}{ma}(a - a_0) = \frac{2k}{m}\left(1 - \frac{a_0}{a}\right)$$

and
$$\text{time period} = \frac{2\pi\sqrt{m/2k}}{\sqrt{1 - a_0/a}} = 2\pi\sqrt{ma/2T_0}.$$

22. *A ball of mass m is connected to rigid walls by means of two wires of lengths l_1 and l_2 (Fig. 1.14). At equilibrium the tension in each wire is T_0. The mass m is displaced slightly from equilibrium in the vertical direction and released. Determine the frequency for small oscillations.*

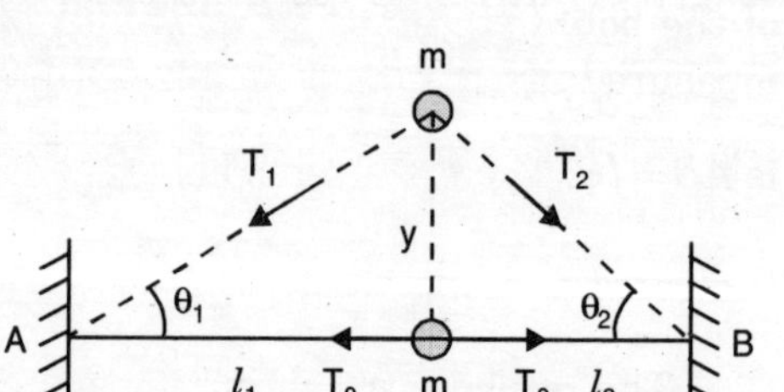

Fig. 1.14

Solution

Restoring force = $T_1 \sin \theta_1 + T_2 \sin \theta_2$.

For small displacements, $T_1 \approx T_0$ and $T_2 \approx T_0$,

$\sin\theta_1 \approx \tan\theta_1 = y/l_1$, $\sin \theta_2 \approx \tan \theta_2 = y/l_2$.

Thus, $$m\ddot{y} = -T_0\left(\frac{y}{l_1}+\frac{y}{l_2}\right) = -T_0\frac{l_1+l_2}{l_1 l_2}y$$

and $$\omega = \left[\frac{T_0(l_1+l_2)}{ml_1l_2}\right]^{1/2}.$$

23. *A vertical spring of length 2L and spring constant k is suspended at one end. A body of mass m is attached to the other end of the spring. The spring is compressed to half its length and then released. Determine the kinetic energy of the body, and its maximum value, in the ensuing motion in the presence of the gravitational field.*

Solution

If the position of the body is measured from the relaxed position of the spring by the coordinate y (positive upward) and if the P.E. V is set equal to zero at $y = 0$, we have

$$V = \frac{1}{2}ky^2 + mgy,$$

$$T + V = T + \frac{1}{2}ky^2 + mgy = E = \text{Total energy},$$

where T is the K.E. of the body.

Now, $T = 0$ when $y = L$, or, $E = \frac{1}{2}kL^2 + mgL.$

Thus, $T = \frac{1}{2}k(L^2 - y^2) + mg(L - y).$

T is maximum when $\frac{dT}{dy} = 0$ or, $y = -\frac{mg}{k}$,

and $T_{max} = \frac{1}{2}k(L + mg/k)^2.$

24. *Find the time period of a simple pendulum.*

Solution

A small bob of mass m is attached to one end of a string of negligible mass and the other end of the string is rigidly fixed at O (Fig. 1.15). OA is the vertical position of the simple pendulum of length l and this is also the equilibrium position of the system. The pendulum can oscillate only in the vertical plane and at any instant of time B is the position of the bob. Let $\angle AOB = \psi$. The displacement of the bob as measured along the perimeter of the circular arc of its path is $AB = l\psi$. The instantaneous tangential velocity is $l\frac{d\psi}{dt}$ and the corresponding tangential acceleration is $l\frac{d^2\psi}{dt^2}$.

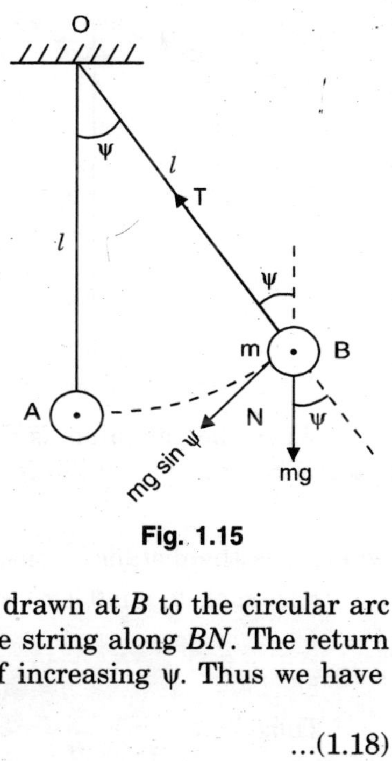

Fig. 1.15

The return force acting on the bob along the tangent BN drawn at B to the circular arc AB is $mg \sin\psi$. There is no component of the tension T of the string along BN. The return force $mg \sin\psi$ acts in a direction opposite to the direction of increasing ψ. Thus we have

$$ml\frac{d^2\psi}{dt^2} = -mg\sin\psi. \quad ...(1.18)$$

Maclaurin's series for $\sin\psi$ is

$$\sin\psi = \psi - \frac{\psi^3}{3!} + \frac{\psi^5}{5!} - \ldots$$

For sufficiently small ψ, $\sin\psi \approx \psi$ (in radians) and we have

$$\frac{d^2\psi}{dt^2} = -\omega^2\psi,$$

with $$\omega^2 = g/l.$$

The motion is simple harmonic and its time period of oscillation is

$$T = \frac{2\pi}{\omega} = 2\pi\sqrt{l/g}.$$

25. *What is the period of small oscillation of an ideal pendulum of length l, if it oscillates in a truck moving in a horizontal direction with acceleration a?*

Solution

Let the equilibrium position be given by the angle ϕ (Fig. 1.16). In this position the force on the mass m along the horizontal axis is equal to ma. The angle ϕ is determined by the equations.

$$T \sin \phi = ma, \; T \cos \phi = mg.$$

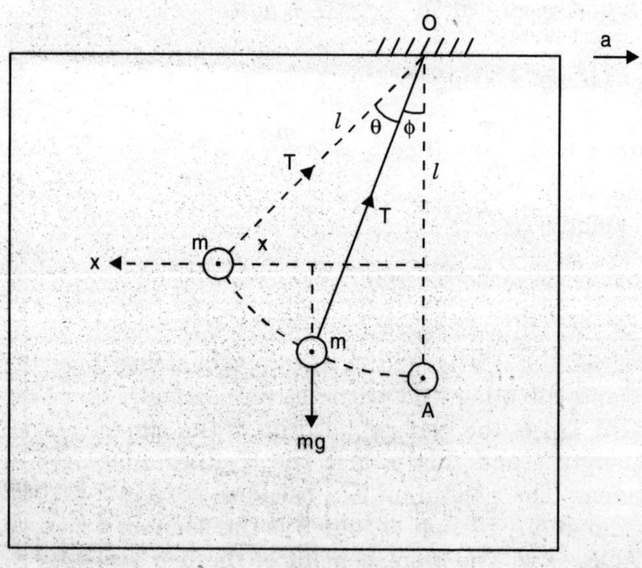

Fig. 1.16

When the pendulum is displaced by a small amount θ, it will perform simple harmonic motion around the equilibrium position. Its equation of motion is

$$m\ddot{x} = -T \sin(\theta + \phi)$$

where x is the distance from the vertical OA.

For small θ, $\sin(\theta + \phi) \approx \theta \cos \phi + \sin \phi$

$$= \theta\frac{mg}{T} + \frac{ma}{T}.$$

Thus $$m\ddot{x} = -ma - mg\theta.$$

θ and l are related geometrically as

$x = l\sin(\theta + \phi) \approx l\,\theta\cos\phi + l\sin\phi$

or $$\ddot{x} = l\cos\phi\,\ddot{\theta}$$

Hence $$l\cos\phi\,\ddot{\theta} = -a - g\theta$$

or $$\ddot{\theta} = -\frac{g}{l\cos\phi}\left(\theta + \frac{a}{g}\right).$$

If we make the following substitution

$$\psi = \theta + \frac{a}{g}$$

we get

$$\ddot{\psi} = -\frac{g}{l\cos\phi}\psi = -\omega^2\psi$$

with time period of oscillation

$$T = \frac{2\pi}{\omega} = 2\pi\left[\frac{l\cos\phi}{g}\right]^{1/2}.$$

Now, $$\cos\phi = \frac{mg}{T} = \frac{mg}{\sqrt{m^2a^2 + m^2g^2}} = \frac{g}{\sqrt{a^2 + g^2}}$$

Thus, $$T = 2\pi\left[\frac{l}{\sqrt{a^2 + g^2}}\right]^{1/2}.$$

26. *A simple pendulum of bob mass m is suspended vertically from O by a massless rigid rod of length L (Fig. 1.17 (a)). The rod is connected to a spring of spring constant k at a distance h form O. The spring has its relaxed length when the pendulum is vertical.*

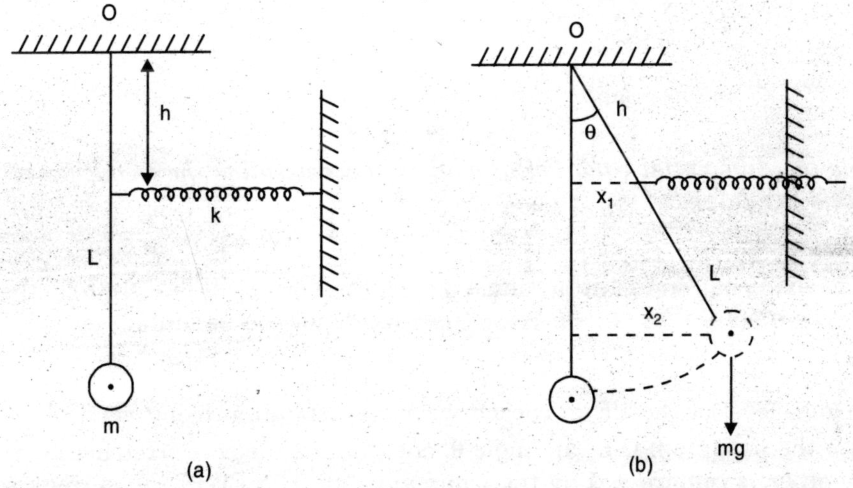

Fig. 1.17

Write the differential equation of motion and determine the frequency for small oscillations of this pendulum.

Solution

Let θ be a small deflection of the pendulum from its equilibrium position. The spring is compressed by x_1 and it exerts a force $F_s = kx_1$ on the rod. We have

$$x_1 = h \sin \theta \text{ and } x_2 = L \sin \theta.$$

Taking the sum of torques about the point O we obtain (for small deflection θ):

$$-F_s h - mgx_2 = mL^2 \ddot{\theta}$$

or

$$mL^2 \ddot{\theta} + (kh^2 + mgL) \sin \theta = 0.$$

Since $\sin \theta \approx \theta$ for small oscillations we get SHM with frequency

$$\omega = \left[\frac{kh^2 + mgL}{mL^2}\right]^{1/2} = \left[\frac{g}{L} + \frac{kh^2}{mL^2}\right]^{1/2}.$$

27. *A simple pendulum of bob mass m is suspended vertically from O by a massless rigid rod of length L. The rod is connected to two identical massless springs on two sides of the rod at a distance a from O (Fig. 1.18). The spring constant of each spring is k. The springs have their relaxed lengths when the pendulum is vertical.*

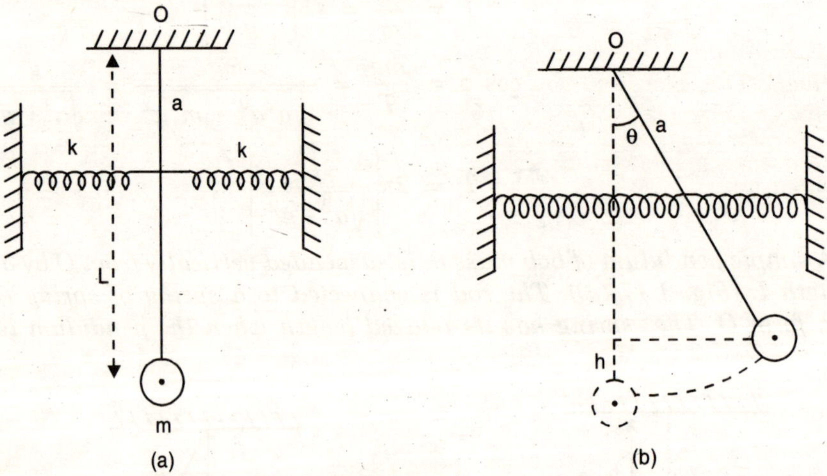

Fig. 1.18

Write the differential equation of motion and determine the frequency of small oscillations of this pendulum.

Solution

When the pendulum is at an angle θ with the vertical [Fig. 1.18(*b*)], the pendulum is raised by the distance $h = L - L \cos \theta$ and the PE of the pendulum is

$$(\text{P.E.})_m = mgh = mgL(1 - \cos \theta).$$

The zero level of the PE is chosen with the pendulum being vertical.

When the pendulum is at an angle θ, one of the springs is stretched by the amount $a\theta$, while the other is compressed by the same amount. The PE of the springs is

$$(\text{P.E.})_s = \frac{1}{2}k(a\theta)^2 \times 2 = ka^2\theta^2.$$

Thus the total P.E. of the system is

$$V = mgL(1 - \cos \theta) + ka^2\theta^2.$$

The kinetic energy is associated only with the mass m. The velocity $v = L\dot{\theta}$ and KE is

$$T = \frac{1}{2}mL^2\dot{\theta}^2$$

Thus the total energy of the system is

$$E = \frac{1}{2}mL^2\dot{\theta}^2 + mg\,L(1 - \cos\theta) + Ka^2\theta^2.$$

Since the total energy of the system is conserved, we have

$$\frac{dE}{dt} = mL^2\dot{\theta}\ddot{\theta} + mgL\,\sin\theta\,\dot{\theta} + 2ka^2\dot{\theta}\,\theta = 0$$

or

$$\ddot{\theta} + \frac{g}{L}\sin\theta + \frac{2ka^2}{mL^2}\theta = 0.$$

Since $\sin\theta \approx \theta$ for small oscillations we get SHM with frequency

$$\omega = \left[\frac{g}{L} + \frac{2ka^2}{mL^2}\right]^{\frac{1}{2}}$$

28. *A simple pendulum is suspended from a peg on a vertical wall. The pendulum is pulled away from the wall to a horizontal position (Fig. 1.19) and released. The ball hits the wall, the coefficient of restitution being $2/\sqrt{5}$. What is the minimum number of collisions after which the amplitude of oscillation becomes less than 60°?* (I.I.T. 1987)

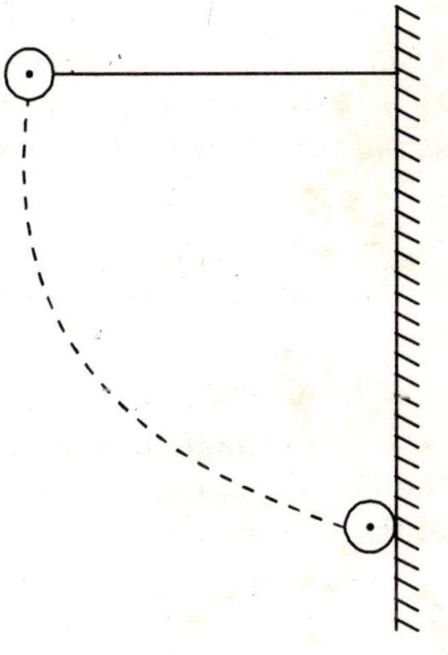

Fig. 1.19

Solution

Let v_0 be the velocity of the bob just before the first collision.

Then

$$\frac{1}{2}mv_0^2 = mgL$$

or

$$v_0 = \sqrt{2gL}.$$

The velocity of the bob just after the 1st collision is

$$v_1 = \frac{2}{\sqrt{5}}v_0 = \frac{2}{\sqrt{5}}\sqrt{2gL}.$$

v_1 will be the velocity of the bob just before 2nd collision. The velocity of the bob just after the second collision is

$$v_2 = \frac{2}{\sqrt{5}}v_1 = \left(\frac{2}{\sqrt{5}}\right)^2\sqrt{2gL}$$

The velocity just after the nth collision is

$$v_n = \left(\frac{2}{\sqrt{5}}\right)^n\sqrt{2gL}.$$

We assume that after n collisions the amplitude of oscillation becomes 60°.

Thus,

$$\frac{1}{2}mv_n^2 = mg\,(L - L\cos 60°) = \frac{1}{2}mgL$$

or

$$v_n^2 = gL.$$

or

$$\left(\frac{2}{\sqrt{5}}\right)^{2n} 2gL = gL$$

or

$$\left(\frac{4}{5}\right)^n = \frac{1}{2}$$

n is slightly greater than 3. In fact

$$n = \frac{\log_{10} 2}{\log_{10} 5 - \log_{10} 4} = 3.1$$

Thus the minimum number of collisions after which the amplitude becomes less than 60° is 4.

29. *A bullet of mass M is fired with a velocity 50 m/s at an angle θ with the horizontal. At the highest point of its trajectory, it collides head-on with a bob of mass 3 M suspended by a massless string of length 10/3 m and gets embedded in the bob. After the collision the string moves through an angle of 120°. Find*

(*i*) *the angle θ*

(*ii*) *the vertical and horizontal coordinates of the initial position of the bob with respect to the point of firing of the bullet. (Take g = 10 m/s²)* (I.I.T 1988)

Solution

(*i*) At the highest point of the bullet the horizontal component of velocity = $u \cos \theta$ and the vertical component of velocity = 0. Let (x, y) be the coordinates of the initial position A of the bob (Fig. 1.20).

We have

$$y = \frac{u^2 \sin^2 \theta}{2g}$$

$$y = \frac{u^2 \sin 2\theta}{2g}.$$

Due to head-on collision of the bullet with the bob at A we have from the conservation of linear momentum

$$Mu \cos \theta = 4Mv$$

where v is the initial velocity of the bob along the x-direction.

Thus,

$$v = \frac{u}{4}\cos \theta.$$

At the highest point (B) of the path of the combined mass let the velocity be v_1. At this position, we have

$$4Mg \cos 60° = \frac{4Mv_1^2}{l}$$

or $$v_1^2 = \frac{gl}{2}$$

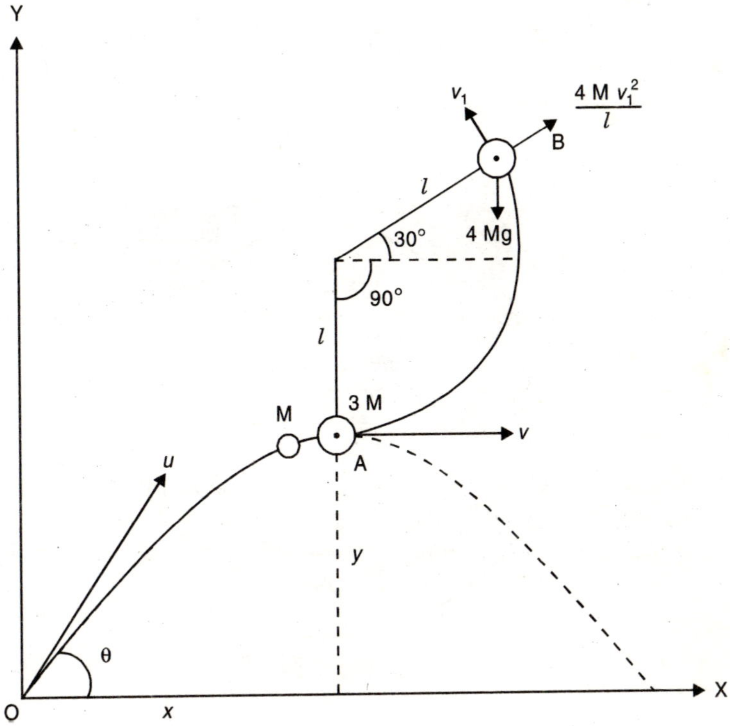

Fig. 1.20

From conservation of mechanical energy of the bob, we get

$$\frac{1}{2}(4M)v^2 = \frac{1}{2}(4M)v_1^2 + 4Mg(l + l \sin 30°)$$

or $$v^2 = v_1^2 + 3gl$$

Thus, $$\frac{u^2}{16}\cos^2\theta = \frac{gl}{2} + 3gl = \frac{7gl}{2}$$

or $$\cos^2\theta = \frac{56gl}{u^2} = \frac{56 \times 10}{50 \times 50} \times \frac{10}{3} = \frac{56}{75}$$

or $$\theta = 30.2°$$

(*ii*) $$y = \frac{u^2 \sin^2\theta}{2g} = \frac{50 \times 50 \times (0.503)^2}{2 \times 10} = 31.6 \text{ m},$$

$$x = \frac{u^2 \sin 2\theta}{2g} = \frac{50 \times 50 \times (0.869)}{2 \times 10} = 108.7 \text{ m}.$$

30. *Two identical balls A and B each of mass 0.1 kg are attached to two identical massless springs. The spring-mass system is constrained to move inside a rigid smooth pipe bent in the form of a circle as shown in Fig. 1.21. The pipe is fixed in a horizontal plane. The centres of the balls can move in a circle of radius 0.06 metre. Each spring has a natural length* *of 0.06 π metre and spring constant 0.1 N/m. Initially, both the balls are displaced by an angle θ = π/6 radian with respect to the diameter PQ of the circle (as shown in the figure) and released from rest.*

(i) *Calculate the frequency of oscillation of ball B.*

(ii) *Find the speed of ball A when A and B are at two ends of the diameter PQ.*

(iii) *What is the total energy of the system?* (I.I.T. 1993)

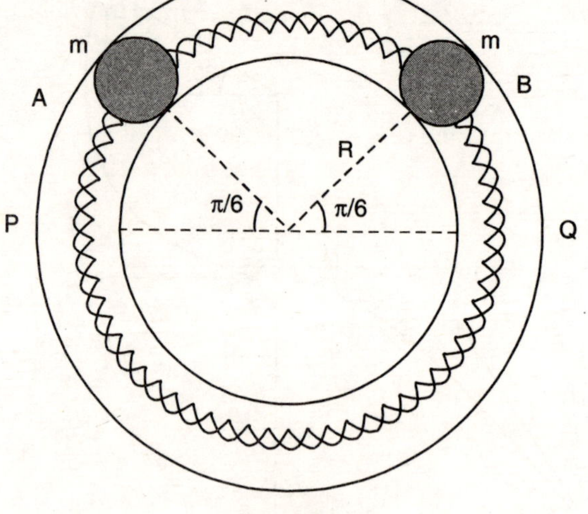

Fig. 1.21

Solution

(*i*) At an angular displacement θ of the balls the compression or extension in respective springs = 2 *R*θ.

Thus, force on *B* due to both springs = 4 *kR*θ, where *k* is the spring constant.

Now, $\frac{d^2\theta}{dt^2}$ = angular acceleration and $R\frac{d^2\theta}{dt^2}$ = linear acceleration.

The equation of motion of the mass *B* is given by

$$mR\frac{d^2\theta}{dt^2} = -4kR\theta$$

or $$\frac{d^2\theta}{dt^2} = -\frac{4k}{m}\theta = -\omega^2\theta,$$ which represents SHM.

Thus, $$\nu = \frac{\omega}{2\pi} = \frac{1}{2\pi}\sqrt{\frac{4k}{m}} = \frac{1}{2\pi}\sqrt{\frac{4\times 0.1}{0.1}} = \frac{1}{\pi}\text{Hz}.$$

(*ii*) $$[\text{K.E.}]_{\theta=0} = [\text{P.E.}]_{\theta=\frac{\pi}{6}}$$

or $$2\times\frac{1}{2}mv^2 = 2\times\frac{1}{2}k(2R\theta)^2$$

or $$v = 2R\theta\sqrt{\frac{k}{m}} = 2\times 0.06\times\frac{\pi}{6} = 0.02\pi\ \text{ms}^{-1}$$

(iii) Total energy $= [\text{P.E.}]_{\theta = \frac{\pi}{6}}$

$$= 4kR^2\theta^2$$

$$= 4 \times 0.1 \times 36 \times 10^{-4} \times \frac{\pi^2}{36}$$

$$= 4\pi^2 \times 10^{-5} \text{ J.}$$

31. *If the earth were a homogeneous sphere of radius R and a straight hole were bored in it through its centre, show that a body dropped into the hole will execute SHM. Find its time period.*

Solution

Suppose AB is a straight hole (Fig. 1.22) passing through the centre O of the earth. A body of mass m is dropped into the hole. At any instant of time the body is at C at a distance x from the centre of the earth. When the body is at C, the force of attraction on the body due to earth is

$$F = -G\frac{4}{3}\pi x^3 \frac{\rho m}{x^2}$$

where ρ, density of the material of earth, is assumed to be uniform everywhere and G, the universal gravitational constant.

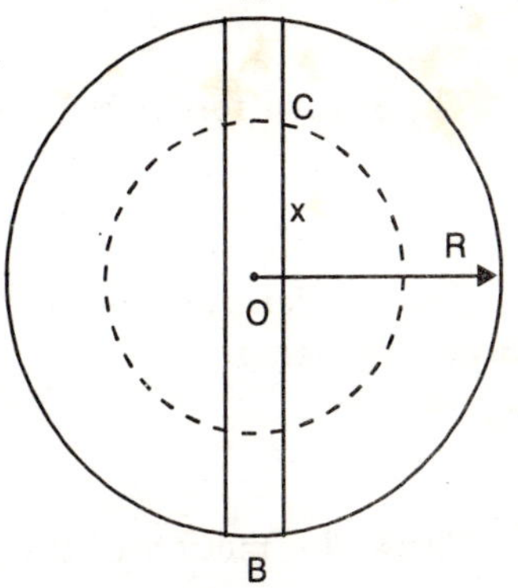

Fig. 1.22

At C, the acceleration of the body towards the centre O is

$$a = \frac{F}{m} = -G\frac{4}{3}\pi\rho x.$$

As $a \propto x$ and it acts opposite to the direction of increasing x, the motion of the body is simple harmonic. We have

$$\omega^2 = G\frac{4}{3}\pi\rho.$$

Now the acceleration of the body on the surface of the earth is

$$g = G\frac{M}{R^2} = G\frac{4}{3}\pi\rho R$$

where M = mass of the earth.

Hence, $$\omega^2 = \frac{g}{R}$$

and the time period $$T = 2\pi\sqrt{\frac{R}{g}}.$$

with $R = 6.4 \times 10^6$ m, $g = 9.8$ m/s^2,

We have $T = 5077.6$ s.

32. *A cylindrical piston of mass M slides smoothly inside a long cylinder closed at one end, enclosing a certain mass of gas. The cylinder is kept with its axis horizontal. If the piston is disturbed from its equilibrium position, it oscillates simple harmonically. Show that the period of oscillation will be (Fig. 1.23)*

$$T = 2\pi\sqrt{\frac{Mh}{PA}} = \frac{2\pi}{A}\sqrt{\frac{MV}{P}}$$ (I.I.T. 1981)

Solution

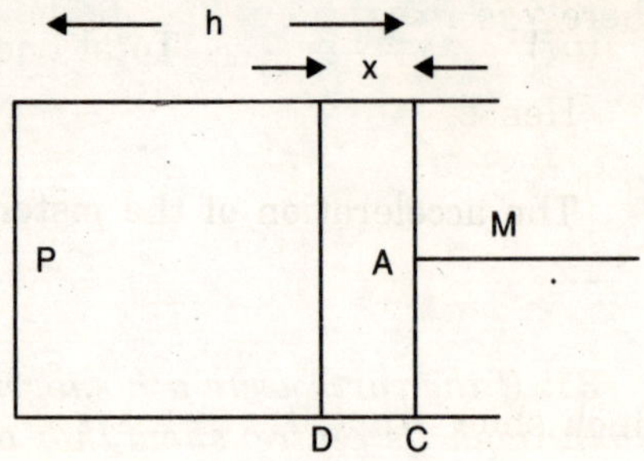

Fig. 1.23

Suppose that the initial pressure of the gas is P and initial volume is $V = Ah$. The piston is moved isothermally from C to D through a distance x (Fig. 1.23). The gas inside the cylinder will be compressed and it will try to push the piston to its original position. When the piston is at D let the pressure of the gas be $P + \delta P$ and volume $= V - \delta V = V - Ax$. Since the process is isothermal, we have

$$PV = (P + \delta P)(V - \delta V) \approx PV - P\delta V + V\delta P$$

or

$$\delta P = \frac{P\delta V}{V} = \frac{PAx}{V}.$$

The return force acting on the piston is

$$A\delta P = \frac{A^2 Px}{V}.$$

The acceleration a of the piston is proportional to x and directed opposite to the direction of increasing x:

$$a = -\frac{A^2 Px}{MV}$$

Thus, the motion of the piston is simple harmonic and its time period is

$$T = 2\pi\sqrt{\frac{MV}{A^2 P}} = \frac{2\pi}{A}\sqrt{\frac{MV}{P}} = 2\pi\sqrt{\frac{Mh}{PA}}.$$

33. *An ideal gas is enclosed in a vertical cylindrical container and supports a freely moving piston of mass M. The piston and the cylinder have equal cross-sectional area A. Atmospheric pressure is P_0 and when the piston is in equilibrium, the volume of the gas is V_0. The piston is now displaced slightly from the equilibrium position. Assuming that the system is completely isolated from its surroundings, show that the piston executes simple harmonic motion and find the frequency of oscillation.* (I.I.T. 1981)

Solution

Since the system is completely isolated from the surroundings, there will be adiabatic change in the container. Let the initial pressure and the volume of the gas be P and V_0 respectively. When the piston is moved down a distance x, the pressure increases to $P + \delta P$ and volume decreases to $V_0 - \delta V$. Thus,

$$PV_0^\gamma = (P + \delta P)(V_0 - \delta V)^\gamma$$

$$= (P + \delta P) V_0^\gamma \left(1 - \frac{\delta V}{V_0}\right)^\gamma$$

$$\approx (P + \delta P) V_0^\gamma \left(1 - \gamma\frac{\delta V}{V_0}\right)$$

$$\approx V_0^\gamma\left(P + \delta P - \frac{\gamma P\delta V}{V_0}\right)$$

where γ is the ratio of specific heats at constant pressure and volume ($\gamma = C_p/C_v$).

Hence,
$$\delta P = \frac{\gamma P \delta V}{V_0} = \frac{\gamma P A x}{V_0}.$$

The acceleration of the piston is given by

$$a = -\frac{\gamma P A x}{V_0} \times \frac{A}{M} = -\frac{\gamma P A^2 x}{M V_0},$$

which shows that the piston executes SHM, with

$$\omega^2 = \frac{\gamma P A^2}{M V_0}$$

Now,
$$P = P_0 + \frac{Mg}{A} = \frac{AP_0 + Mg}{A}$$

The frequency of oscillation is given by

$$f = \frac{\omega}{2\pi} = \frac{1}{2\pi}\left[\frac{\gamma A(AP_0 + Mg)}{MV_0}\right]^{1/2}.$$

34. *Two non-viscous, incompressible and immiscible liquids of densities ρ and 1.5 ρ are poured into two limbs of a circular tube of radius R and small cross-section kept fixed in a vertical plane as shown in Fig. 1.24. Each liquid occupies one-fourth the circumference of the tube. (a) Find the angle θ that the radius vector to the interface makes with the vertical in equilibrium position. (b) If the whole liquid is given a small displacement from its equilibrium position, show that the resulting oscillations are simple harmonic. Find the time period of these oscillations.* (I.I.T. 1991)

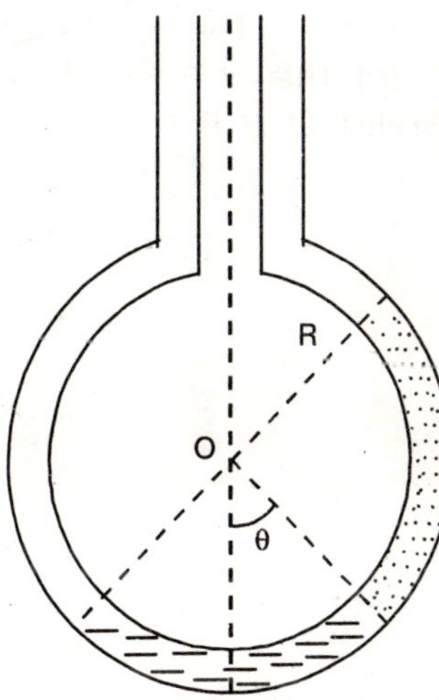

Fig. 1.24

Solution

(*a*) Since each liquid occupies one-fourth the circumference of the tube, $\angle AOC = 90° = \angle BOC$ [Fig. 1.25 (*a*)].

The pressure P_1 at D due to liquid on the left limb is

$$P_1 = (R - R\sin\theta)\, 1.5\, \rho g$$

The pressure P_2 at D due to liquid on the right limb is

$$P_2 = (R - R\cos\theta)\, 1.5\, \rho g + (R\sin\theta + R\cos\theta)\rho g$$

At equilibrium $P_1 = P_2$. Thus, we have

$$(1 - \sin\theta)\, 1.5 = (1 - \cos\theta)\, 1.5 + \sin\theta + \cos\theta$$

Solving this equation, we get $2.5 \sin\theta = 0.5\cos\theta$,

or
$$\tan\theta = \frac{0.5}{2.5} = 0.2$$

or
$$\theta = 11.3°.$$

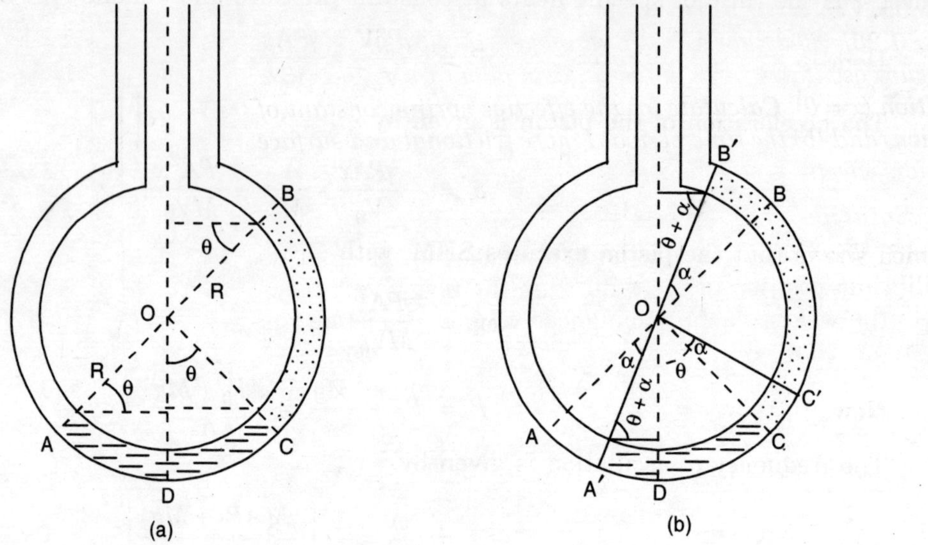

Fig. 1.25

(*b*) When the liquid is given a small upward displacement $y = BB'$ in the right limb [Fig. 1.25 (*b*)], then $y = R\alpha$ where $\alpha = \angle B'OB$, and A goes to A' and C goes to C'. The pressure difference at D is

$$
\begin{aligned}
dP &= P_2' - P_1' \\
&= [R - R\cos(\theta + \alpha)\ 1.5\ \rho g + [R\sin(\theta + \alpha) \\
&\qquad + [R\cos(\theta + \alpha)]\rho g - [R - R\sin(\theta + \alpha)]\ 1.5\ \rho g \\
&= R\rho g\ \{2.5\sin(\theta + \alpha) - 0.5\cos(\theta + \alpha)\} \\
&\approx R\rho g\ \{2.5\ [\sin\theta + \alpha\cos\theta] - 0.5[\cos\theta - \alpha\sin\theta]\} \\
&= R\rho g\ \{2.5\alpha\cos\theta + 0.5\alpha\sin\theta\} \\
&= y\rho g\ \{2.5\cos\theta + 0.5\sin\theta\} \\
&= 2.55\rho g y
\end{aligned}
$$

Thus, Restoring force $= -\,2.55\ \rho g y \times A,$

where A is the area of cross-section of the tube.

Mass of the liquid in the tube is

$$m = \frac{2\pi R}{4}A\rho + \frac{2\pi R}{4}A \times 1.5\rho = 1.25\ \pi R A\rho.$$

The acceleration of the liquid column is

$$a = -\frac{2.55\rho g\ yA}{1.25\pi R A\rho} = -2.04\left(\frac{g}{\pi R}\right)y$$

which shows that the motion is simple harmonic.

The time period of oscillations is given by

$$T = 2\pi\sqrt{\frac{\pi R}{2.04g}} = 2.49\sqrt{R}\ \text{s}\cdot$$

35. *Ten kg of mercury are poured into a glass U tube [Fig. 1.26]. The tube's inner diameter is 1.0 cm and the mercury oscillates freely up and down about its equilibrium position (x = 0). Calculate (a) the effective spring constant of motion, and (b) the time period. Ignore frictional and surface tension effects.*

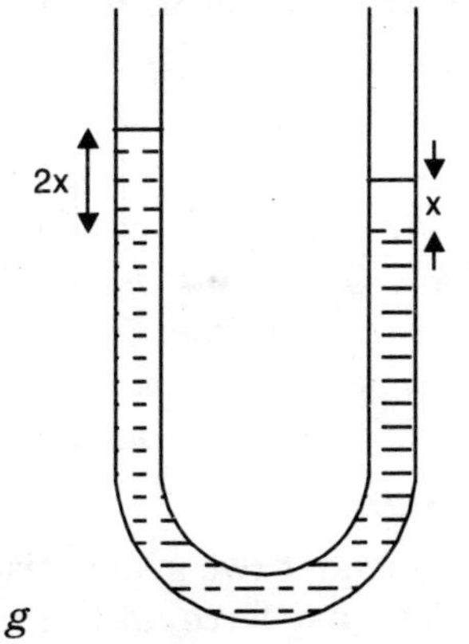

Fig. 1.26

Solution

(*a*) When the mercury is displaced x metres from its equilibrium position in the right arm, the restoring force is due to the weight of the unbalanced column of mercury of weight $2x$. Now,

$$\text{Weight} = \text{Volume} \times \text{Density} \times g = (\pi r^2 2x) \times \rho \times g$$

where $\rho = 13.6 \text{ g/cm}^3 = \dfrac{13.6 \times 10^{-3} \text{kg}}{10^{-6} \text{m}^3} = 13.6 \times 10^3 \text{ kg/m}^3.$

Thus, the restoring force = $-(2\pi r^2 \rho g)\, x$, and Hooke's law is valid, thereby we see that the effective spring constant for the system is

$$k = 2\pi r^2 \rho g = 2\pi(0.005)^2 (13.6 \times 10^3) \times 9.8 = 20.94 \text{ N/m}$$

(*b*) $$T = 2\pi\sqrt{\frac{m}{k}} = 2\pi\sqrt{\frac{10}{20.94}} = 4.34 \text{ s.}$$

36. *Two identical positive point charges + Q each, are fixed at a distance of 2a apart. A point charge + q lies midways between the fixed charges. Show that for a small displacement along the line joining the fixed charges, the charge + q executes simple harmonic motion. Find the frequency of oscil-lations.*

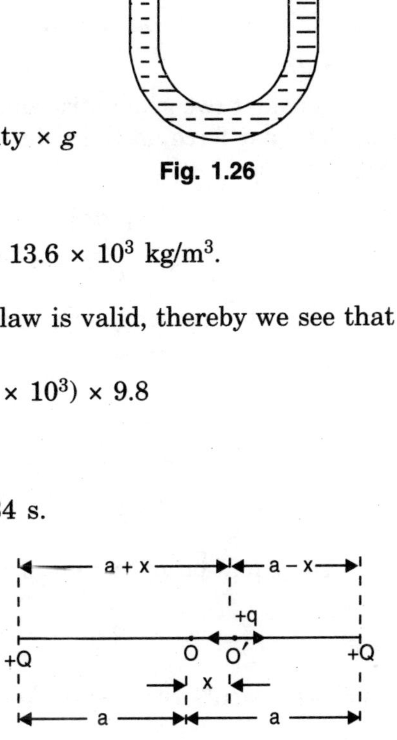

Fig. 1.27

Solution

Let the charge + q be displaced through a distance x to the right (Fig. 1.27). Restoring force on charge + q is

$$F = \frac{Qq}{4\pi \epsilon_0 (a+x)^2} - \frac{Qq}{4\pi \epsilon_0 (a-x)^2}$$

$$= -\frac{4aQq\,x}{4\pi \epsilon_0 \left(a^2 - x^2\right)^2}$$

$$\approx -\frac{Qq\,x}{\pi \epsilon_0 a^3} \quad \text{since } x \ll a,$$

where $\epsilon_0 = 8.85 \times 10^{-12} \text{ C}^2/\text{N.m}^2$.

The equation of motion of the charge + q of mass m *is*

$$m\ddot{x} = -\frac{Qq\,x}{\pi \epsilon_0 a^3}$$

which represents SHM with frequency

$$\nu = \frac{1}{2\pi}\sqrt{\frac{Qq}{\pi \in_0 a^3 m}}.$$

37. *A thin ring of radius R carries uniformly distributed charge + Q. A point charge – q is placed on the axis of the ring at a distance x from the centre of the ring and released from rest. Show that the motion of the charged particle is approximately simple harmonic. Find the frequency of oscillations.*

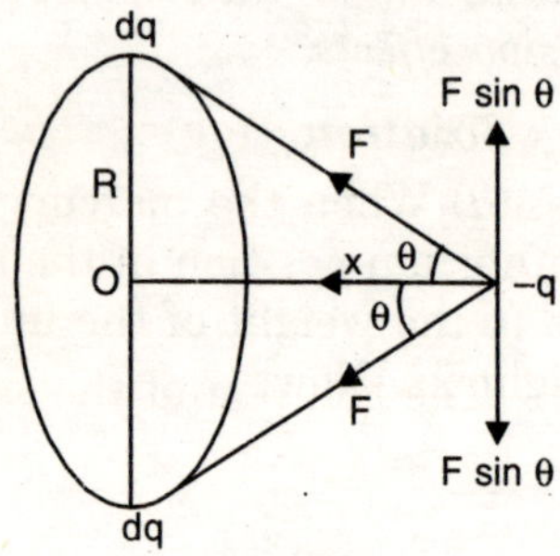

Fig. 1.28

Solution

Consider two symmetric small charge elements dq of the ring. The net force on the point charge along the x-direction (Fig. 1.28) is

$$-\frac{(2dq)q}{4\pi \in_0 \left(R^2 + x^2\right)} \cos\theta = -\frac{(2dq)q\,x}{4\pi \in_0 \left(R^2 + x^2\right)^{3/2}}.$$

Thus the net force on $(-q)$ due to the total charge $(+Q)$ on the ring is

$$F = -\frac{Q\,q\,x}{4\pi \in_0 \left(R^2 + x^2\right)^{3/2}}$$

$$\approx -\frac{Qqx}{4\pi \in_0 R^3} \quad \text{since } x \ll R.$$

The equation of motion of the point charge is

$$m\ddot{x} = -\frac{Qqx}{4\pi \in_0 R^3}$$

which represents SHM with frequency

$$\nu = \frac{1}{2\pi}\sqrt{\frac{Qq}{4\pi \in_0 mR^3}}.$$

38. *An object of 98 N weight suspended from the end of a vertical spring of negligible mass stretches the spring by 0.1 m. (a) Determine the position of the object at any time if initially it is pulled down 0.05 m and then released. (b) Find the amplitude, period and frequency of the motion.*

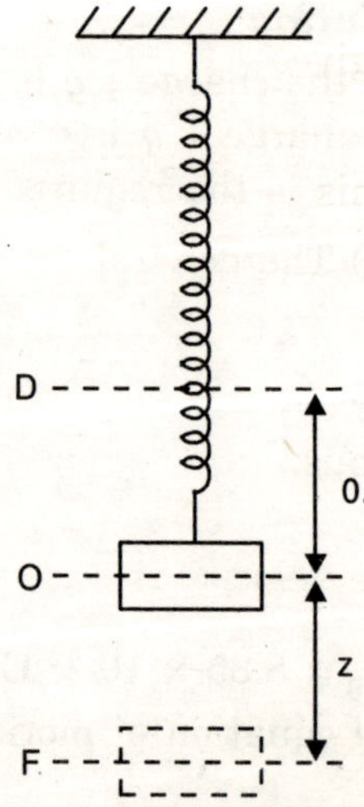

Fig. 1.29

Solution

(*a*) Let D and O represent the position of the end of the spring before and after the object is put (Fig. 1.29). Position O is the equilibrium position of the object. The positive z-axis is downward with origin at the equilibrium position O. When the elongation of the spring is 0.1 m, the force on it is 98 N.

When the elongation is $(0.1 + z)$ m, the force on it is $\frac{98}{0.1} \times (0.1 + z)$ N. Thus, when the object is released at F, there is an upward force acting on it of magnitude $\frac{98}{0.1} \times (0.1 + z)$ N and a downward force due to its weight of magnitude 98 N. Hence, we can write

$$\frac{98}{9.8}\frac{d^2z}{dt^2}\hat{k} = 98\,\hat{k} - \frac{98}{0.1}(0.1+z)\hat{k} = -\frac{98}{0.1}z\hat{k}.$$

The resulting motion is simple harmonic with angular frequency

$$\omega = \sqrt{\frac{9.8}{0.1}} = 7\sqrt{2}\ \text{s}^{-1}.$$

The solution of the differential equation is (see Eqn. 1.6)

$$z = C\cos 7\sqrt{2}\,t + D\sin 7\sqrt{2}\,t.$$

At $t = 0$, $z = 0.05$ m, $\frac{dz}{dt} = 0$ which give $C = 0.05$ and $D = 0$. Thus, the position of the object at any time is given by

$$z = 0.05\cos 7\sqrt{2}\,t$$

(*b*) Amplitude = 0.05 m, period $= \frac{\sqrt{2}\pi}{7}$ s and frequency $= \frac{7\sqrt{2}}{2\pi}$ Hz.

39. *A particle of mass 3 units moves along the x-axis attracted toward origin by a force whose magnitude is numerically equal to 27x. If it starts from rest at x = 8 units, find (a) the differential equation describing the motion of the particle (b) the position and velocity of the particle at any time and (c) The amplitude and period of the vibration.*

Solution

(*a*) Let $\vec{r} = x\hat{i}$ be the position vector of the particle. The force acting on the particle is

$$3\ddot{x}\hat{i} = -27x\hat{i}$$

which gives

$$\ddot{x} + 9x = 0.$$

This is the required differential equation.

(*b*) The general solution of the differential equation is

$$x = C\cos 3t + D\sin 3t.$$

The initial conditions are $x = 8$, $\dot{x} = 0$ at $t = 0$, which give $C = 8$ and $D = 0$.

Thus, $$x = 8\cos 3t$$

and the velocity is $$\frac{dx}{dt}\hat{i} = -24\sin 3t\,\hat{i}$$

(*c*) Amplitude = 8 units, period $= \frac{2\pi}{3}$ s.

40. *Work the previous problem if the particle is initially at $x = 8$ units but is moving (a) to the right with speed 18 units, (b) to the left with speed 18 units. Find the amplitude, frequency and the phase angle in each case. Are the two motions (a) and (b) 180° out of phase with each other?*

Solution

(*a*) $x = C\cos 3t + D\sin 3t.$

At $t = 0$, $x = 8$ and $\dot{x} = 18$, which give $C = 8$ and $D = 6$.

Thus,

$$x = 8\cos 3t + 6\sin 3t$$

$$= \sqrt{8^2 + 6^2}\,\cos(3t - \phi)$$

$$= 10\cos(3t - \phi)$$

where

$$\cos\phi = \frac{8}{10},\ \sin\phi = \frac{6}{10} \text{ and } \tan\phi = \frac{3}{4}.$$

The angle ϕ is called the phase angle which is in the first quadrant: $\phi = 36.87°$.

Amplitude = 10 and frequency = $\dfrac{3}{2\pi}$ Hz.

(*b*) At $t = 0$, $x = 8$ and $\dot{x} = -18$, which give $C = 8$ and $D = -6$ so that

$$x = 8\cos 3t - 6\sin 3t$$

$$= 10\cos(3t - \psi)$$

with $\cos\psi = \dfrac{8}{10}$, $\sin\psi = -\dfrac{6}{10}$ and $\tan\psi = -\dfrac{3}{4}$.

The phase angle ψ is in the fourth quadrant:

$$\psi = 323.13°.$$

The amplitude and frequency are the same as in part (*a*). The only difference is in the phase angle. Here we have $\sin(\phi + \psi) = 0$ and $\cos(\phi + \psi) = 1$ and $\psi + \phi = 2\pi$. The two motions are not 180° out of phase with each other since $\psi - \phi \neq 180°$.

41. *A pail of water, at the end of a rope of length r, is whirled in a horizontal circle at constant speed v. A distant ground-level spot light casts a shadow of the pail onto a vertical wall which is perpendicular to the spotlight beam. Show that the shadow executes SHM with angular frequency $\omega = v/n$.*

Solution

The figure gives a top view of the set-up (Fig. 1.30).

Let θ (*t*) denote the instantaneous angular position in radians of the pail, measured counter clockwise from the + ve *x*-axis. Then $\omega = \pm\dfrac{v}{r}$, with the sign depending upon which way the pail is whirled. Letting $\theta(t = 0) = \theta_0$, the angular position

$$\theta(t) = \pm\frac{v}{r}\,t + \theta_0.$$

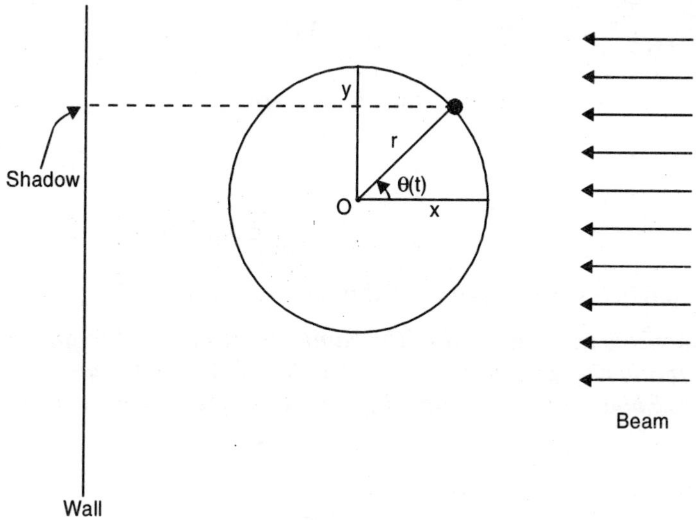

Fig. 1.30

Now,
$$y(t) = r \sin \theta(t) = r \sin \left(\pm \frac{vt}{r} + \theta_0 \right)$$
$$= \pm r \sin \left(\frac{vt}{r} \pm \theta_0 \right)$$

So,
$$\ddot{y} = \mp \frac{v^2}{r} \sin \left(\frac{vt}{r} \pm \theta_0 \right)$$

and the differential equation for the motion of the shadow is

$$\ddot{y} + \frac{v^2}{r^2} y = 0$$

or
$$\ddot{y} + \omega^2 y = 0$$

i.e., we have SHM of amplitude $|\pm r| = r$ and angular frequency $\omega = \frac{v}{r}$.

42. *Two particles oscillate in simple harmonic motion along a common straight line segment of length A. Each particle has a period of 1.5 s but they differ in phase by 30°. (a) How far apart are they (in terms of A) 0.5 s after the lagging particle leaves one end of the path? (b) Are they moving in the same direction, towards each other, or away from each other at this time?*

Solution

(*a*) Let the equations of two particles be

$$x_1 = \frac{A}{2} \cos\left(\frac{4\pi}{3} t + \frac{\pi}{6} \right)$$
$$x_2 = \frac{A}{2} \cos\left(\frac{4\pi}{3} t \right)$$

x_2 reaches one end of the path when $t = 0$ and at that time

$$x_1 = \frac{A}{2} \cdot \frac{\sqrt{3}}{2}.$$

When $t = 0.5$ s, $x_2 = -\frac{A}{4}$ and $x_1 = -\frac{A\sqrt{3}}{4}$, and $|x_2 - x_1| = 0.183\ A$.

(*b*) At $t = 0.5$ s, velocities of the particles are

$$\dot{x}_1 = -\frac{A}{2}\frac{4\pi}{3}\sin 150° = -\text{ ve},$$

$$\dot{x}_2 = -\frac{A}{2}\frac{4\pi}{3}\sin 120° = -\text{ ve}.$$

Thus, the particles are moving in the same direction.

43. *Two particles execute SHM of the same amplitude and frequency along the same line. They pass one another when going in opposite directions each time their displacement is half their amplitude. Show that the phase difference between them is 120°.*

Solution

Let the equations be

$$x_1 = A\sin(\omega t + \phi_1),$$
$$x_2 = A\sin(\omega t + \phi_2)$$

with $\phi_1 \neq \phi_2$.

Let the particles cross each other at $t = 0$, so that

$$x_1 = x_2 = \frac{A}{2} \text{ at } t = 0$$

which give $\phi_1 = 30°$ and $\phi_2 = 150°$.

At $t = 0$, $\dot{x}_1 = A\omega\cos\phi_1 = +$ ve and $\dot{x}_2 = A\omega\cos\phi_2 = -$ ve which show that the particles are moving in opposite directions.

Phase difference $= \phi_2 - \phi_1 = 120°$.

44. *Show that the phase-space diagram (p_x versus x curve) of SHM is an ellipse with area equal to $\frac{E}{\nu}$ where E = Total energy and ν = frequency of oscillations.*

Solution

For a particle executing SHM, we have

$$x = A\cos\omega t$$

which gives

$$p_x = m\dot{x} = -m\omega A\sin\omega t$$

and

$$\frac{x^2}{A^2} + \frac{p_x^2}{m^2\omega^2 A^2} = 1.$$

Since

$$E = \frac{1}{2}m\omega^2 A^2, \text{ we have}$$

$$\frac{x^2}{\frac{2E}{m\omega^2}} + \frac{p_x^2}{2mE} = 1$$

which is the equation of an ellipse in the xp_x–plane with semi-major axis $= a = \sqrt{2E/(m\omega^2)}$ and semi-minor axis $= b = \sqrt{2mE}$.

Area of the ellipse = $\pi ab = \dfrac{2\pi E}{\omega} = \dfrac{E}{\nu}$.

45. *Show that the force $\vec{F} = -kx\hat{i}$ acting on a simple harmonic oscillator is conservative.*

Solution

We have
$$\vec{\nabla} \times \vec{F} = \begin{vmatrix} \hat{i} & \hat{j} & \hat{k} \\ \dfrac{\partial}{\partial x} & \dfrac{\partial}{\partial y} & \dfrac{\partial}{\partial z} \\ -kx & 0 & 0 \end{vmatrix} = 0.$$

Hence $\vec{F}$ is conservative.

46. *Solve the differential equation*

$$\frac{1}{2}m\left(\frac{dx}{dt}\right)^2 + \frac{1}{2}kx^2 = E$$

and show that x(t) represents simple harmonic motion. What is the frequency and amplitude of vibration of the motion?

Solution

We have
$$\frac{dx}{dt} = \omega\sqrt{A^2 - x^2}$$

where
$$\omega = \sqrt{\frac{k}{m}} \text{ and } A^2 = \frac{2E}{k}.$$

After integration, we get

$$x = A\sin(\omega t + \phi)$$

which represents a simple harmonic motion. Here ϕ is the phase angle.

$$\text{Frequency} = \frac{\omega}{2\pi} = \frac{1}{2\pi}\sqrt{\frac{k}{m}} \text{ and Amplitude} = A = \sqrt{\frac{2E}{k}}.$$

47. *A spring of mass M and spring constant k is hanged from a rigid support. A mass m is suspended at the lower end of the spring. If the mass is pulled down and released then it will execute SHM. Find the frequency of oscillations.*

Solution

Let l be the length of the coiled wire of the spring. Let the suspended mass m be at a distance z from the equilibrium position. We consider an element of length dx of the spring situated between points x and $x + dx$ measured along the spring from the point of support. Since the stretching is assumed to be uniform, the element, distant x from the fixed end, will experience a displacement $(x/l)\,z$ and will have velocity $(x/l)\,\dot{z}$. Since the mass of the element dx is $(M/l)\,zdx$, the kinetic energy of this element is

$$\frac{1}{2}\left(\frac{M}{l}\right)dx\left[\frac{xdz}{l\,dt}\right]^2 = \frac{M}{2l^3}\left[\frac{dz}{dt}\right]^2 x^2\,dx.$$

Thus at that instant total kinetic energy of the spring is

$$\frac{M}{2l^3}\left[\frac{dz}{dt}\right]^2\int_0^1 x^2\,dx = \frac{M}{6}\left(\frac{dz}{dt}\right)^2,$$

and the kinetic energy of the mass m is $\frac{1}{2}m\left(\frac{dz}{dt}\right)^2$.

The potential energy of the system when the spring is elongated by z is $\left(\frac{1}{2}\right)kz^2$. Hence the total energy of the system is

$$\left(\frac{1}{2}\right)\left(m+\frac{M}{3}\right)\left(\frac{dz}{dt}\right)^2 + \frac{1}{2}kz^2 = \text{constant}. \quad ...(1.19)$$

Due to finite mass of the spring the effective mass of the system becomes $\left(m+\frac{M}{3}\right)$. From problem (46), we find that

$$\omega^2 = \frac{k}{m+\frac{M}{3}} \quad ...(1.20)$$

Differentiating Eqn. (1.19) with respect to t, we obtain

$$\frac{1}{2}\left(m+\frac{M}{3}\right)2z\ddot{z} + \frac{1}{2}k\,2z\,\dot{z} = 0$$

or

$$\left(m+\frac{M}{3}\right)\ddot{z} + kz = 0$$

which shows that the motion is simple harmonic with ω^2 given by Eqn. (1.20).

48. *Show that if the assumption of small vibration (see problem 24) is not made, then the time period of a simple pendulum is given by*

$$T = 4\sqrt{\frac{l}{g}}\int_0^{\frac{p}{2}}\frac{df}{\sqrt{1-k^2\sin^2 f}}$$

where $k = \sin\left(\frac{\psi_0}{2}\right)$, ψ_0 being the maximum angle made by the string with the vertical. The initial conditions are $\psi = \psi_0$ and $\dot{\psi} = 0$ at time $t = 0$.

Solution

The equation of motion for a simple pendulum, if small vibrations are not assumed, is:

$$\frac{d^2\psi}{dt^2} = -\frac{g}{l}\sin\psi \quad ...(1.21)$$

We put $u = \frac{d\psi}{dt}$ so that

$$\frac{d^2\psi}{dt^2} = \frac{du}{dt} = \frac{du}{d\psi}\frac{d\psi}{dt} = u\frac{du}{d\psi}$$

and Eqn. (1.21) becomes

$$u\frac{du}{d\psi} = -\frac{g}{l}\sin\psi.$$

On integration, we get

$$\frac{u^2}{2} = \frac{g}{l}\cos\psi + C.$$

When $\psi = \psi_0,\ \dot{\psi} = u = 0$, so that

$$C = -\frac{g}{l}\cos\psi_0.$$

Thus, we have $$u = \pm\left[\frac{2g}{l}(\cos\psi - \cos\psi_0)\right]^{\frac{1}{2}}.$$

We consider that part of the motion where the bob goes from $\psi = \psi_0$ to $\psi = 0$ which represents a time equal to $\frac{T}{4}$. In this case ψ is decreasing so that ψ is negative:

$$\frac{d\psi}{dt} = -\sqrt{\frac{2g}{l}}(\cos\psi - \cos\psi_0)^{\frac{1}{2}}.$$

Integrating from $\psi = \psi_0$ to $\psi = 0$, we get

$$\frac{T}{4} = -\sqrt{\frac{l}{2g}}\int_{\psi_0}^{0}\frac{d\psi}{(\cos\psi - \cos\psi_0)^{1/2}}$$

$$= \sqrt{\frac{l}{2g}}\int_0^{\psi_0}\frac{d\psi}{\sqrt{2}\left(\sin^2\frac{\psi_0}{2} - \sin^2\frac{\psi}{2}\right)^{1/2}}.$$

Let $\sin\frac{\psi}{2} = \sin\frac{\psi_0}{2}\sin\phi$, so that

$$\frac{1}{2}\cos\frac{\psi}{2}d\psi = \sin\frac{\psi_0}{2}\cos\phi\, d\phi.$$

Putting $k = \sin\frac{\psi_0}{2}$, we get

$$T = 4\sqrt{\frac{l}{g}}\int_0^{\frac{\pi}{2}}\frac{d\phi}{\sqrt{1 - k^2\sin^2\phi}} \qquad ...(1.22)$$

This integral is called an elliptical integral. For small vibrations $k \approx 0$ and

$$T = 2\pi\sqrt{\frac{l}{g}}.$$

49. *Show that the time period given in problem 48 can be written as*

$$T = 2p\sqrt{\frac{l}{g}}\left[1 + \left(\frac{1}{2}\right)^2 k^2 + \left(\frac{1.3}{2.4}\right)^2 k^4 + ...\right]$$

Solution

Since $k^2 \sin^2 \phi < 1$, we make binomial expansion of $\left(1 - k^2 \sin^2 \phi\right)^{-1/2}$ and integrate term by term. We finally find

$$T = 2\pi\sqrt{\frac{l}{g}}\left[1 + \left(\frac{1}{2}\right)^2 k^2 + \left(\frac{1.3}{2.4}\right)^2 k^4 + ...\right]$$

where we have made use of the following integration formula

$$\int_0^{\pi/2} \sin^{2n} \phi d\phi = \frac{1.3.5....(2n-1)}{2.4.6....2n}\frac{\pi}{2}.$$

50. *A particle of mass m is located in a one-dimensional potential field where the potential energy of the particle depends on the coordinates x as $V(x) = V_0 (1 - \cos ax)$. Find the period of small oscillations that the particle performs about the equilibrium position.*

Solution

$$\frac{dV}{dx} = 0 \text{ when } \sin ax = 0 \text{ or } x = 0.$$

Again $x = 0$ is the point of minimum of $V(x)$ since $\frac{d^2V}{dx^2} > 0$ at $x = 0$. The force acting on the particle is

$$F_x = -\frac{\partial V}{\partial x} = -V_0 a \sin ax.$$

For small values of x we have

$$m\ddot{x} = -V_0 a(ax) = -V_0 a^2 x.$$

The time period of small oscillations is

$$T = \frac{2\pi}{a}\sqrt{\frac{m}{V_0}}.$$

51. *A bead of mass m slides on a frictionless wire of nearly parabolic shape (Fig. 1.31). Let the point P be the point at the bottom of the wire. Show that the bead will oscillate about P if displaced slightly from P and released.*

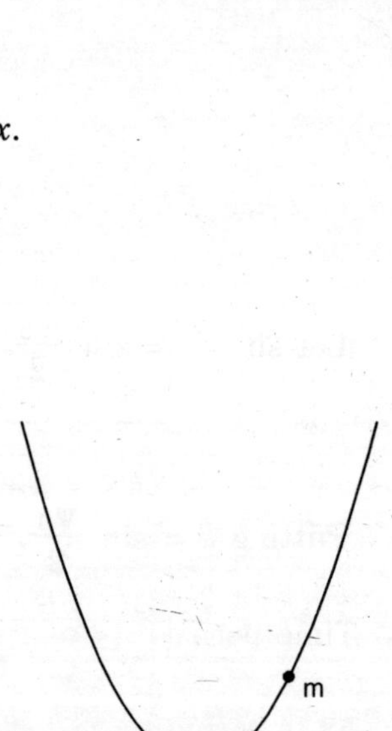

Fig. 1.31

Solution

Since the shape of the wire near P is a parabola, the potential energy of the bead is given by $V = cx^2$ in the neighbourhood of P, where x is measured from P and c is a constant. Now, $F_x = -\frac{\partial V}{\partial x} = -2cx$, *i.e.*, $F_x \propto x$ and directed opposite to the direction of increasing x. So, the bead executes SHM and its time period is given by $T = 2\pi\left(\frac{m}{2c}\right)^{1/2}$.

Note that this line of reasoning leads to a general result: Any conservative system will oscillate with SHM about a minimum in its potential energy curve provided the oscillation amplitude is small enough.

52. *The potential energy of a particle of mass m is given by*

$$V(x) = \frac{a}{x^2} - \frac{b}{x}$$

where a and b are positive constants. Find the minimum of V(x) and expand V(x) about the point of minimum of V(x). Find the period of small oscillations that the particle performs about the position of minimum of V(x).

Solution

$\frac{dV}{dx} = 0$ when $x = \frac{2a}{b}$ and $\frac{d^2V}{dx^2} > 0$ at $x = \frac{2a}{b}$. Hence $x = \frac{2a}{b}$ is the point of minimum of $V(x)$.

Now,
$$V\left(\frac{2a}{b}\right) = -\frac{b^2}{4a}$$

$$V'\left(\frac{2a}{b}\right) = 0$$

$$V''\left(\frac{2a}{b}\right) = \frac{b^4}{8a^3}$$

Thus the expansion of $V(x)$ about $x = \frac{2a}{b}$ is given by

$$V(x) = -\frac{b^2}{4a} + \frac{\left(x - \frac{2a}{b}\right)^2}{2!}\frac{b^4}{8a^3} + \cdots$$

If we put $y = x - \left(\frac{2a}{b}\right)$, the equation of motion about the position of minimum of $V(x)$ is

$$m\ddot{y} = -\frac{b^4}{8a^3}y.$$

Time period
$$T = 2\pi/\left(\frac{b^4}{8a^3m}\right)^{\frac{1}{2}} = \frac{4\pi a}{b^2}\sqrt{2ma}.$$

53. *A thin rod of length 10 cm and mass 100 g is suspended at its midpoint from a long wire. Its period T_a of angular SHM is measured to be 2 s. An irregular object, which we call object X, is then hung from the same wire, and its period T_x is found be 3 s. What is the rotational inertia of the object X about its suspension axis?*

Solution

The moment of inertia of the thin rod about a perpendicular axis through its midpoint is

$$I_a = \frac{1}{12}mL^2 = \frac{1}{12} \times 0.1 \times (0.1)^2 = \frac{1}{12} \times 10^{-3} \text{ kg.m}^2$$

We know $T_a = 2\pi\sqrt{\frac{I_a}{C}}$ and $T_x = 2\pi\sqrt{\frac{I_x}{C}}$

Thus, $I_x = I_a \frac{T_x^2}{T_a^2} = \frac{1}{12} \times 10^{-3} \times \frac{9}{4}$

$= 1.875 \times 10^{-4}$ kg.m^2.

54. *A uniform disc of radius R and mass M is attached to the end of a uniform rigid rod of length L and mass m. When the disc is suspended from a pivot as shown in Fig. 1.32, what will be the period of motion?*

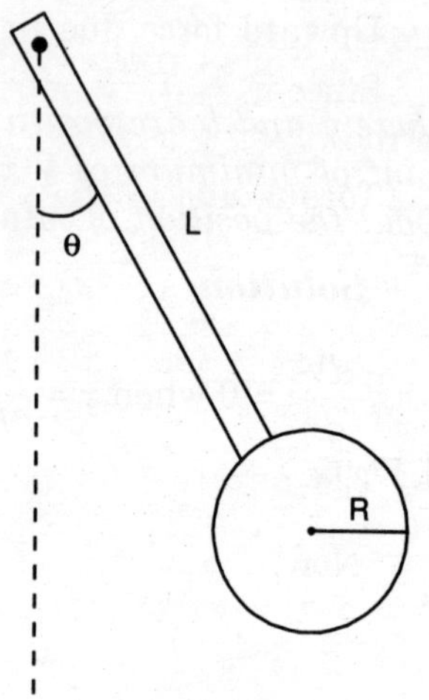

Fig. 1.32

Solution

The equation of motion is $\Gamma = I\alpha$, where Γ is the external torque, I is the moment of inertia, α is the angular acceleration, and both Γ and I are about the pivot point. Let θ be a small angular displacement from the vertical. Now, external torque comes both from the rod and the disc:

$$\Gamma = -mg\frac{L}{2}\sin\theta - Mg(R+L)\sin\theta$$

$$\approx -\left[\frac{1}{2}mgL + Mg(R+L)\right]\theta = -C\theta$$

Where we put $\sin\theta \approx \theta$.

Now, $I = I_{rod} + I_{disc} = \frac{1}{3}mL^2 + \left[\frac{1}{2}MR^2 + M(R+L)^2\right]$.

The time period T is given by

$$T = 2\pi\sqrt{\frac{I}{C}}$$

$$= 2\pi\left(\frac{L}{g}\right)^{\frac{1}{2}} \frac{\left[\frac{a}{3} + \frac{b^2}{2} + (1+b)^2\right]^{1/2}}{\left[1 + b + \frac{a}{2}\right]^{1/2}}$$

Where $a = \frac{m}{M}$ and $b = \frac{R}{L}$.

55. *A thin rod of length L and area of cross-section S is pivoted at its lowest point P inside a stationary, homogeneous and non-viscous liquid (Fig. 1.33). The rod is free to rotate in a vertical plane about a horizontal axis passing through P. The density d_1 of the material of the rod is smaller than the density d_2 of the liquid. The rod is displaced by a small angle θ from its equilibrium position and then released. Show that the motion of the rod is simple harmonic and determine its angular frequency in terms of the given parameters.* (I.I.T. 1996)

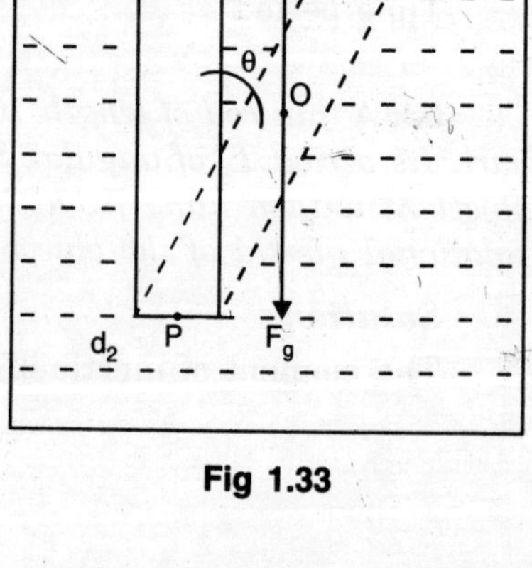

Fig 1.33

Solution

Volume of the rod, $V = SL$ and its mass, $M = Vd_1 = SLd_1$. In the slightly displaced position two forces are acting on the rod:

Downward force due to gravity, $F_g = Mg = SLd_1g$.

Upward force due to buoyancy, $F_B = Vd_2g = SLd_2g$.

Since $d_2 > d_1$, a net upward force of magnitude $F_B - F_g$ acts at O (the middle point of the rod).

Torque due to the net upward force about P is

$$\Gamma = -SLg(d_2 - d_1)\frac{L}{2}\sin\theta$$

$$\approx -\frac{1}{2}SL^2g(d_2 - d_1)\theta = -C\theta.$$

The negative sign is due to the fact that the torque acts in the opposite direction of increasing θ. Now, the moment of inertia of the rod about the pivot point P is

$$I = \frac{1}{3}ML^2 = \frac{1}{3}Sd_1L^3.$$

The equation of motion of the rod is

$$\Gamma = I\frac{d^2\theta}{dt^2}$$

or

$$\frac{d^2\theta}{dt^2} = -\frac{C}{I}\theta = -\omega^2\theta.$$

The motion of the rod is simple harmonic. The angular frequency is

$$\omega = \sqrt{\frac{C}{I}} = \left[\frac{3g(d_2 - d_1)}{2d_1L}\right]^{1/2}.$$

56. *A thin light beam of uniform cross-section A is clamped at one end and loaded at free end by placing a mass M. [Such a beam is called a loaded cantilever.] If the loaded free end of the beam is slightly displaced from its equilibrium position, it starts executing SHM. Find an expression for the time period of vibration of the loaded light cantilever.*

Solution

We shall assume that the bar is not subjected to any tension and the amplitude of motion is so small that the rotatory effect can be neglected. The x-axis is taken along the length of the bar and the transverse vibration is taking place in the y-direction. The radius of curvature R is given by

$$\frac{1}{R} = \frac{d^2y}{dx^2}\Bigg/\left[1+\left(\frac{dy}{dx}\right)^2\right]^{3/2}.$$

For a small transverse vibration $\frac{dy}{dx} \ll 1$ and

$$\frac{1}{} \quad \frac{d^2y}{}$$

In the bent position of the rod we consider a cross-section ABCD of a small segment of the rod of length δx measured along the central line PQ so that $EQ = EP = R$ = radius of curvature of the central filament PQ (Fig. 1.34). The filaments above PQ are extended whereas the filaments below PQ are contracted. PQ is the neutral filament and $PQ = \delta x$. Let us consider a filament MN above PQ at a distance r from PQ. Let Δ be the extension of this filament so that $MN = \delta x + \Delta$. From Fig. 1.34 we have

$$\phi = \frac{\delta x}{R} = \frac{\delta x + \Delta}{R + r}$$

or

$$\Delta = \frac{r\delta x}{R}$$

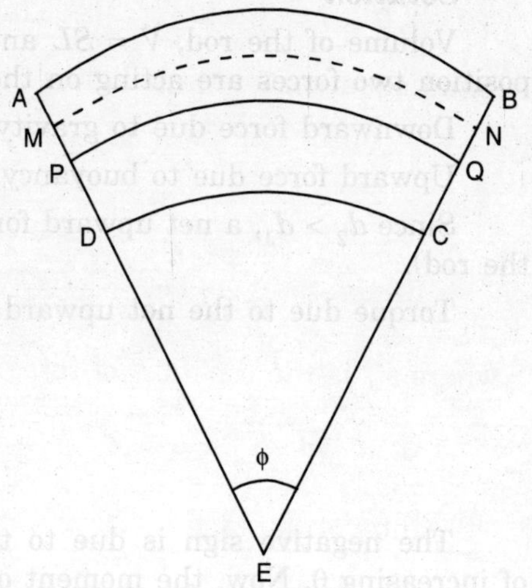

Fig. 1.34

Thus, longitudinal strain $= \dfrac{\Delta}{\delta x} = \dfrac{r}{R}.$

If Y denotes the Young's modulus of the material of the beam, the longitudinal stress is $\dfrac{Yr}{R}$. Hence the force acting on the filament is $\alpha \dfrac{Yr}{R}$, where α is the area of cross-section of the filament MN. The total bending moment of the bar is

$$\Gamma = \frac{Y}{R}\Sigma r^2\alpha \approx Y\frac{d^2y}{\delta x^2}.I_g$$

where $\Sigma r^2\alpha = AK^2 = I_g$ is known as geometrical moment of inertia of the cross-section of the rod about the neutral axis, and K is the radius of gyration of the section about the neutral axis.

Let OG be the cantilever of length l clamped at the end O and loaded at the free end G (Fig. 1.35). We neglect the weight of the beam. We are interested in finding the depression at any point F (x, y) of the cantilever. The bending moment at F is

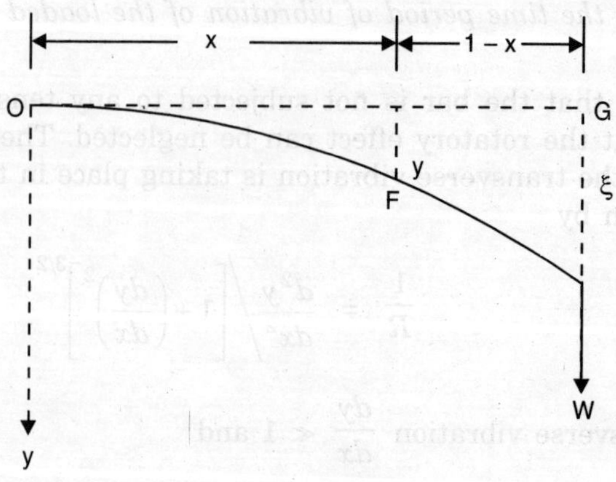

Fig. 1.35

$W(l-x)$ where $W = Mg$. This equals the resisting moment $YI_g \dfrac{d^2y}{dx^2}$. Thus the differential equation for the bending of the beam is

$$\frac{d^2y}{dx^2} = \frac{W}{YI_g}(l-x).$$

Integrating, we get
$$\frac{dy}{dx} = \frac{W}{YI_g}(lx - x^2/2) + C_1$$

where C_1 is the constant of integration.

Now, $\dfrac{dy}{dx} = 0$ when $x = 0$ so that $C_1 = 0$.

Integrating again, we get

$$y = \frac{W}{YI_g}(lx^2/2 - x^3/6) + C_2.$$

We have the boundary condition, $y = 0$ when $x = 0$ so that $C_2 = 0$. The depression of the loaded end ($x = l$) is

$$\xi = \frac{W}{YI_g}\left(\frac{l^3}{3}\right) = \frac{l^3}{3YI_g}(Mg)$$

or
$$Mg = \frac{3YI_g}{l^3}\xi.$$

The restoring force in magnitude on the mass M is $\left(\dfrac{3YI_g}{l^3}\right)\xi$ which is proportional to ξ. The restoring force is opposite to the direction of increasing ξ. The mass M executes SHM which may be written as

$$M\frac{d^2\xi}{dt^2} + \frac{3YI_g}{l^3}\xi = 0.$$

The time period of vibration of the mass is

$$T = 2\pi\left[\frac{Ml^3}{3YI_g}\right]^{1/2}.$$

57. *A rectangular light beam of breadth b, thickness d and length l is clamped at one end and loaded at free end by placing a mass M. Show that the time period of vibration of the mass is*

$$T = 2\pi\left[\frac{4Ml^3}{Ybd^3}\right]^{1/2}$$

Solution

Let us consider the cross-section of the beam (Fig. 1.36). PQ is the neutral line. We consider the strip ST of thickness dx at a distance x from the neutral axis PQ. The geometrical moment of inertia of the cross-section of the beam about the neutral axis is

$$I_g = 2\int_0^{d/2} bdx \cdot x^2 = \frac{bd^3}{12}.$$

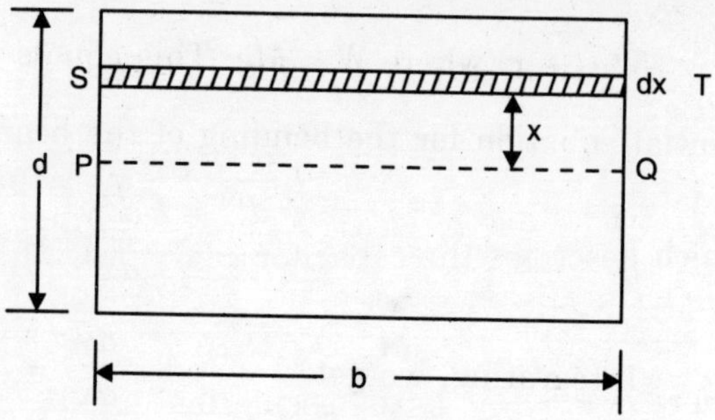

Fig. 1.36

Thus, $$T = 2\pi\left[\frac{4Ml^3}{Ybd^3}\right]^{1/2}.$$

58. *A light beam of circular cross-section of radius a and length l is clamped at one end and loaded at free end by placing a mass M. Show that the time period of vibration of this rod is*

$$T = \frac{2\pi}{a^2}\left[\frac{4Ml^3}{3\pi Y}\right]^{1/2}.$$

Solution

We consider an elementary area $dr\, rd\theta$ of the circular cross-section of the rod (Fig. 1.37). The geometrical moment of inertia of the cross-section of the rod about the neutral line PQ is

$$I_g = 2\int_0^a\int_0^\pi (r\,dr\,d\theta)\,(r\sin\theta)^2 = \frac{\pi a^4}{4}.$$

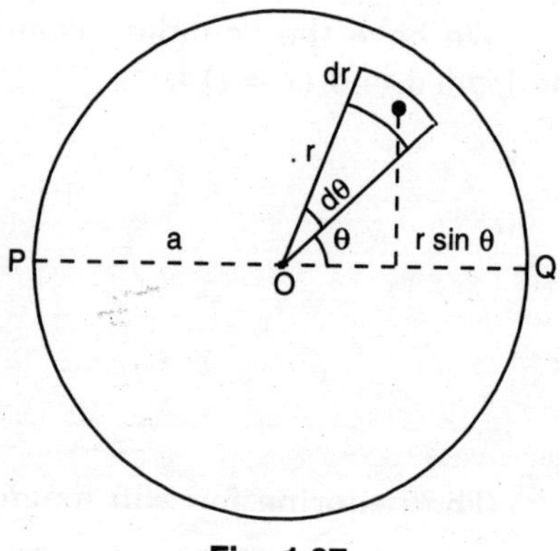

Hence, $$T = \frac{2\pi}{a^2}\left[\frac{4Ml^3}{3\pi Y}\right]^{1/2}.$$

Fig. 1.37

59. *We would like to make an LC circuit that oscillates at 440 Hz. If we have a 2 H inductor, what value of capacitance should we use? If the capacitor is initially charged to 5 V, what will be the peak charge on the capacitor? What is the total energy in the circuit ?*

Solution

The total energy in the circuit is the sum of the magnetic and electric energy:

$$E = E_B + E_E = \frac{1}{2}LI^2 + \frac{1}{2}\frac{q^2}{C}$$

where I = Current and q = Capacitor charge.

Since the total energy does not change, $\frac{dE}{dt} = 0$. Thus we have

$$LI\frac{dI}{dt} + \frac{q}{C}\frac{dq}{dt} = 0.$$

Substituting $I = \dfrac{dq}{dt}$ and $\dfrac{dI}{dt} = \dfrac{d^2q}{dt^2}$, we obtain

$$L\frac{d^2q}{dt^2} + \frac{1}{C}q = 0$$

which describes the capacitor charge as a function of time. The solution of this equation is

$$q = q_0 \cos(\omega t + \phi)$$

where $\omega = \dfrac{1}{\sqrt{LC}}$ is the oscillation angular frequency. Thus,

$$C = \frac{1}{\omega^2 L} = \frac{1}{4\pi^2\nu^2 L} = 0.065\ \mu\text{F}.$$

Peak charge, $q_0 = CV_0 = (0.065\ \mu\text{F})\ (5.0\ \text{V}) = 0.33\ \mu\text{C}.$

Now, $I = \dfrac{dq}{dt} = -\omega q_0 \sin(\omega t + \phi)$

Peak current, $I_0 = \omega q_0 = 2\pi\nu q_0 = 0.91$ m A

$$\text{Total energy} = \frac{1}{2}\frac{q_0^2}{C} = \frac{0.33 \times 0.33}{2 \times 0.065}\ \mu\ \text{J}$$
$$= 0.84\ \mu\ \text{J}.$$

60. *Two particles of mass m each are tied at the ends of a light string of length 2a. The whole system is kept on a frictionless horizontal surface with the string held tight so that each mass is at a distance a from the centre P (as shown in the Figure 1.38). Now the mid-point of the string is pulled vertically upwards with a small but constant force F. As a result, the particles move towards each other on the surface.*

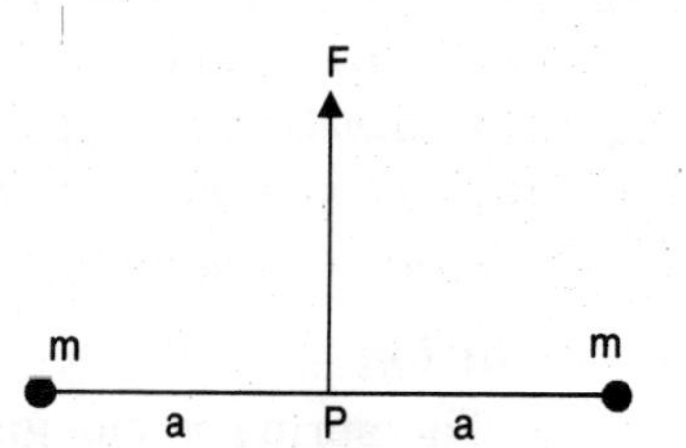

Fig. 1.38

The magnitude of acceleration, when the separation between them becomes 2x, is

(*a*) $\dfrac{F}{2m}\dfrac{a}{\sqrt{a^2 - x^2}}$

(*b*) $\dfrac{F}{2m}\dfrac{x}{\sqrt{a^2 - x^2}}$

(*c*) $\dfrac{F}{2m}\dfrac{x}{a}$

(*d*) $\dfrac{F}{2m}\dfrac{\sqrt{a^2 - x^2}}{x}$

(I.I.T. 2007)

Solution

From Fig. 1.39, we have

$$F = 2T \sin\theta$$

and $$T \cos\theta = mf$$

where f is the acceleration

Thus, $$f = \frac{T\cos\theta}{m} = \frac{F}{2\sin\theta}\cdot\frac{\cos\theta}{m}$$
$$= \frac{F}{2m}\frac{x}{\sqrt{a^2 - x^2}}$$

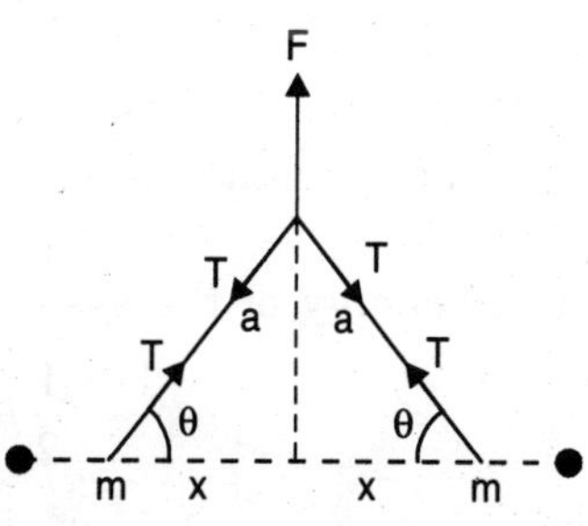

Fig. 1.39

Correct choice : (*b*)

61. *A student performs an experiment for determination of g* $\left(=\frac{4\pi^2 l}{T^2}\right)$, $l \approx 1m$, *and he commits an error of* Δl. *For T, he takes the time of n oscillations with the stop watch of least count* ΔT *and he commits a human error of 0.1 s. For which of the following data, the measurement of g will be most accurate?*

	Δl	ΔT	n	*Amplitude of oscillation*
(a)	5 mm	0.2s	10	5 mm
(b)	5 mm	0.2s	20	5 mm
(c)	5 mm	0.1s	20	1 mm
(d)	1 mm	0.1s	50	1 mm

(I.I.T. 2006)

Solution

The error in T decreases with increase in the number of oscillations (n). The amplitude should be small for SHM of the simple pendulum. In (D), we have minimum error in l (Δl = 1 mm) and T (ΔT = 0.1s).

Correct choice: (*d*).

62. (*a*) *A small body attached to one end of vertically hanging spring is performing SHM about its mean position with angular frequency* ω *and amplitude a. If at a height* y^* *from the mean position the body gets detached from the spring, calculate the value of* y^* *so that the height H attained by the mass is maximum. The body does not interact with the spring during its subsequent motion after the detachment* ($a\omega^2 > g$).

(*b*) *Find the maximum value of H.* (I.I.T. 2005)

Solution

(*a*) The spring is elongated by a distance l due to the weight mg. Thus, we have

$$kl = mg \text{ or, } l = \frac{mg}{k} = \frac{g}{\omega^2} < a$$

where k is the spring constant and $\omega^2 = (k/m)$ The amplitude of oscillation a is greater than l. Now if the mass is pulled down through a distance from the equilibrium position A (Fig. 1.40) and released from rest it executes SHM about the mean position. When the mass is moving up, suppose, it is at the position C at a distance y^* from the mean position. At this position P.E.

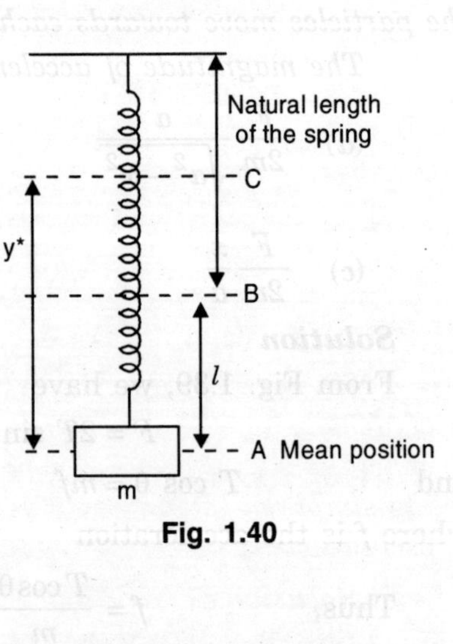

Fig. 1.40

stored in the spring $= \frac{1}{2}k\,(y^* - l)^2$

Gravitational P.E. $= mgy^*$

[Zero of Gravitational P.E. is taken at level A (mean position)]

Total energy of the system is

$$E = \frac{1}{2}k\,(y^* - l)^2 + mgy^* + \text{K.E. of the mass at } C = \text{Constant}$$

when $\quad y^* = a$, K.E. of the mass $= 0$

Thus, $$E = \frac{1}{2}k(a-l)^2 + mga$$

$$= \frac{1}{2}k(a^2+l^2).$$

Suppose, when the mass is moving up at C, it gets detached from the spring, and due to its K.E. It goes further up by a height h so that the K.E. of the mass at $C = mgh$.

Thus, we have

$$E = \frac{1}{2}k(y^*-l)^2 + mgy^* + mgh = \frac{1}{2}k(a^2+l^2)$$

or $$h = \left[\frac{1}{2}k\left(a^2+l^2\right) - \frac{1}{2}k\left(y^*-l\right)^2 - mgy^*\right]\Big/mg$$

We have to find a condition so that $y^* + h = H$ is maximum.

$$H = y^* + h = \left[\frac{1}{2}k\left(a^2+l^2\right) - \frac{1}{2}k\left(y^*-l\right)^2\right]\Big/mg$$

$$\frac{dH}{dy^*} = -\frac{k(y^*-l)}{mg} = 0$$

or $$y^* = l = \frac{mg}{k} = \frac{g}{\omega^2}$$

and $$\frac{d^2H}{dy^{*2}} = -\frac{k}{mg} = -\text{ve}$$

Thus H attains its maximum value when $y^* = l$ [at the position B]. The spring has its natural length at this position.

(*b*) The maximum value of H is

$$H_{\max} = \frac{1}{2}k\left(a^2 + \frac{m^2g^2}{k^2}\right)\Big/mg$$

$$= \frac{1}{2}\frac{ka^2}{mg} + \frac{1}{2}\frac{mg}{k}$$

$$= \frac{1}{2}\frac{\omega^2a^2}{g} + \frac{1}{2}\frac{g}{\omega^2}.$$

63. *A solid sphere of radius R is floating in a liquid of density ρ with half of its volume submerged. If the sphere is slightly pushed and released, it starts performing simple harmonic motion. Find the frequency of these oscillations.* (I.I.T. 2004)

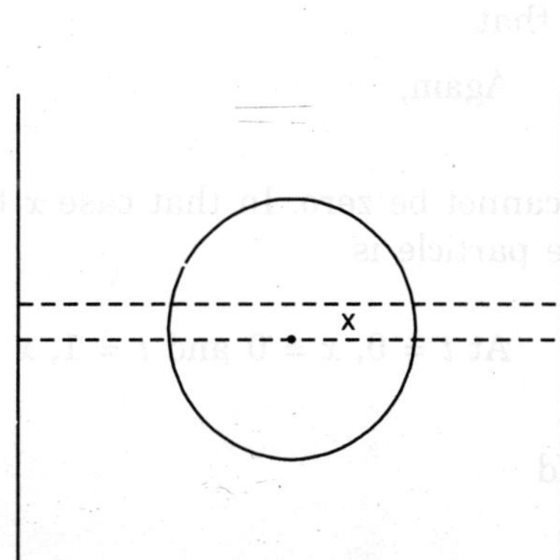

Fig. 1.41

Solution

Initially at equilibrium, mass of the solid sphere = Mass of the displaced liquid

or $$\frac{4}{3}\pi R^3\rho_1 = \frac{1}{2}\left(\frac{4}{3}\pi R^3\right)\rho$$

where ρ_1 = Density of the solid sphere

Thus, we have $2\rho_1 = \rho$.

Now, the sphere is pushed downward slightly by a distance x inside the liquid (Fig. 1.41).

Net downward force on the sphere is

$$\frac{4}{3}\pi R^3\rho_1 g - \left(\frac{1}{2}\cdot\frac{4}{3}\pi R^3 + \pi R^2 x\right)\rho g = -\pi R^2 x\rho g$$

Thus, the restoring force $= -\pi R^2\rho g x = m\ddot{x}$

or
$$\ddot{x} = -\frac{\pi^2 R^2\rho g}{m}x = -\omega^2 x.$$

The motion is simple harmonic with

$$\omega^2 = \frac{\pi R^2\rho g}{\frac{4}{3}\pi R^3\rho_1} = \frac{3}{2}\frac{g}{R}$$

The frequency of oscillation is

$$\nu = \frac{\omega}{2\pi} = \frac{1}{2\pi}\sqrt{\frac{3g}{2R}}.$$

64. *A particle of mass m moves on the x-axis as follows: it starts from rest at t = 0 from the point x = 0 and comes to rest at t = 1 at the point x = 1. No other information is available about its motion at intermediate times (0 < t < 1). If α denotes the instantaneous acceleration of the particle, then*

(*a*) *α cannot remain positive for all t in the interval (0 ≤ t ≤ 1).*

(*b*) *|α| cannot exceed 2 at any point in its path.*

(*c*) *|α| must be ≥ 4 at some point or points in its path.*

(*d*) *α must change sign during the motion, but no other assertion can be made with the information given.* (I.I.T. 1993)

Solution

We may consider a motion of the type

$$x = x_0 + A\cos\omega t$$

so that
$$\dot{x} = -A\omega\sin\omega t = 0 \text{ at time } t = 0$$

Again,
$$\dot{x} = 0 \text{ at time } t = 1$$

or
$$\sin\omega = 0 \text{ or } \omega = \pi, 2\pi, 3\pi, ...$$

ω cannot be zero. In that case x becomes independent of t. Thus the equation of motion of the particle is

$$x = x_0 + A\cos n\pi t,\ n = 1, 2, 3, ...$$

At $t = 0$, $x = 0$ and $t = 1$, $x = 1$,

$$x_0 = -A$$

and
$$1 = x_0(1 - \cos n\pi)$$

or
$$x_0 = \frac{1}{1-\cos n\pi},\ n \neq 2, 4, 6, ...$$

$$n = 1, 3, 5, ...$$

or $$x_0 = \frac{1}{2}$$

Thus, the equation of the particle satisfying all the condition is

$$x = \frac{1}{2}(1 - \cos n\pi t),\ n = 1, 3, 5$$

Acceleration $$\alpha = \ddot{x} = \frac{1}{2}(n\pi)^2 \cos n\pi t$$

$\cos n\pi t$ changes sign when t varies from 0 to 1.

Maximum value of $|\alpha|$ is $\dfrac{n^2\pi^2}{2} > 4$.

Correct choice : (*a*) and (*c*).

65. *A spring of force constant k is cut into two pieces such that one piece is n times the length of the other. Find the force constant of the long piece.*

Solution

If the spring is divided into $(n + 1)$ equal parts then each has a spring constant $(n + 1)\,k$. The long piece has n such springs which are in series. The equivalent spring constant K of the long piece is given by

$$\frac{1}{K} = \frac{1}{(n+1)k} + \frac{1}{(n+1)k} + \text{.........} n \text{ terms}$$

$$= \frac{n}{(n+1)k}$$

or $$K = \frac{(n+1)\,k}{n}.$$

66. *A particle free to move along the x-axis has potential energy given by*

$$U(x) = k\,[1 - \exp(-x^2)],\ -\alpha \le x \le \alpha$$

where k is a positive constant dimensions. Then

(*a*) *At points away from the origin, the particle is in unstable equilibrium.*

(*b*) *For any finite non-zero value of x, there is a force directed away from the origin.*

(*c*) *If its total mechanical energy is k/2, it has its minimum kinetic energy at the origin.*

(*d*) *For small displacement from x = 0, the motion is simple harmonic.* (I.I.T. 1999)

Solution

$$\text{Force} = -\frac{dU}{dx} = -\,2kx\,e^{-x^2}$$

The –ve sign indicates that the force is directed towards the origin.

For small x, Force = $-\,2kx$, the motion is simple harmonic.

For small x, $U(x) \approx k\,[1 - 1 + x^2] = kx^2$.

Minimum P.E. is at $x = 0$ and thus the maximum K.E. is at $x = 0$.

Far away from the origin $U(x) \approx k$ and the force = $-\dfrac{dU}{dx} = 0$ [stable equilibrium]

Correct choice : (*d*).

67. *A particle of mass m is executing oscillations about the origin on the x-axis. Its potential energy $U(x) = kx^3$ where k is a positive constant. If amplitude of oscillation is a, then its time period T is*

(a) Proportional to $\frac{1}{\sqrt{a}}$ *(b) Independent of a*

(c) Proportional to $\sqrt{a}$ *(d) Proportional to* $a^{3/2}$. (I.I.T. 1998)

Solution

For $x > 0$

Total energy $= E = \frac{1}{2}mv^2 + kx^3 = ka^3$ from conservation of energy.

Thus, $$v = \pm\left(\frac{2k}{m}\right)^{1/2}\sqrt{a^3 - x^3} = \frac{dx}{dt}$$

or $$dt = \left(\frac{m}{2k}\right)^{1/2}\frac{dx}{\sqrt{a^3 - x^3}}.$$

We consider +ve velocity.

Integrating from $x = 0$ to $x = a$, we have

$$\int_0^{T/4} dt = \left(\frac{m}{2k}\right)^{1/2}\int_0^a \frac{dx}{\sqrt{a^3 - x^3}}.$$

We put $x = a \sin\theta$ so that $dx = a\cos\theta\, d\theta$

Thus, $$\frac{T}{4} = \left(\frac{m}{2k}\right)^{1/2}\int_0^{\pi/2}\frac{a\cos\theta\, d\theta}{\sqrt{a^3 - a^3\sin^3\theta}}$$

$$= \left(\frac{m}{2k}\right)^{1/2}\frac{a}{a^{3/2}}\int_0^{\pi/2}\frac{\cos\theta\, d\theta}{\sqrt{1-\sin^3\theta}}.$$

The integral is a constant

$$T = \text{Const.}\ \frac{1}{\sqrt{a}}$$

or $$T \propto \frac{1}{\sqrt{a}}$$

Correct choice : (*a*).

68. *Two blocks A and B each of mass m are connected by a massless spring of natural length L and spring constant K. The blocks are initially resting on a smooth horizontal floor with the spring at its natural length as shown in Fig. 1.42.*

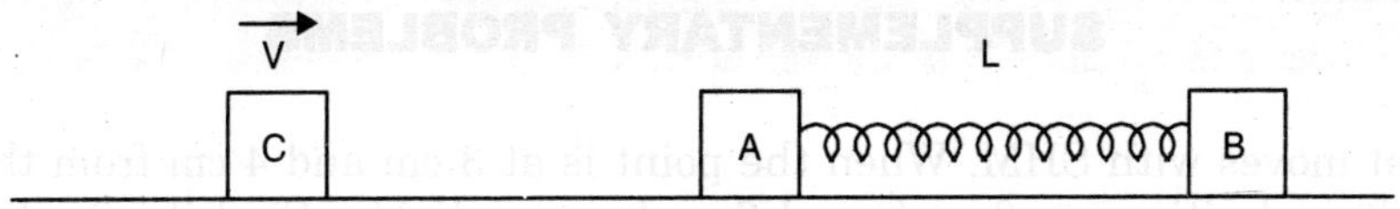

Fig. 1.42

A third identical block C also of mass m, moves on the floor with a speed V along the line joining A and B and collides with A. Then

(a) the kinetic energy of the A–B system at maximum compression of the spring, is zero,

(b) the kinetic energy of the A–B system at maximum compression of the spring is $mV^2/4$.

(c) the maximum compression of the spring is $V\sqrt{m/K}$.

(d) the maximum compression of the spring is $V\sqrt{m/2K}$. (I.I.T. 1993)

Solution

The block C will come to rest after colliding with the block A and its energy will be partly converted to the K.E. of the A-B system and the remaining energy goes into the internal energy of the A-B system.

Suppose, V' = Velocity of the A-B system after the collision. From the principle of conservation of momentum, we get

$$mV = 2mV' \quad \text{or,} \quad V' = \frac{V}{2}$$

At the maximum compression of the spring the internal energy is the potential energy of the spring. The A-B system moves with velocity V' after the collision. Thus, the kinetic energy of the A-B system is

$$\frac{1}{2}\,(2m)V'^2 = m\left(\frac{V}{2}\right)^2 = \frac{mV^2}{4}$$

The P.E. of the A-B system is

$$\text{P.E.} = \frac{1}{2}mV^2 - \frac{mV^2}{4} = \frac{mV^2}{4}$$

If x is the maximum compression of the spring, then

$$\text{P.E.} = \frac{1}{2}Kx^2 = \frac{mV^2}{4}$$

or

$$x = V\sqrt{\frac{m}{2K}}$$

Correct Choice : (*b*) and (*d*).

SUPPLEMENTARY PROBLEMS

1. A point moves with SHM. When the point is at 3 cm and 4 cm from the centre of its path, its velocities are 8 cm/s and 6 cm/s respectively. Find its amplitude and time period. Find its acceleration when it is at the greatest distance from the centre.

2. A particle is moving with SHM in a straight line. When the distances of the particle from the equilibrium position are x_1 and x_2, the corresponding values of the velocity are u_1 and u_2. Show that the period is

$$T = 2\pi\left[\left(x_2^2 - x_1^2\right) / \left(u_1^2 - u_2^2\right)\right]^{\frac{1}{2}}.$$

3. A particle of mass 0.005 kg is vibrating 15 times per second with an amplitude of 0.08 m. Find the maximum velocity and its total energy.

4. A particle moves with SHM. If its acceleration at a distance d from the mean position is a, show that the time period of motion is $2\pi\sqrt{d/a}$.

5. At the moment $t = 0$ a body starts oscillating along the x-axis according to the law

$$x = A \sin \omega t.$$

Find (*a*) the mean value of its velocity $<v>$ and (*b*) the mean value of the modulus of the velocity $<|v|>$ averaged over 3/8 of the period after the start.

6. Plot (*dP*/*dx*) of problem 6 (page 9) as a function of x. Find the probability of finding the particle within the interval from – (*A*/2) to + (*A*/2).

7. A particle is executing SHM. Show that, average K.E. over a cycle = average P.E. over a cycle = Half of the total energy.

8. A particle moves with simple harmonic motion in a straight line. Its maximum speed is 4 m/s and its maximum acceleration is 16 m/s^2. Find (*a*) the time period of the motion, (*b*) the amplitude of the motion.

9. A loudspeaker produces a musical sound by the oscillation of a diaphragm. If the amplitude of oscillation is limited to 9.8×10^{-4} mm, what frequency will result in the acceleration of the diaphragm exceeding g?

10. A small body is undergoing SHM of amplitude A. While going to the right from the equilibrium position, it takes 0.5 s to move from $x = +(A/2)$ to $x = +A$. Find the period of the motion.

11. A block is on a piston that is moving vertically with SHM. (*a*) At what amplitude of motion will the block and piston separate if time period = 1 s? (*b*) If the piston has an amplitude of 4.0 cm, what is the maximum frequency for which the block and piston will be in contact continuously?

12. The piston in the cylindrical head of a locomotive has a stroke of 0.8 m. What is the maximum speed of the piston if the drive wheels make 180 rev/min and the piston moves with simple harmonic motion?

$$\left[\textbf{Hints: } \nu = \frac{180}{60} = 3 \text{ Hz and } A = 0.4 \text{ m}\right].$$

13. A 40 g mass hangs at the end of a spring. When 25 g more is added to the end of the spring, it stretches 7.0 cm more. (*a*) Find the spring constant and (*b*) if 25 g is now removed, what will be the time period of the motion?

14. Two bodies M and N of equal masses are suspended from two separate massless springs of spring constants k_1 and k_2 respectively. If the two bodies oscillate vertically such that their maximum velocities are equal, the ratio of the amplitude of M to that of N is

(*a*) $\frac{k_1}{k_2}$ (*b*) $\sqrt{\left(\frac{k_1}{k_2}\right)}$ (*c*) $\frac{k_2}{k_1}$ (*d*) $\sqrt{\left(\frac{k_2}{k_1}\right)}$ (*I.I.T. 1988*)

15. A block whose mass is 700g is fastened to a spring whose spring constant k is 63 N/m. The block is pulled a distance 10 cm from its equilibrium position and released from rest. (*a*) Find the time period of oscillation of the block, (*b*) what is the mechanical energy of the oscillator? (*c*) What are the potential energy and kinetic energy of this oscillator when the particle is halfway to its end point? [Neglect gravitational P.E.]

16. A cubical block vibrates horizontally in SHM with an amplitude of 4.9 cm and a frequency of 2 Hz. If a smaller block sitting on it is not to slide, what is the minimum value that the coefficient of static friction between the two blocks can have?

[**Hints:** Maximum force on the smaller body = $m\omega^2 A = \mu mg$]

17. The vibration frequencies of atoms in solids at normal temperatures are of the order of 10^{13} Hz. Imagine the atoms to be connected to one another by "springs". Suppose that a single silver atom vibrates with this frequency and that all the other atoms are at rest. Compute the effective spring constant. One mole of silver has a mass of 108 g and contains 6.023×10^{23} atoms. [**Hints:** $k = \omega^2 m = 4\pi^2\nu^2 m$]

18. Suppose that in Fig. 1.5 the 100 g mass initially moves to the left at a speed of 10 m/s. It strikes the spring and becomes attached to it. (*a*) How far does it compress the spring? (*b*) If the system then oscillates back and forth, what is the amplitude of the oscillation?

$$\left[\textbf{Hints: } \frac{1}{2}(0.1\text{ kg})(10\text{ m/s})^2 = \frac{1}{2}\times(500\text{ N/m})x_0^2\right]$$

19. Suppose that in Fig. 1.5 the 100 g mass compresses the spring 10 cm and is then released. After sliding 50 cm along the flat table from the point of release the mass comes to rest. How large a friction force opposes the motion?

20. A mass of 200 g placed at the lower end of a vertical spring stretches it 20 cm. When it is in equilibrium the mass is hit upward and due to this it goes up a distance of 8 cm before coming down. Find (*a*) the magnitude of the velocity imparted to the mass when it is hit, (*b*) the period of motion.

21. With a 100 g mass at its end a spring executes SHM with a frequency of 1 Hz. How much work is done in stretching the spring 10 cm from its unstretched length?

22. A popgun uses a spring for which k = 30 N/cm. When cocked the spring is compressed 2 cm. How high can the gun shoot a 4 g projectile?

23. A block of mass M, at rest on a horizontal frictionless table, is attached to a rigid support by a spring of spring constant k. A bullet of mass m and velocity v strikes the block as shown in Fig. 1.38. The [illegible] remains embedded in the block. Determine (*a*) the velocity of the block immediately after the collisions and (*b*) the amplitude of the resulting simple harmonic motion.

$$\left[\textbf{Hints: } mv = (M+m)V\,;\ \frac{1}{2}(M+m)V^2 = \frac{1}{2}kA^2\right]$$

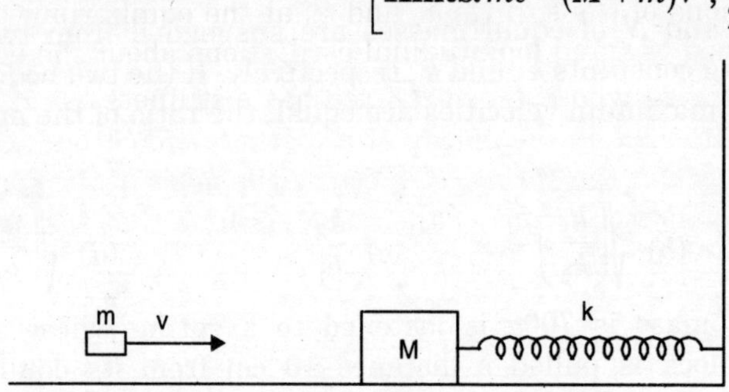

Fig. 1.43

24. A 500 g mass at the end of a Hookean spring vibrates up and down in such a way that it is 2 cm above the table top at its lowest point and 12 cm above the table top at its highest point. Its period is 5s. Find (*a*) the spring constant, (*b*) the amplitude of vibration, (*c*) the speed and acceleration of the mass when it is 10 cm above the table top.

25. A thin metallic wire of length L and area of cross-section A is suspended from free end which stretches it through a distance l. Show that the vertical oscillations of the system are simple harmonic in nature and its time period is given by

$$T = 2\pi\sqrt{l/g} = 2\pi\sqrt{\frac{mL}{(AY)}}$$

where Y is the Young's modulus of the material of the wire.

26. There are two spring systems (*a*) and (*b*) of Fig. 1.10 with k_1 = 5 kN/m and k_2 = 10 kN/m. A 100 kg block is suspended from each system. If the block is constrained to move in the vertical direction only, and is displaced 0.01 m down from its equilibrium position, determine for each spring system: (1) The equivalent single spring constant, (2) Time period of vibration, (3) The maximum velocity of the block, and (4) The maximum acceleration of the block.

27. A 10 kg electric motor is mounted on four vertical springs, each having a spring constant of 20 N/cm. Find the frequency with which the motor vibrates vertically.

28. A spring of force constant k is cut into three equal parts, the force constant of each part will be ..... . *(I.I.T. 1978)*

29. A horizontal spring system of mass M executes SHM. When the block is passing through its equilibrium position, an object of mass m is put on it and the two move together. Show that the new frequency and the new amplitude in terms of old frequency and old amplitude are given by

$$\omega' = \omega\sqrt{\frac{M}{(M+m)}},\ A' = A\sqrt{\frac{M}{(M+m)}}.$$

30. Find the period of small oscillations in the vertical plane performed by a ball of mass m = 50 g fixed at the middle of a horizontally stretched string l = 1.0 m in length. The tension of the string is assumed to the constant and equal to T = 10 N.

31. A body of mass m on a horizontal frictionless plane is attached to two horizontal springs of spring constants k_1 and k_2 and equal relaxed lengths L. Now the free ends of the springs are pulled apart and fastened to two fixed walls a distance $3L$ apart. Find the elongations of the springs k_1 and k_2 at the equilibrium position of the body and the time period of small longitudinal oscillations about the equilibrium position.

32. A non-deformed spring whose ends are fixed has a stiffness $k = 12$ N/m. A small body of mass 12 g is attached on the spring at a distance 1/3 l from one end of the spring where l is the length of the spring. Neglecting the mass of the spring find the period of small longitudinal oscillations of the body. Assume that the gravitational force is absent.

[**Hints:** The spring of length $\frac{1}{3}l$ has stiffness $k_1 = \frac{lk}{\frac{1}{3}l} = 3k$ and the spring of length $\frac{2}{3}l$ has stiffness $k_2 = \frac{lk}{\frac{2}{3}l} = \frac{3}{2}k.$]

33. A uniform spring whose unstretched length is L has a force constant k. The spring is cut into two pieces of unstretched lengths L_1 and L_2, with $L_1 = nL_2$. What are the corresponding force constants k_1 and k_2 in terms of n and k?

34. Two bodies of masses m_1 and m_2 are interconnected by a weightless spring of stiffness k and placed on a smooth horizontal surface. The bodies are drawn closer to each other and released simultaneously. Show that the natural oscillation frequency of the system is

$$\omega = \sqrt{k/\mu} \qquad \text{where } \mu = \frac{m_1 m_2}{m_1 + m_2}.$$

35. A particle executes SHM with an amplitude A. At what displacement will the K.E. be equal to twice the P.E.?

36. A body of mass 0.1 kg is connected to three identical springs of spring constant $k = 1$ N/m and in their relaxed state the springs are fixed to three corners of an equilateral triangle ABC (Fig. 1.44). Relaxed length of each spring is 1m. The mass m is displaced from the initial position O to the point D, the mid-point of BC and then released from rest. What will be the kinetic energy of m if it returns to the point O? What will be the speed of the body at O?

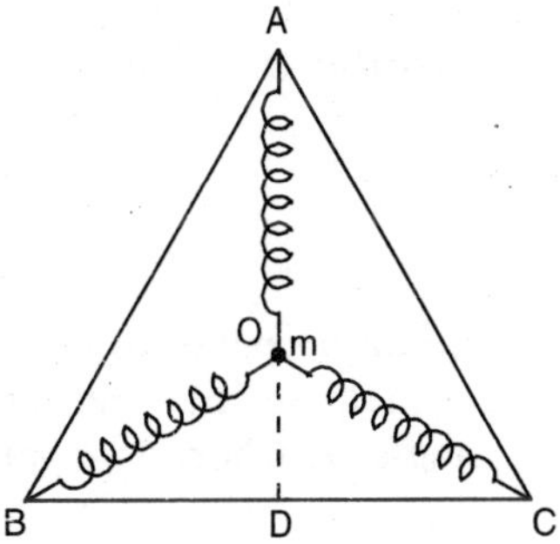

Fig. 1.44

37. Find the length of a second pendulum ($T = 2$ s) at a place where $g = 9.8$ m/s^2.

38. Compare the period of the simple pendulum at the surface of the earth to that at the surface of the moon.

39. The time periods of a simple pendulum on the earth's surface and at a height h from the earth's surface are T and T' respectively. Show that the radius (R) of the earth is given by

$$R = \frac{Th}{T' - T}.$$

40. A simple pendulum of length L and mass (bob) M is oscillating in a plane about a vertical line between angular limits $-\phi$ and $+\phi$. For an angular displacement θ ($|\theta| < \phi$) the tension in the string and the velocity of the bob are T and v respectively. The following relations hold under the above conditions [Tick the correct relations] :

(*a*) $T \cos \theta = Mg$.

(*b*) $T - Mg \cos \theta = Mv^2/L$

(*c*) The magnitude of the tangential acceleration of the bob $|a_T| = g \sin \theta$.

(*d*) $T = Mg \cos \theta$. (*I.I.T. 1986*)

41. A simple pendulum of length l and mass m is suspended in a car that is travelling with a constant speed v around a circular path of radius R. If the pendulum executes small oscillations about the equilibrium position, what will be its time period of oscillation?

42. A simple pendulum of length l and having a bob of mass m and density ρ is completely immersed in a liquid of density σ ($\rho > \sigma$). Find the time period of small oscillation of the bob in the liquid.

43. Solve problem 27 (Fig. 1.18) by summing the torques about the point O.

44. The mass and diameter of a planet are twice those of the earth. What will be the period of oscillation of a pendulum on this planet if it is a second's pendulum on the earth? (*I.I.T. 1973*)

45. One end of a long metallic wire of length L is tied to the ceiling. The other end is tied to a massless spring of spring constant k. A mass m hangs freely from the free end of the spring. The area of cross-section and the Young's modulus of the wire are A and Y respectively. If the mass is slightly pulled down and released show that it will oscillate with a time period T equal to

$$2\pi\sqrt{\frac{m(YA+kL)}{(YAk)}}.$$ (*I.I.T. 1993*)

[**Hints:** If x_1 and x_2 are elongations of metallic wire and spring due to force F, then

$$F = -AYx_1/L = -kx_2$$

and $$x = x_1 + x_2 = -F\left(\frac{L}{AY}+\frac{1}{k}\right).$$

46. A simple pendulum of mass M is suspended by a thread of length l when a bullet of mass m hits the bob horizontally and sticks in it. The system is deflected by an angle α, where $\alpha < 90°$. Show that the speed of the bullet is

$$\frac{2(M+m)}{m}\sin\left(\frac{\alpha}{2}\right)\sqrt{gl}.$$

47. A cylinder having axis vertical floats in a liquid of density ρ. It is pushed down slightly and released. Find the period of oscillations if the cylinder has weight W and cross-sectional area A.

48. A vertical U-tube of uniform cross-section contains a liquid of total mass M. The mass of the liquid per unit length is m. When disturbed the liquid oscillates back and forth from arm to arm. Calculate the time period if the liquid on one side is depressed and then released. Compute the effective spring constant of the motion.

49. Two identical positive point charge + Q each, are fixed at a distance of $2a$ apart. A point negative charge $(-q)$ of mass m lies midway between the fixed charges. Show that for a small displacement perpendicular to the line joining the fixed charges, the charge $(-q)$ executes SHM and the frequency of oscillations is

$$\frac{1}{2\pi}\sqrt{\frac{Qq}{2\pi\epsilon_0 a^3 m}}$$

50. A thin fixed ring of radius 1 m has a positive charge of 1×10^{-5}C uniformly distributed over it. A particle of mass 0.9 g and having a negative charge 1×10^{-6}C is placed on the axis at a distance of 1 cm from the centre of the ring. Show that the motion of the negative charged particle is approximately simple harmonic. Calculate the time period of oscillations. *(I.I.T. 1982)*

51. A simple pendulum consists of a small sphere of mass m suspended by a thread of length l. The sphere carries a positive charge q. The pendulum is placed in a uniform electric field of strength E directed vertically upwards. With what period the pendulum oscillates if the electrostatic force acting on the sphere is less than the gravitational force?

Assume that the oscillations are small) *(I.I.T. 1977)*

[**Hints:** Net downward force acting on the pendulum is $ma = mg - Eq$]

52. A 2.0 g particle at the end of a spring moves according to the equation

$$y = 0.1 \sin 2\pi t \text{ cm}$$

where t is in seconds. Find the spring constant and the position of the particle at time $t = \frac{1}{\pi}$ s.

53. A particle moves according to the equation

$$y = \frac{1}{\sqrt{2}}\sin 10\sqrt{2}\,t + \frac{1}{10}\cos 10\sqrt{2}\,t.$$

Find the amplitude of the motion.

54. A particle vibrates about the origin of the coordinates along the y-axis with a frequency of 15 Hz and an amplitude of 3.0 cm. The particle is at the origin at time $t = 0$. Find its equation of motion.

55. A particle of mass m moves along the x-axis, attracted toward the origin O by a force proportional to the distance from O. Initially the particle is at distance x_0 from O and is given a velocity of magnitude v_0 (a) away from O (b) toward O. Find the position at any time, the amplitude and maximum speed in each case.

56. An object of mass 2 kg moves with SHM on the x-axis. Initially $(t = 0)$ it is located at a distance 2 m away from the origin $x = 0$, and has velocity 4 m/s and acceleration 8 m/s^2 directed toward $x = 0$. Find (a) the position at any time (b) the amplitude and period of oscillations, (c) the force on the object when $t = \pi/8$ s.

57. A point particle of mass 0.1 kg is executing SHM of amplitude 0.1 m. When the particle passes through the mean position, its kinetic energy is 8×10^{-3} J. Obtain the equation of motion of the particle if the initial phase of oscillation is 45°. *(Roorkee 1991)*

58. Retaining terms up to k^2 in problem 49 (page 35) show that the time period of the pendulum is given approximately by

$$T = 2\pi\sqrt{\frac{l}{g}}\left(1+\frac{\psi_0^2}{16}\right)$$

where ψ_0 is the maximum angle made by the string with the vertical.

59. The potential energy of a particle of mass m is given by

$$V(x) = (1 - ax)\exp(-ax), \quad x \ge 0$$

where a is a positive constant. Find the location of the equilibrium point(s), the nature of the equilibrium, and the period of small oscillations that the particle performs about the equilibrium position.

60. An engineer wants to find the moment of inertia of an odd-shaped object about an axis passing through its centre of mass. The object is supported with a wire through its centre of mass along the desired axis. The wire has a torsional constant $C = 0.50$ Nm. The engineer observes that this torsional pendulum oscillates through 20 complete cycles in 50s. What value of moment of inertia is obtained?

61. A 90 kg solid sphere with a 10 cm radius is suspended by a vertical wire attached to the ceiling of a room. A torque of 0.20 Nm is required to twist the sphere through an angle of 0.85 rad. What is the period of oscillation when the sphere is released from this position?

62. Compare the time periods of vibrations of two loaded light cantilevers made of the same material and having the same length and weight at the free end with the only difference that while one has a circular cross-section of radius a, the other has a square cross-section, each side of which is equal to a.

63. A long horizontal wire AB, which is free to move in a vertical plane and carries a steady current of 20 A, is in equilibrium at a height of 0.01 m over another parallel long wire CD, which is fixed in a horizontal plane and carries a steady current of 30 A, as shown in Fig. 1.45. Show that when AB is slightly depressed, it executes simple harmonic motion. Find the period of oscillations. *(I.I.T. 1994)*

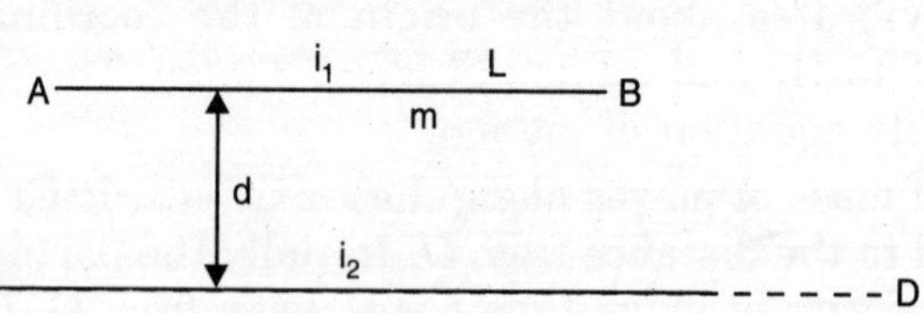

Fig. 1.45

[**Hints:** $\dfrac{\mu_0 i_1 i_2 L}{2\pi d} = mg$. If d is changed to $d - x$, then the restoring force is

$$F = -\frac{\mu_0 i_1 i_2 L}{2\pi(d-x)} + mg \approx -\frac{\mu_0 i_1 i_2 L}{2\pi d^2}x = -\frac{mgx}{d}$$]

64. You have a 2.0 mH inductor and wish to make an LC circuit whose resonant frequency can be tuned across the AM radio band (550 kHz to 1600 kHz). What range of capacitance should your variable capacitor cover?

65. An object of mass 0.2 kg executes simple harmonic oscillations along the x-axis with a frequency of $(25/\pi)$ Hz. At the position $x = 0.04$ m, the object has kinetic energy 0.5 J and potential energy 0.4 J. Find the amplitude of oscillations. (*I.I.T. 1994*)

$$\left[\textbf{Hints: } \frac{\text{K.E.}}{\text{P.E.}} = \frac{(A^2 - x^2)}{x^2}\right]$$

66. T_1 is the time period of a simple pendulum. The point of suspension moves vertically upwards according to $y = kt^2$ where $k = 1$ m/s^2. Now the time period is T_2. Then $\frac{T_1^2}{T_2^2}$ is ($g = 10$ m/s^2)

(*a*) $\frac{4}{5}$ (*b*) $\frac{6}{5}$

(*c*) $\frac{5}{6}$ (*d*) 1 (*I.I.T. 2005*)

[**Hints :** Upward acceleration of the point of suspension is $a = 2k = 2$ m/s^2 and in this case the effective g is (10 + 2) m/s^2]

67. A simple pendulum has a time period T_1 when on the earth's surface and T_2 when taken to a height R above the earth's surface where R is the radius of the earth. Show that the value of $(T_2/T_1)\frac{T_2}{T_1}$ is 2.

$$\left[\textbf{Hints: } mg = G\frac{Mm}{r^2}, T_1 = 2\pi\sqrt{l\Big/\frac{GM}{R^2}}, T_2 = 2\pi\sqrt{l\Big/\frac{GM}{4R^2}}\right]$$

68. A particle executes simple harmonic motion between $x = -A$ to $x = +A$. The time taken for it to go from 0 to $A/2$ is T_1 and to go from $A/2$ to A is T_2. Show that $T_2/T_1 = 2$.

$$\left[\textbf{Hints: } x = A\sin\omega t,\ \omega T_1 = \frac{\pi}{6},\ \omega(T_1 + T_2) = \frac{\pi}{2}\right]$$

2 Superposition Principle and Coupled Oscillations

2.1 DEGREES OF FREEDOM

Number of independent coordinates required to specify the configuration of a system completely is known as degrees of freedom.

2.2 SUPERPOSITION PRINCIPLE

For a linear homogeneous differential equation, the sum of any two solutions is itself a solution.

Consider a linear homogeneous differential equation of degree n:

$$a_n \frac{d^n y}{dt^n} + a_{n-1} \frac{d^{n-1} y}{dt^{n-1}} + ... + a_1 \frac{dy}{dt} + a_0 y = 0.$$

If y_1 and y_2 are two solutions of this equation then $y_1 + y_2$ is also a solution, which can be proved by direct substitution.

2.3 SUPERPOSITION PRINCIPLE FOR LINEAR INHOMOGENEOUS EQUATION

Consider a driven harmonic oscillator

$$m \frac{d^2 x}{dt^2} = -kx + F(t)$$

where $F(t)$ is the external force which is independent of x. Suppose that a driving force $F_1(t)$ produces an oscillation $x_1\ (t)$ and another driving force $F_2(t)$ produces an oscillation $x_2\ (t)$ [when $F_2(t)$ is the only driving force]. When the total driving force is $F_1(t) + F_2(t)$, the corresponding oscillation is given by $x(t) = x_1(t) + x_2(t)$.

2.4 SUPERPOSITION OF SIMPLE HARMONIC MOTIONS ALONG A STRAIGHT LINE

If a number of simple harmonic motions along the x-axis

$$x_i = a_i \sin (\omega_i t + \phi_i),\ i = 1, 2, .., N \qquad ...(2.1)$$

are superimposed on a particle simultaneously, the resultant motion is given by

$$X = \sum_i x_i = \sum_i a_i \sin(\omega_i t + \phi_i). \qquad ...(2.2)$$

2.5 SUPERPOSITION OF TWO SIMPLE HARMONIC MOTIONS AT RIGHT ANGLES TO EACH OTHER

If two simple harmonic motions

$$x = a \sin \omega_1 t, \qquad ...(2.3)$$

$$y = b \sin(\omega_2 t + \phi) \qquad ...(2.4)$$

act on a particle simultaneously perpendicular to each other the particle describes a path known as Lissajous figure when ω_1 and ω_2 are in simple ratio. The equation of the path is obtained by eliminating t from these two equations. The position of the particle in the xy plane is given by

$$\vec{r} = x\hat{i} + y\hat{j} \qquad ...(2.5)$$

SOLVED PROBLEMS

1. *Two simple harmonic motions of same angular frequency* ω

$$x_1 = a_1 \sin \omega t,$$

$$x_2 = a_2 \sin (\omega t + \phi)$$

act on a particle along the x-axis simultaneously. Find the resultant motion.

Solution

The resultant displacement is

$$X = x_1 + x_2 = \sin \omega t\, [a_1 + a_2 \cos \phi] + \cos \omega t\, [a_2 \sin \phi].$$

We put

$$R \cos \theta = a_1 + a_2 \cos \phi, \qquad ...(2.6)$$

$$R \sin\theta = a_2 \sin \phi \qquad ...(2.7)$$

so that

$$R^2 = a_1^2 + a_2^2 + 2a_1a_2 \cos \phi \qquad ...(2.8)$$

and

$$\tan \theta = \frac{a_2 \sin \phi}{a_1 + a_2 \cos \phi} \qquad ...(2.9)$$

The resultant displacement is

$$X = R \sin(\omega t + \theta) \qquad ...(2.10)$$

which is also simple harmonic along the x-axis with the same angular frequency ω. The amplitude R and the phase angle θ of the resultant motion are given by Eqns. (2.8) and (2.9) respectively.

Special Cases

(*i*) $\phi = \pm 2n\pi$, $n = 0, 1, 2, ...$ or, the two SHMs x_1 and x_2 are in phase,

$$R = a_1 + a_2$$

(*ii*) $\phi = \pm (2n + 1)\pi$, $n = 0, 1, 2, ...$or, the two SHMs x_1 and x_2 are in opposite phase,

$$R = a_1 \sim a_2.$$

In this case, the resultant amplitude is zero when $a_1 = a_2$ and one motion is destroyed by the other.

2. *Find the resultant motion due to superposition of a large number of simple harmonic motions of same amplitude and same frequency along the x-axis but differing progressively in phase.*

Solution

The simple harmonic motions are given by

$$x_1 = a \sin \omega t,$$
$$x_2 = a \sin(\omega t + \phi),$$
$$x_3 = a \sin(\omega t + 2\phi),$$
$$\vdots \qquad \vdots$$
$$x_N = a \sin[\omega t + (N - 1)\phi].$$

The resultant displacement is

$$X = \sum_i x_i = a \sin \omega t \; [1 + \cos \phi + \cos 2\phi + ... + \sin (N - 1) \phi],$$
$$+ a \cos \omega t \; [0 + \sin \phi + \sin 2\phi + ... + \sin (N - 1) \phi],$$
$$= R \sin (\omega t + \theta) \qquad ...(2.11)$$

where $R \cos \theta = a \; [1 + \cos \phi + \cos 2\phi + ... + \cos (N - 1) \phi],$

$R \sin \theta = a \; [0 + \sin \phi + \sin 2\phi + ... + \sin (N - 1) \phi]$

Now, $$e^{i\phi} + e^{2i\phi} + ... + e^{i(N-1)\phi} = \frac{e^{i\phi}(e^{(N-1)\phi} - 1}{e^{i\phi} - 1}$$

$$= e^{\frac{iN\phi}{2}} \frac{\sin (N-1)\phi/2}{\sin \phi/2}$$

Equating the real and imaginary parts, we get

$$\cos \phi + \cos 2\phi + ... + \cos (N - 1) \phi = \frac{\cos N \phi/2 \sin (N-1)\phi/2}{\sin \phi/2}$$

$$\sin \phi + \sin 2\phi + ... + \sin (N - 1)\phi = \frac{\sin N \phi/2 \sin(N-1) \phi/2}{\sin \phi/2}$$

Thus, we write

$$1 + \cos \phi + \cos 2\phi + ... + \cos (N - 1) \phi = 1 + \frac{\cos N\phi/2 \sin(N-1)\phi/2}{\sin \phi/2}$$

$$= \frac{\sin\{N - (N-1)\}\phi/2 \cos N\phi/2 \sin(N-1)\phi/2}{\sin \phi/2}$$

$$= \frac{\sin N\phi/2 \cos (N-1)\phi/2}{\sin \phi/2}$$

The resultant motion of Eqn. (2.11) is simple harmonic with amplitude and phase angle given by

$$R = a\frac{\sin(N\phi/2)}{\sin(\phi/2)} \qquad ...(2.12)$$

$$\theta = (N-1)\phi/2 \qquad ...(2.13)$$

When N is large and ϕ is small, we may write

$$\theta \approx N\phi/2, \qquad ...(2.14)$$

$$R \approx Na\frac{\sin\theta}{\theta} \qquad ...(2.15)$$

and the phase difference between the first component vibration x_1 and Nth component vibration x_N is nearly equal to 2θ.

The resultant amplitude may be obtained by the vector polygon method (Fig. 2.1). The polygon OABCD is drawn with each side of length a and making an angle ϕ with the neighbouring side. The resultant has the amplitude OD with the phase angle = $\angle DOA$ with respect to the first vibration.

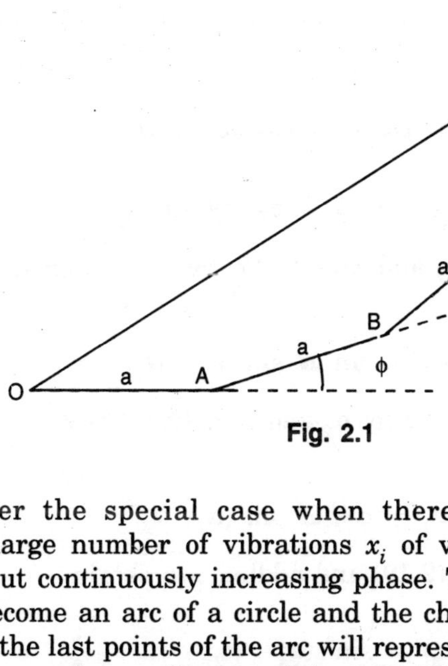

Fig. 2.1

Special Cases

(*i*) We consider the special case when there is superposition of a large number of vibrations x_i of very small amplitude a but continuously increasing phase. The polygon will then become an arc of a circle and the chord joining the first and the last points of the arc will represent the amplitude of the resultant vibration (Fig. 2.2). When the last component vibration is at A, the first and the last component vibration are in opposite phase and the amplitude of the resultant vibration = OA = diameter of the circle. When the last component vibration is at B, the first and the last component vibrations are in phase, the polygon becomes a complete circle and the amplitude of the resultant vibration is zero.

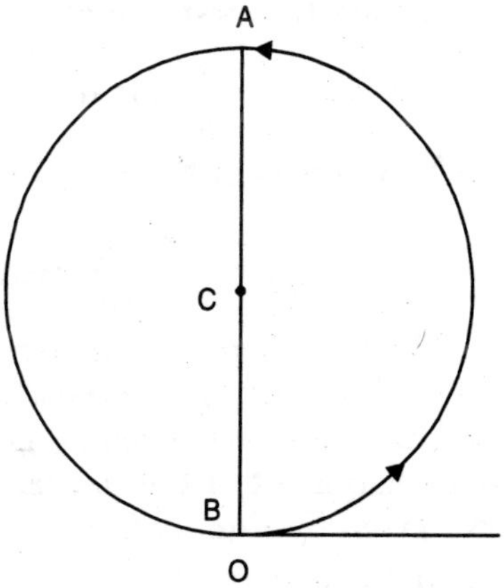

Fig. 2.2

(*ii*) When the successive amplitudes of a large number of component vibrations decrease slowly and the phase angles increase continuously the polygen becomes a spiral converging asymptotically to the centre of the first semicircle.

3. *The displacement y of a particle executing periodic motion is given by*

$$y = 4\cos^2\left(\frac{1}{2}t\right)\sin(1000\,t).$$

show that this expression may be considered to be a result of the superposition of three independent harmonic motions. (I.I.T. 1992)

Solution

$$y = 4\cos^2\left(\frac{1}{2}t\right)\sin(1000t)$$
$$= 2\,[\cos t + 1]\,\sin(1000t)$$
$$= [\sin(1000+1)t + \sin(1000-1)t] + 2\sin 1000t$$
$$= \sin 1001t + \sin 999t + 2\sin 1000t.$$

4. *Two simple harmonic motions of same frequency ω but having displacements in two perpendicular directions act simultaneously on a particle:*

$$x = a\sin(\omega t + \alpha_1)$$

and

$$y = b\sin(\omega t + \alpha_2).$$

Find the resultant motion for various values of the phase difference $\delta = \alpha_1 - \alpha_2$.

Solution

We have

$$\frac{x}{a} = \sin\omega t\cos\alpha_1 + \cos\omega t\sin\alpha_1, \quad ...(2.16)$$

$$\frac{y}{b} = \sin\omega t\cos\alpha_2 + \cos\omega t\sin\alpha_2 \quad ...(2.17)$$

Multiplying Eqn. (2.16) by $\sin\alpha_2$ and Eqn. (2.17) by $\sin\alpha_1$ and subtracting the second from the first, we get

$$\frac{x}{a}\sin\alpha_2 - \frac{y}{b}\sin\alpha_1 = \sin\omega t\sin(\alpha_2 - \alpha_1) \quad ...(2.18)$$

Similarly multiplying Eqn. (2.16) by $\cos\alpha_2$ and Eqn. (2.17) by $\cos\alpha_1$ and subtracting the second from the first, we obtain

$$\frac{x}{a}\cos\alpha_2 - \frac{y}{b}\cos\alpha_1 = \cos\omega t\sin(\alpha_1 - \alpha_2) \quad ...(2.19)$$

Now squaring Eqns. (2.18) and (2.19) and adding, we obtain

$$\frac{x^2}{a^2} + \frac{y^2}{b^2} - \frac{2xy}{ab}\cos(\alpha_1 - \alpha_2) = \sin^2(\alpha_1 - \alpha_2) \quad ...(2.20)$$

This represents the general equation of an ellipse. Thus, due to superposition of two simple harmonic vibrations at right angles to each other, the displacement of the particle will be along a curve given by Eqn. (2.20).

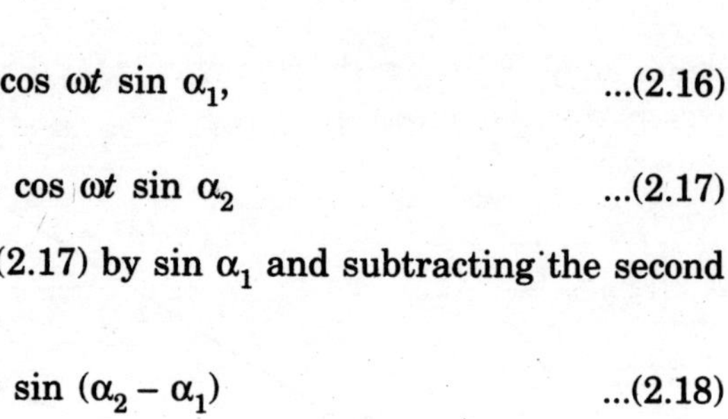

Fig. 2.3

Special Cases

(*i*) $\delta = \alpha_1 - \alpha_2 = 0, 2\pi, 4\pi, ...$

$\cos\delta = 1$, $\sin\delta = 0$ and

$$\left(\frac{x}{a}-\frac{y}{b}\right)^2 = 0 \text{ or, } y = \frac{b}{a}x.$$

The particle vibrates simple harmonically along the straight line BD (Fig. 2.3).

(*ii*) $\delta = \pi, 3\pi, 5\pi,...$

We have

$$y = -\frac{b}{a}x$$

This equation represents a straight line with slope = $-b/a$. The particle vibrates along the straight line AC (Fig. 2.4)

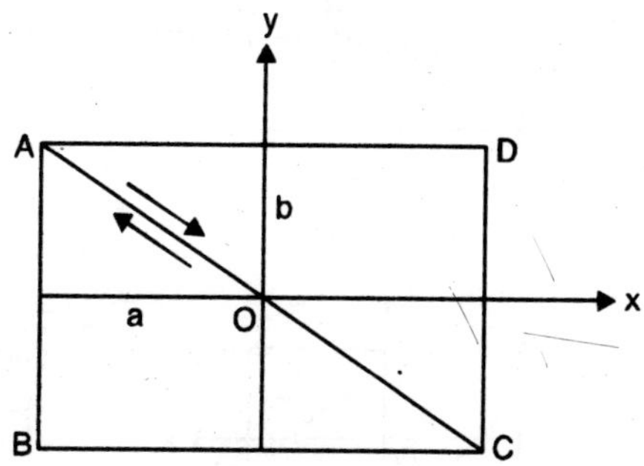

Fig. 2.4

(*iii*) $\delta = \frac{\pi}{2}, \frac{3\pi}{2}, \frac{5\pi}{2}, ...$

We have $\frac{x^2}{a^2}+\frac{y^2}{b^2} = 1$

which is an ellipse with semimajor and semiminor axes a and b, coinciding with the x- and y-axes, respectively (Fig. 2.5). If $a = b$, we get the equation of the circle $x^2 + y^2 = a^2$ with radius a.

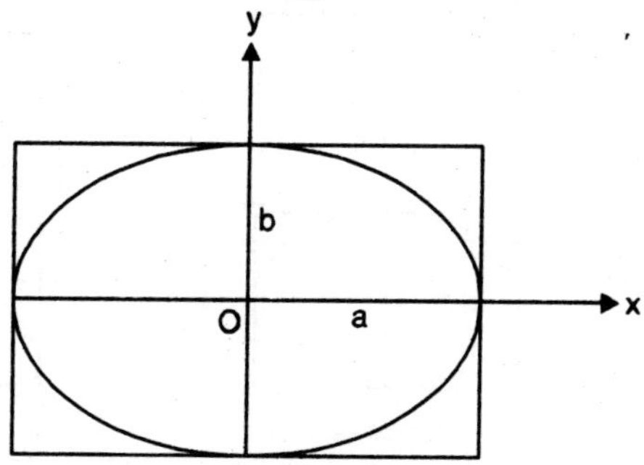

Fig. 2.5

(*iv*) $\delta = \frac{\pi}{4}, \frac{7\pi}{4}, \frac{9\pi}{4}, \frac{15\pi}{4},....$

We have $$\frac{x^2}{a^2}+\frac{y^2}{b^2}-\frac{\sqrt{2}xy}{ab}=\frac{1}{2}.$$

This is an oblique ellipse (Fig. 2.6).

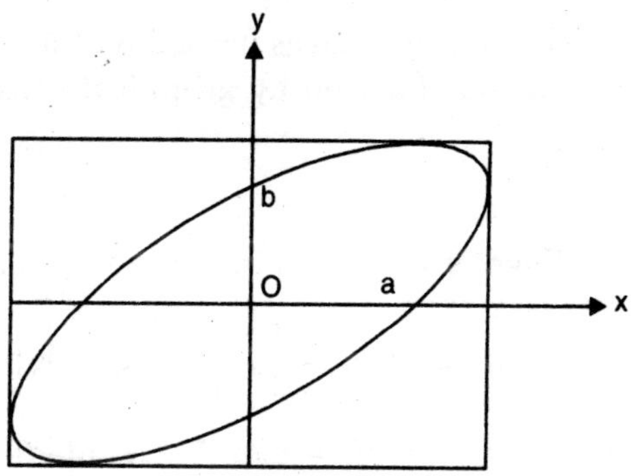

Fig. 2.6

(*v*) $\delta = \frac{3\pi}{4}, \frac{5\pi}{4}, \frac{11\pi}{4}, \frac{13\pi}{4},....$

We have now $$\frac{x^2}{a^2}+\frac{y^2}{b^2}+\frac{\sqrt{2}xy}{ab}=\frac{1}{2}$$

We get the oblique ellipse (Fig. 2.7).

The direction of rotation (clockwise or anticlockwise) of the particle may be obtained form the x- and y-motions of the particle when t is increased gradually. How the path of the particle with direction changes as δ is increased gradually is shown in Fig. 2.8.

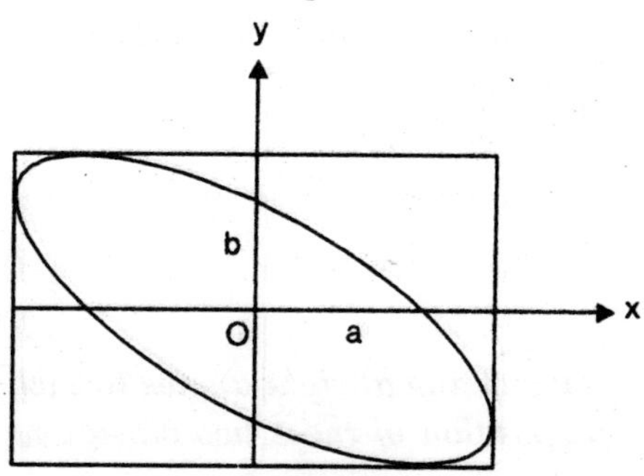

Fig. 2.7

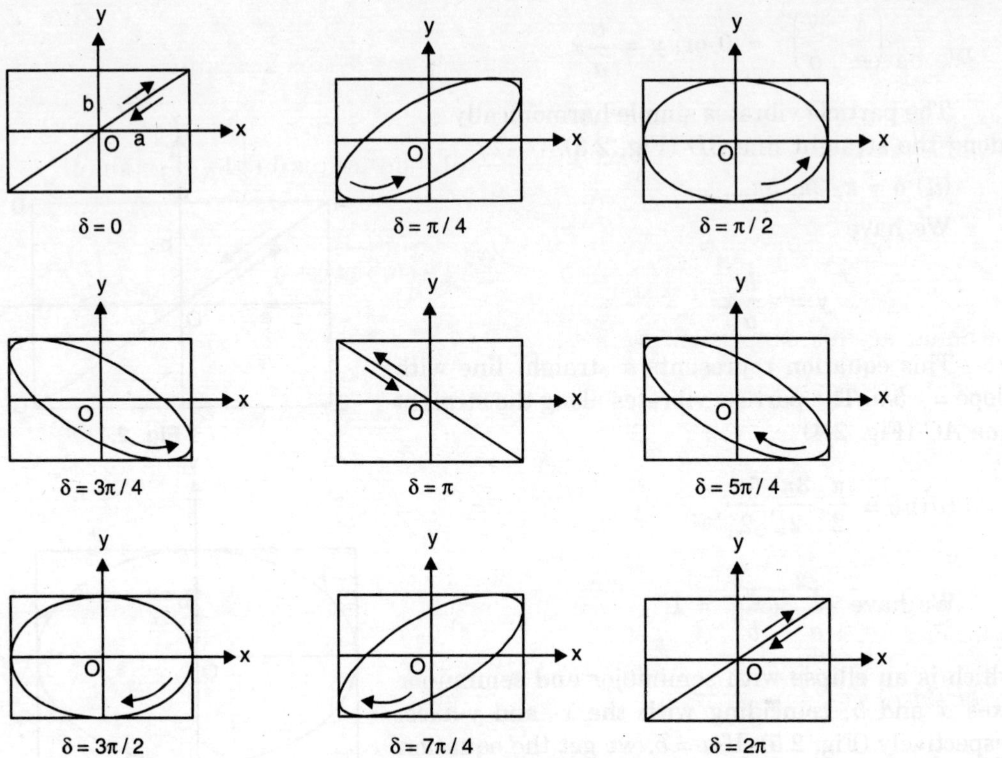

Fig. 2.8

That the two cases $\delta = \pi/2$ and $\delta = 3\pi/2$, although giving the same path, are physically different may be seen by graphical constructions. When $\delta = \pi/2$, we have

$$x = a \cos (\omega t + \alpha_2)$$

$$y = b \sin (\omega t + \alpha_2)$$

When $t = 0$ $x = a \cos \alpha_2$ and $y = b \sin \alpha_2$.

When $\omega t + \alpha_2 = \dfrac{\pi}{2}$, $x = 0$ and $y = b$.

When $\omega t + \alpha_2 = \pi$, $x = -a$ and $y = 0$.

Thus in this case the particle moves in the anticlockwise direction.

Similarly, it can be shown that the particle moves in the clockwise direction when $\delta = \dfrac{3\pi}{2}$.

5. *A particle is subjected to two SHMs represented by the following equations*

$$x = a_1 \sin \omega t,$$

$$y = a_2 \sin (2\omega t + \delta)$$

in a plane acting at right angles to each other. Discuss the formation of Lissajous' figures due to superposition of these two vibrations.

Solution

We have
$$\frac{y}{a_2} = \sin 2\omega t \cos \delta + \cos 2\omega t \sin \delta$$
$$= \frac{2x}{a_1}\sqrt{1-\sin^2 \omega t}\cos \delta + \left(1-\frac{2x^2}{a_1^2}\right)\sin \delta$$

or
$$\frac{y}{a_2} - \left(1-\frac{2x^2}{a_1^2}\right)\sin \delta = \frac{2x}{a_1}\sqrt{1-\sin^2 \omega t}\cos \delta$$

Squaring this expression, we get
$$\left(\frac{y}{a_2} - \sin\delta\right)^2 = \frac{4x^2}{a_1^2}\left(1-\frac{x^2}{a_1^2}-\frac{y}{a_2}\sin\delta\right)$$

or
$$\frac{4x^2}{a_1^2}\left(\frac{x^2}{a_1^2}+\frac{y}{a_2}\sin\delta - 1\right)+\left(\frac{y}{a_2}-\sin\delta\right)^2 = 0 \quad ...(2.21)$$

This gives the general equation of the resultant motion for any phase difference and amplitudes.

Special Cases

(*i*) When $\delta = \frac{\pi}{2}$, Eqn. (2.21) reduces to
$$\left(\frac{y}{a_2} - 1 + \frac{2x^2}{a_1^2}\right)^2 = 0$$

where represents two coincident parabolas:
$$x^2 = \frac{a_1^2}{2a_2}(a_2 - y) \quad ...(2.22)$$

The curve given by Eqn. (2.22) is shown in Fig. 2.9.

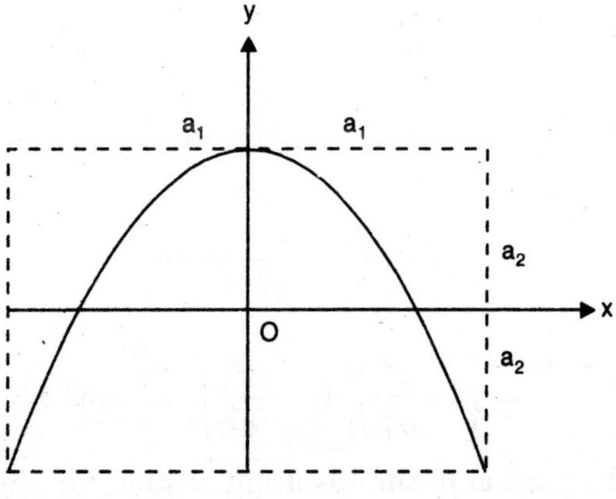

Fig. 2.9

(*ii*) When δ = 0, Eqn. (2.21) reduces to

$$\frac{y^2}{a_2^2} + \frac{4x^2}{a_1^2}\left(\frac{x^2}{a_1^2}-1\right) = 0 \qquad \text{...(2.23)}$$

This is an equation of 4th degree in x and it represents a curve having two loops (Fig. 2.10)

In Fig. 2.10,

$$y = 0 \text{ when } x = 0, \pm a_1$$

and

$$y = \pm a_2 \text{ when } x = \pm \frac{a_1}{\sqrt{2}}.$$

As the phase difference is changed gradually, the shape of the loop also changes gradually. Fig. 2.10 give the Lissajous' figure for two simple harmonic vibrations in phase (δ = 0) with a frequency ratio of 1:2 [frequency of x-vibration: frequency of y-vibration = 1 : 2].

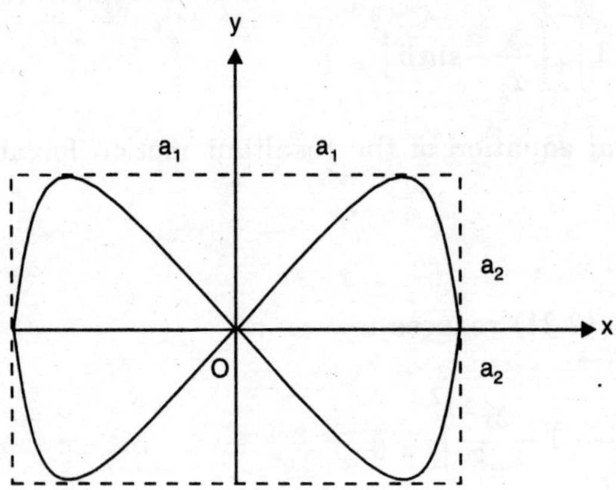

Fig. 2.10

6. *Two vibrations of frequencies in the ratio 1 : 3 and initial phase difference δ, given by*

$$x = a_1 \sin \omega t,$$

$$y = a_2 \sin (3\omega t + \delta)$$

act simultaneously on a particle at right angles to each other. Find the equation of the figure traced by the particle.

Solution

We have

$$\frac{y}{a_2} = (3 \sin \omega t - 4 \sin^3 \omega t) \cos \delta + (4 \cos^3 \omega t - 3 \cos \omega t) \sin \delta$$

or

$$\left[\frac{y}{a_2} - \left(\frac{3x}{a_1} - \frac{4x^3}{a_1^3}\right)\cos\delta\right]^2$$

$$= \left(1-\frac{x^2}{a_1^2}\right)\left[4\left(1-\frac{x^2}{a_1^2}\right)-3\right]^2 \sin^2\delta \qquad \text{...(2.24)}$$

This gives the general equation of the resultant motion for any phase difference δ and amplitudes a_1 and a_2.

Special Cases

(*i*) When $\delta = 0$, Eqn. (2.24) reduces to

$$\left[\frac{y}{a_2} - \left(\frac{3x}{a_1} - \frac{4x^3}{a_1^3}\right)\right]^2 = 0 \qquad \text{...(2.25)}$$

which gives two coincident cubic curves (Fig. 2.11)

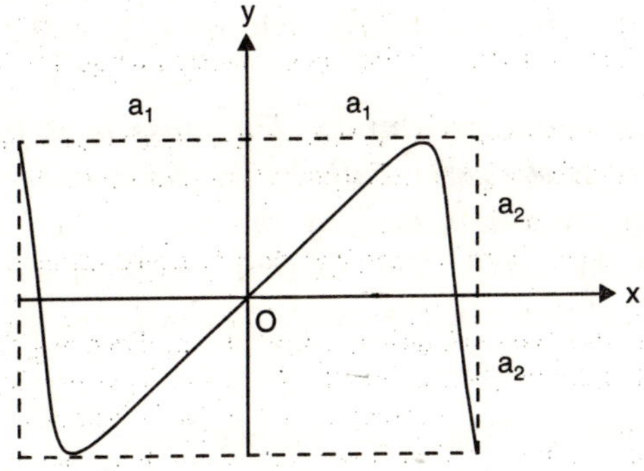

Fig. 2.11

(*ii*) When $\delta = \frac{\pi}{2}$, Eqn. (2.24) reduces to

$$\frac{y^2}{a_2^2} = \left(1 - \frac{x^2}{a_1^2}\right)\left(1 - \frac{4x^2}{a_1^2}\right)^2. \qquad \text{...(2.26)}$$

This is an equation of sixth degree giving a curve of three loops (Fig. 2.12).
It should be noted from Eqn. (2.26) that

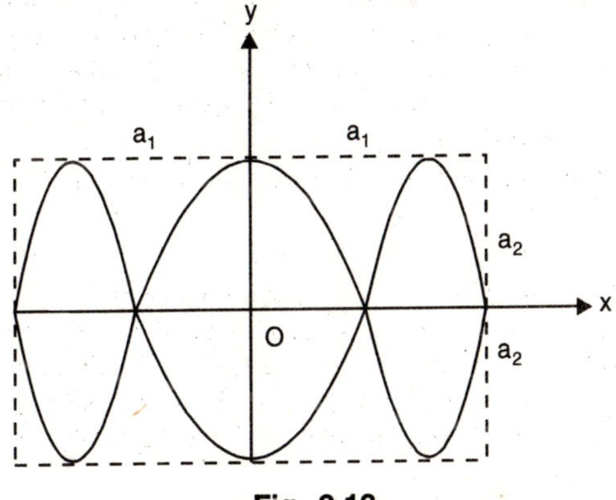

Fig. 2.12

when $\quad x = \pm\frac{a_1}{2}, \pm a_1, y = 0;$

when $\quad x = 0, y = \pm a_2;$

when $\quad x = \pm \frac{\sqrt{3}}{2} a_1, y = \pm a_2.$

It is clear that Eqn. (2.26) gives three loops. In general, if the frequencies are in the ratio 1 : *N*, the curve will have *N* loops.

7. *In an experiment to obtain Lissajous' figures, one tuning fork is of 250 Hz and a circular figure occurs after five seconds. What deductions may be made about the frequency of the other tuning fork?*

Solution

The Lissajous' figures are repeated in 5s *i.e.* the second tuning fork gains or loses one vibration over the first one in 5s. Thus, the difference in frequencies = 1/5 = 0.2 Hz. Hence, the possible frequencies of the second tuning fork are (250 + 0.2) = 250.2 Hz or (250 – 0.2) = 249.8 Hz.

8. *Two-dimensional harmonic oscillation: The mass m is free to move in the xy plane (Fig. 2.13) (a). It is connected to the rigid walls by two unstretched massless springs of spring constant, k_1 oriented along the x-axis and by the unstretched massless springs of spring constant k_2 oriented along the y-axis. The relaxed length of each spring is a. Discuss the motion of the mass in the xy plane in the small oscillation approximation.*

Solution

In the general configuration [Fig. 2.13 (*b*)], we have

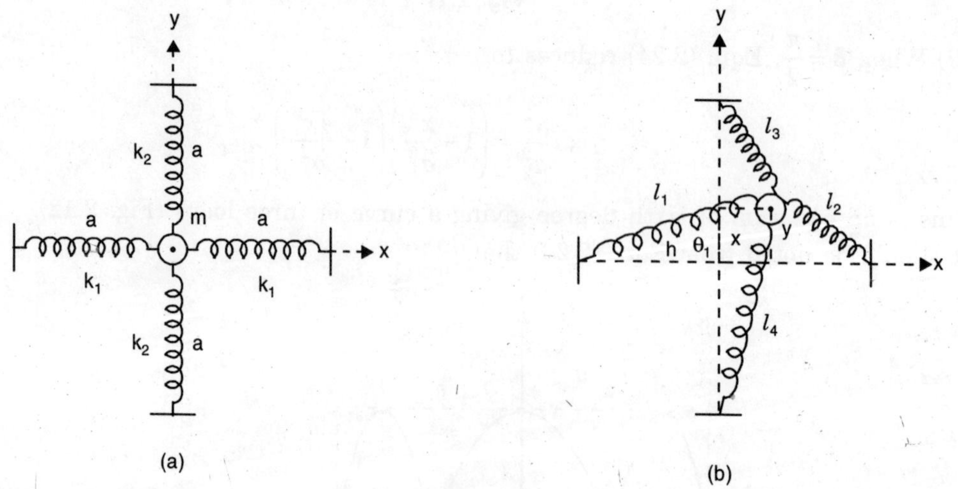

Fig. 2.13

$$l_1^2 = (a + x)^2 + y^2,$$

$$l_2^2 = (a - x)^2 + y^2,$$

$$l_3^2 = (a - y)^2 + x^2,$$

$$l_4^2 = (a + y)^2 + x^2.$$

Thus,

$$l_1 = a\left[1+\frac{x^2+2ax+y^2}{a^2}\right]^{1/2}$$

$$\approx a\left[1+\frac{2ax}{2a^2}\right] = a + x$$

where we have neglected x^2/a^2, y^2/a^2 and xy/a^2 terms in comparison with 1 in the small oscillation approximation. Similarly we may write

$$l_2 \approx a - x,$$
$$l_3 \approx a - y,$$
$$l_4 \approx a + y.$$

The tension in the first spring of length l_1 is $T_1 \approx k_1 (a + x - a) = k_1 x$. Magnitude of x-component of this tension is $T_1 \cos \theta_1 \approx T_1 = kx_1$ since the angle θ_1 made by l_1 with the x-axis is small. The y-component of this tension is $T_1 \sin \theta_1 \approx 0$. Thus we find that the x-component of the return force is entirely due to the two springs of lengths l_1 and l_2:

$$F_x = -2\, k_1 x.$$

The y-component of the return force is also entirely due to the two springs of lengths l_3 and l_4;

$$F_y = -2\, k_2 y.$$

Thus we get two uncoupled differential equations for the mass m along the x- and y-directions:

$$m\ddot{x} = -2\, k_1 x,$$
$$m\ddot{y} = -2\, k_2 y.$$

The solutions of these two equations are

$$x = A \cos(\omega_1 t + \phi_1) \quad \text{with } \omega_1^2 = 2k_1/m,$$

$$y = B \cos(\omega_2 t + \phi_2) \quad \text{with } \omega_2^2 = 2k_2/m.$$

The complete motion can be thought of as the superposition of the motions $x\hat{i} + y\hat{j}$. The position of the mass in the xy plane is given by

$$\vec{r} = x\,\hat{i} + y\,\hat{j}.$$

9. *The spherical pendulum: Consider a simple pendulum of length l. At equilibrium the string is vertical along the z-axis and the bob is at x = 0, y = 0. Find the motion of the bob for small oscillations (x and y are small).*

Solution

Suppose A is the position of the bob at any instant of time (Fig. 2.14). From A we drop a perpendicular AB on the z-axis. We have

$$AB^2 = l^2 - (l - z)^2$$
$$= 2lz - z^2$$

Again, $$z^2 + AB^2 = OA^2 = x^2 + y^2 + z^2$$

Thus, $$2lz = x^2 + y^2 + z^2.$$

Since z is a small quantity, we may write

$$z \approx \frac{x^2 + y^2}{2l}$$

which shows that z is a small quantity of second order. The potential energy of the bob at A is

$$V = mgz \approx \frac{mg}{2l}(x^2 + y^2).$$

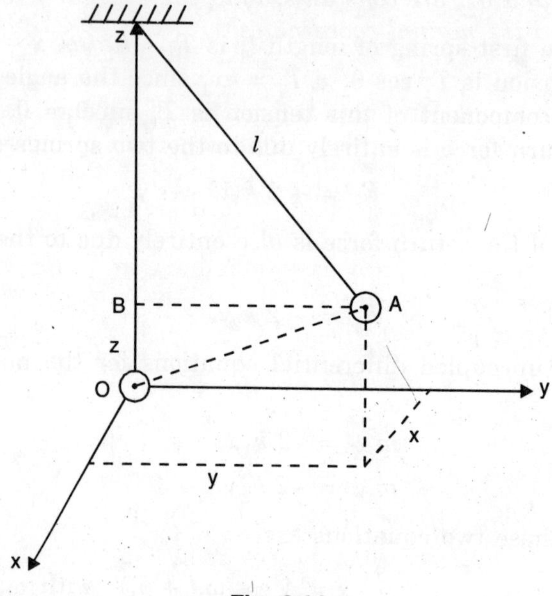

Fig. 2.14

Force on the bob along the x-direction is $F_x = -\frac{\partial V}{\partial x} = -\frac{mg}{l}x$ and the force along the y-direction is $F_y = -\frac{\partial V}{\partial y} = -\frac{mg}{l}y$. Therefore, we have two uncoupled differential equations along x-and y-directions.

$$m\ddot{x} = -\frac{mg}{l}x,$$

$$m\ddot{y} = -\frac{mg}{l}y.$$

These are simple harmonic motions. These equation can be solved independently:

$$x = A_1 \cos(\omega t + \phi_1),$$

$$y = A_2 \cos(\omega t + \phi_2)$$

with $\omega^2 = g/l$.

The constants A_1, A_2, ϕ_1 and ϕ_2 are determined by the initial conditions of displacement and velocity in the x-and y-directions. The complete motion can be thought of as a superposition of the motions $\hat{i}x$ and $\hat{j}y$ when we neglect the motion in the z-direction. Depending on the phase relationship between ϕ_1 and ϕ_2 we get an ellipse or a straight line for the path of the bob. For the x- and y-modes of vibrations we have the same frequency ω; the two modes are then said to be 'degenerate'.

10. *Consider two coupled first order linear homogeneous differential equations*

$$\ddot{x}_1 = -a_{11}x_1 - a_{12}x_2, \quad ...(2.27)$$

$$\ddot{x}_2 = -a_{21}x_1 - a_{22}x_2 \quad ...(2.28)$$

where a_{11}, a_{12}, a_{21} and a_{22} are constants. Find the angular frequencies for the normal modes of oscillations and the normal coordinates.

Solution

For normal modes of oscillations, both the degrees of freedom, namely x_1 and x_2 oscillate with the same frequency and they oscillate in phase or out of phase with one another:

$$x_1 = A \exp(i\omega t), \quad ...(2.29)$$

$$x_2 = B \exp(i\omega t) \quad ...(2.30)$$

where A and B are in general complex to take account of the possibility that x_1 and x_2 might oscillate out of phase with one another.

Substituting Eqns. (2.29) and (2.30) into Eqns. (2.27) and (2.28), we get

$$(a_{11} - \omega^2) A + a_{12}B = 0, \quad (2.31)$$

$$a_{21}A + (a_{22} - \omega^2) B = 0. \quad (2.32)$$

For non-trivial solutions, we have

$$(a_{11} - \omega^2)(a_{22} - \omega^2) - a_{12}a_{21} = 0. \quad (2.33)$$

This is a quadratic equation in the variable ω^2. It has two solutions in general, which we call ω_1^2 and ω_2^2. From Eqn. (2.31), we find the ratio B/A as

$$\frac{B}{A} = \frac{\omega^2 - a_{11}}{a_{12}} \quad ...(2.34)$$

For normal mode 1 with $\omega^2 = \omega_1^2$, we may write

$$x_1(t) = A_1 \exp(i\omega_1 t)$$

$$x_2(t) = A_1 \frac{\omega_1^2 - a_{11}}{a_{12}} \exp(i\omega_1 t),$$

For normal mode 2 with $\omega^2 = \omega_2^2$, we have

$$x_1(t) = A_2 \exp(i\omega_2 t),$$

$$x_2(t) = A_2 \frac{\omega_2^2 - a_{11}}{a_{12}} \exp(i\omega_2 t).$$

Due to superposition of the two modes the general solution is given by

$$x_1(t) = A_1 \exp(i\omega_1 t) + A_2 \exp(i\omega_2 t),$$

$$x_2(t) = \frac{1}{a_{12}}[A_1(\omega_1^2 - a_{11}) \exp(i\omega_1 t) + A_2(\omega_2^2 - a_{11}) \exp(i\omega_2 t)]$$

where A_1 and A_2 are arbitrary constants and ω_1 and ω_2 are the normal mode frequencies.

Normal coordinates: If the differential equations are coupled, we have to search for new variables which satisfy uncoupled differential equations. The new variables are then called normal coordinates. Suppose X and Y are the normal coordinates satisfying the differential equations

$$\ddot{X} = -\omega_1^2 X, \quad ...(2.35)$$

$$\ddot{Y} = -\omega_2^2 Y. \quad ...(2.36)$$

If X and Y are the normal coordinates, then any constant multiple of X and Y also satisfy Eqns. (2.35) and (2.36). Suppose X and Y are obtained from the linear combinations of x_1 and x_2 so that we may write

$$X = x_1 + \alpha x_2, \quad ...(2.37)$$

$$Y = x_1 + \beta x_2 \quad ...(2.38)$$

where α and β are constants. By solving x_1 and x_2, we get

$$x_1 = \frac{\alpha Y - \beta X}{\alpha - \beta}, \quad ...(2.39)$$

$$x_2 = \frac{X - Y}{\alpha - \beta}. \quad ...(2.40)$$

We substitute Eqns. (2.39) and (2.40) into Eqns. (2.27) and (2.28) and separate the uncoupled differential equations for X and Y, which give

$$\omega_1^2 = \frac{\beta a_{11} - a_{12}}{\beta} = a_{22} - \beta a_{21}, \quad ...(2.41)$$

$$\omega_2^2 = \frac{\alpha a_{11} - a_{12}}{\alpha} = a_{22} - \alpha a_{21}. \quad ...(2.42)$$

Thus α and β satisfy the same quadratic equation having the roots

$$\alpha, \beta = \frac{1}{2a_{21}}\left[a_{22} - a_{11} \pm [(a_{22} - a_{11})^2 + 4a_{12}a_{21}]^{1/2}\right].$$

When $a_{22} = a_{11}$ and $a_{12} = a_{21}$, $\alpha, \beta = \pm 1$ and the normal coordinates are $x_1 + x_2$ and $x_1 - x_2$.

11. *Longitudinal oscillations of two coupled masses: Two bodies of masses m_1 and m_2 are attached to each other and to two fixed points by three identical light springs of relaxed length a. The whole arrangement rests on a smooth horizontal table. Find the angular frequencies of the normal modes for longitudinal oscillations of small amplitude. Describe the motions of the two bodies for each normal mode. Find normal coordinates when $m_1 = m_2$.*

Solution

Let the spring constant of each spring be k. We consider the forces acting on m_1 and m_2 when m_1 is displaced from its equilibrium position by x_1 and m_2 is displaced by x_2 from its equilibrium position (Fig. 2.15). The first spring is stretched by amount x_1 and the force

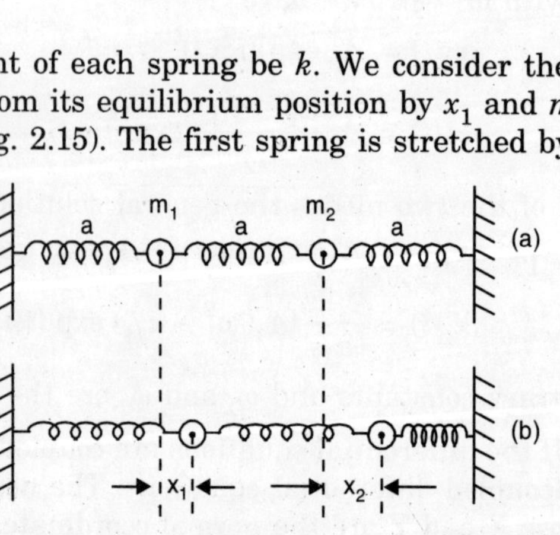

Fig. 2.15

on the mass m_1 due to the first spring = $-kx_1$. Length of the second spring = $(2a + x_2) - (a + x_1) = a + x_2 - x_1$. The second spring is stretched by $x_2 - x_1$. The force on the mass m_1 due to the second spring = $k(x_2 - x_1)$ and the force on the mass m_2 due to the second spring = $-k(x_2 - x_1)$. The third spring is compressed by amount x_2. Force on mass m_2 due to the third spring = $-kx_2$. So the equations of motion of the two bodies are

$$m_1\ddot{x}_1 = -kx_1 + k(x_2 - x_1), \quad \text{...(2.43)}$$

$$m_2\ddot{x}_2 = -k(x_2 - x_1) - kx_2. \quad \text{...(2.44)}$$

For normal modes of oscillations we put

$$x_1 = A\exp(i\omega t), \quad \text{...(2.45)}$$

$$x_2 = B\exp(i\omega t). \quad \text{...(2.46)}$$

Substituting the solutions (2.45) and (2.46) into (2.43) and (2.44), we get

$$-m_1\omega^2 A = -2kA + kB, \quad \text{...(2.47)}$$

$$-m_2\omega^2 B = kA - 2kB \quad \text{...(2.48)}$$

or, in the matrix notation we may write

$$\begin{pmatrix} m_1\omega^2 - 2k & k \\ k & m_2\omega^2 - 2k \end{pmatrix}\begin{pmatrix} A \\ B \end{pmatrix} = \begin{pmatrix} 0 \\ 0 \end{pmatrix}$$

For non-trivial solutions we have

$$(m_1\omega^2 - 2k)(m_2\omega^2 - 2k) - k^2 = 0.$$

This is a quadratic equation in ω^2 which can be solved to give

$$\omega^2 = \frac{k}{m_1 m_2}\left[m_1 + m_2 \pm \sqrt{m_1^2 + m_2^2 - m_1 m_2}\right] \quad \text{...(2.49)}$$

From Eqn. (2.48) we find the ratio A/B as

$$\frac{A}{B} = \frac{2k - m_2\omega^2}{k} \quad \text{...(2.50)}$$

Mode 1: The lower frequency solution of (2.49) has the ratio

$$\frac{A}{B} = \frac{m_1 - m_2 + \sqrt{m_1^2 + m_2^2 - m_1 m_2}}{m_1}$$

which is real and positive. The two bodies oscillate in phase.

Mode 2: The higher frequency solution of (2.49) has the ratio

$$\frac{A}{B} = \frac{m_1 - m_2 - \sqrt{m_1^2 + m_2^2 - m_1 m_2}}{m_1}$$

which is real and negative. The two bodies oscillate in antiphase.

Let us take the simple case $m_1 = m_2 = m$. The mode with lower frequency ($\omega_1^2 = k/m$) has the ratio $A/B = 1$ and therefore $x_1 = x_2$ for all time. The central spring has the same length as it had at equilibrium, so that the central spring exerts no force on either mass. When the left hand mass goes to the right, the right hand mass also goes to the right; when the left hand mass goes to the left, the right hand mass also goes to the left, the length of the central spring remaining unchanged always.

When $m_1 = m_2 = m$, the mode with higher frequency ($\omega_2^2 = 3\ k/m$) has the ratio $A/B = -1$ so that $x_1 = -x_2$ for all time. When the left hand mass goes to the right, the right hand mass goes to the left [the central spring is compressed]. When the left hand mass goes to the left, the right hand mass to the right [the central spring is elongated]. The two masses move in opposite directions (antiphase). When the central spring is compressed, the side springs are elongated, and when the central spring is elongated, the side springs are compressed.

Normal coordinates: When $m_1 = m_2 = m$, the lower frequency $\omega_1 = \sqrt{k/m}$ and higher frequency $\omega_2 = \sqrt{3k/m}$. By adding Eqns. (2.43) and (2.44) and subtracting Eqn. (2.44) from Eqn. (2.43), we get

$$\ddot{x}_1 + \ddot{x}_2 = -\omega_1^2 (x_1 + x_2)$$

$$\ddot{x}_1 - \ddot{x}_2 = -\omega_2^2 (x_1 - x_2)$$

which show that the normal coordinates are $x_1 + x_2$ and $x_1 - x_2$ having frequencies ω_1 and ω_2 respectively.

12. *Two bodies of masses m_1 and m_2 are attached to each other and to two fixed points by three identical light springs of relaxed length a_0. Find the angular frequencies of the normal modes for transverse oscillations. Describe the motions of the two bodies for each normal mode. Find the normal coordinates when $m_1 = m_2 = m$.*

Solution

At equilibrium let the length of each spring be a and the spring constant be k. Let y_1 and y_2 be the vertical displacements of the masses m_1 and m_2 from the initial positions at any instant of time (Fig. 2.16). In this position the lengths of the springs are given by

$$l_1^2 = y_1^2 + a^2,$$

$$l_2^2 = (y_2 - y_1)^2 + a^2,$$

$$l_3^2 = y_2^2 + a^2.$$

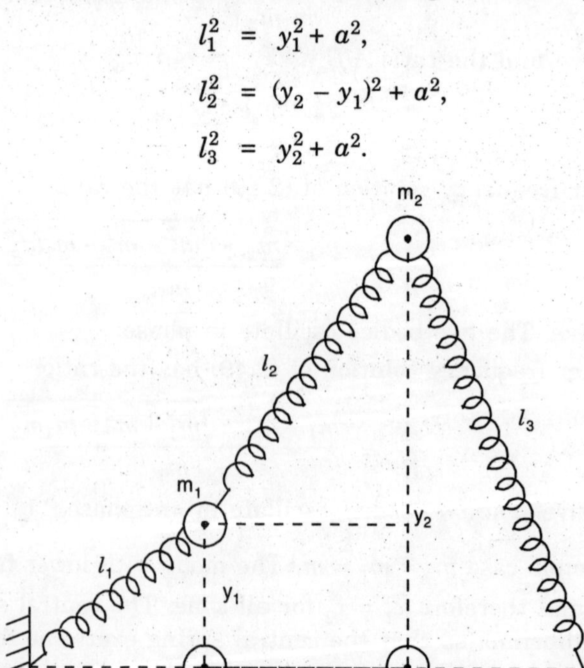

Fig. 2.16

Elongation of the left hand spring = $l_1 - a_0$ and the component of the force due to this spring on mass m_1 along the y-direction = $-k(l - a_0)\, y_1/l_1$. The central spring is elongated by $(l_2 - a_0)$ and the component of the force due to the central spring on mass m_1 along the y-direction is $k\,(l_2 - a_0)\,(y_2 - y_1)/l_2$. The component of the force along the y-direction for mass m_2 due to the central and the right hand springs is

$$-k(l_2 - a_0)\,(y_2 - y_1)/l_2 - k(l_3 - a_0)y_2/l_3.$$

Thus, the equations of motion of the two masses are

$$m_1\ddot{y}_1 = -ky_1\left(1-\frac{a_0}{l_1}\right) + k(y_2 - y_1)\left(1-\frac{a_0}{l_2}\right), \qquad \text{...(2.51)}$$

$$m_2\ddot{y}_2 = -k(y_2 - y_1)\left(1-\frac{a_0}{l_2}\right) - ky_2\left(1-\frac{a_0}{l_3}\right). \qquad \text{...(2.52)}$$

(*i*) *Slinky approximation:* a_0 is small and a_0/l_1, a_0/l_2, $a_0/l_3 << 1$. Under this approximation Eqns. (2.51) and (2.52) reduce to

$$m_1\ddot{y}_1 = -2ky_1 + ky_2, \qquad \text{...(2.53)}$$

$$m_2\ddot{y}_2 = ky_1 - 2ky_2, \qquad \text{...(2.54)}$$

which are identically same as Eqns. (2.43) and (2.44) of problem 11. Thus the normal mode frequencies are given by Eqn. (2.49). When $m_1 = m_2 = m$, the two normal mode frequencies are $\sqrt{k/m}$ and $\sqrt{3k/m}$ with normal coordinates $y_1 + y_2$ and $y_1 - y_2$. For the mode with lower frequency we get $y_1 = y_2$ *i.e.* the centre spring is never compressed or elongated. For the mode with higher frequency we have $y_1 = -y_2$ *i.e.* the two masses move in opposite directions.

(*ii*) *Small oscillation approximation*: When y_1 and y_2 are small in comparison with a, we have

$$l_1 \approx a\left(1+\frac{y_1^2}{2a^2}\right) \approx a,$$

$$l_2 \approx a\left(1+\frac{(y_2 - y_1)^2}{2a^2}\right) \approx a,$$

$$l_3 \approx a\left(1+\frac{y_2^2}{2a^2}\right) \approx a.$$

Under this approximation Eqns. (2.51) and (2.52) become

$$m_1\ddot{y}_1 = -2k\left(1-\frac{a_0}{a}\right)y_1 + k\left(1-\frac{a_0}{a}\right)y_2,$$

$$m_2\ddot{y}_1 = k\left(1-\frac{a_0}{a}\right)y_1 - 2k\left(1-\frac{a_0}{a}\right)y_2,$$

which are of the same form as Eqns. (2.53) and (2.54) if k is replaced by $k\left(1-\frac{a_0}{a}\right)$. The normal mode frequencies are $\left[\frac{k}{ma}(a-a_0)\right]^{1/2}$ and $\left[\frac{3k}{ma}(a-a_0)\right]^{1/2}$ when $m_1 = m_2 = m$.

13. *Two identical simple pendulums having the same length l and same bob mass m are suspended by strings of negligible mass. The bobs are connected by a spring of relaxed length a. At the equilibrium position the spring has its relaxed length. Assuming small-oscillation amplitudes in the vertical plane, find the normal modes of oscillations of the system.*

Solution

At the general configuration (Fig. 2.17) the bobs are displaced by x_1 and x_2 from their equilibrium positions, where x_1 and x_2 are small. If the spring were not present, we have for the x-component of motion of the spherical pendulums.

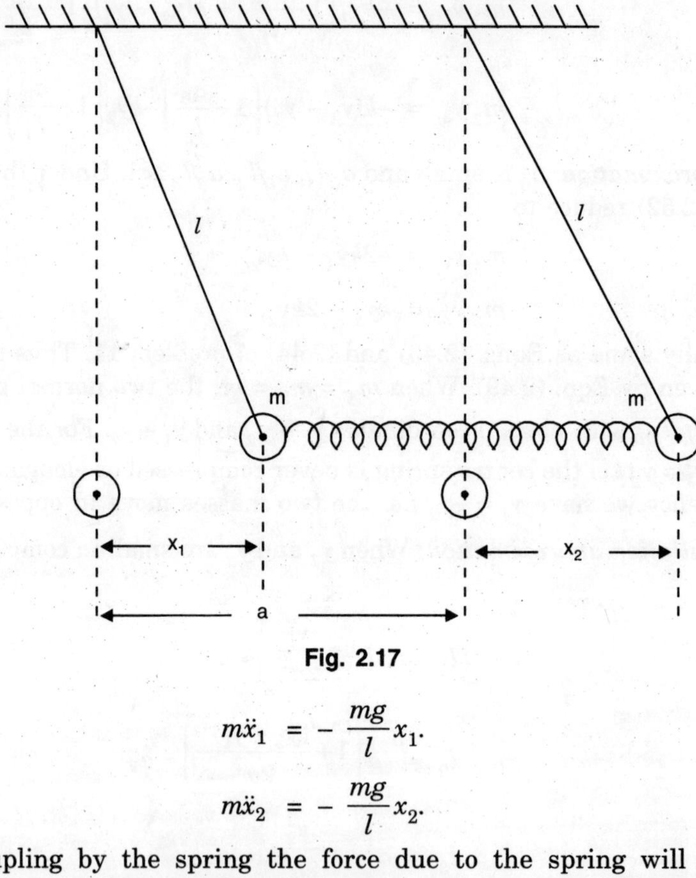

Fig. 2.17

$$m\ddot{x}_1 = -\frac{mg}{l}x_1.$$

$$m\ddot{x}_2 = -\frac{mg}{l}x_2.$$

Due to coupling by the spring the force due to the spring will act on the bobs. Length of the spring in the general configuration = $(a + x_2 - x_1)$. Elongation of the spring = $(a + x_2 - x_1) - a = x_2 - x_1$. Thus the equations of motion for the coupled pendulums are

$$m\ddot{x}_1 = -\frac{mg}{l}x_1 + k(x_2 - x_1),$$

$$m\ddot{x}_2 = -\frac{mg}{l}x_2 - k(x_2 - x_1).$$

The normal mode frequencies are given by $\omega_1^2 = g/l$ and $\omega_2^2 = g/l + 2k/m$ with the normal coordinates $x_1 + x_2$ and $x_1 - x_2$. For the mode with lower frequency ω_1 we get $x_1 = x_2$ *i.e.* the spring is never compressed or elongated and the whole system moves with the pendulum frequency $\sqrt{g/l}$. For the mode with the higher frequency ω_2 we have $x_1 = -x_2$ *i.e.* the two bobs move in the opposite directions and the spring is alternately compressed and elongated.

14. *Discuss the formation of beats due to superposition of two harmonic oscillations ψ_1 and ψ_2 having slightly different angular frequencies ω_1 and ω_2:*

$$\psi_1 = A \cos \omega_1 t,$$
$$\psi_2 = A \cos \omega_2 t.$$

How does the modulation amplitude vary with time?

Solution

We define
$$\omega_{av} \equiv \frac{1}{2}(\omega_1 + \omega_2),$$

$$\omega_{\text{mod}} \equiv \frac{1}{2}(\omega_1 - \omega_2), \; \omega_1 > \omega_2$$

where ω_{av} is the average angular frequency and ω_{mod} is the modulation angular frequency. Due to superposition of ψ_1 and ψ_2, we get

$$\psi = \psi_1 + \psi_2 = A[\cos(\omega_{av} + \omega_{\text{mod}})t + \cos(\omega_{av} - \omega_{\text{mod}})t]$$

$$= A_{\text{mod}}(t) \cos \omega_{av} t \quad ...(2.55)$$

where
$$A_{\text{mod}}(t) = 2A \cos \omega_{\text{mod}} t. \quad ...(2.56)$$

Equation (2.55) represents an oscillation at angular frequency ω_{av} with an amplitude A_{mod} that varies with time according to Eqn. (2.56). If $\omega_1 \approx \omega_2$, $\omega_{\text{mod}} \ll \omega_{av}$. In this case the modulation amplitude $A_{\text{mod}}(t)$ varies only slightly during many fast oscillations of $\cos \omega_{av} t$ and therefore Eqn. (2.55) represents almost harmonic oscillation at angular frequency ω_{av} with a slowly varying amplitude (Fig. 2.18).

When the two waves are in phase, the resultant ($\psi = \psi_1 + \psi_2$) is maximum and A_{mod} has its maximum value. When the two waves are out of phase the resultant is minimum and A_{mod} has zero amplitude. There are two maximum amplitude positions for A_{mod} (one positive and one negative) in a $\frac{1}{2}$-modulation cycle. Number of beats per second is twice the modulation frequency.

$$\omega_{\text{beat}} = 2\omega_{\text{mod}} = \omega_1 - \omega_2 \quad ...(2.57)$$

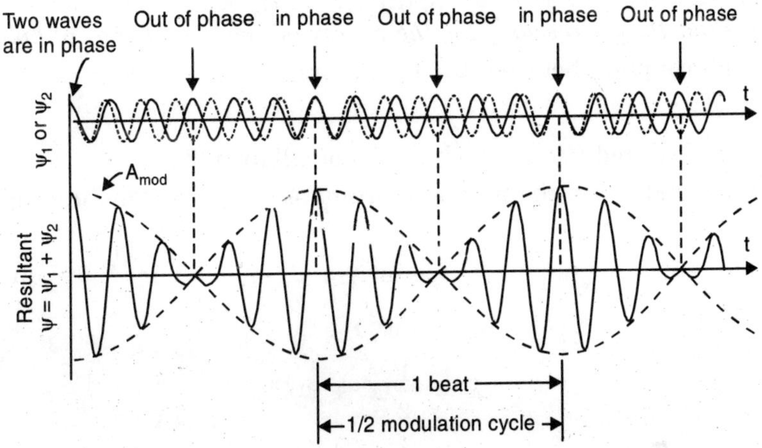

Fig. 2.18

Equation (2.56) gives the variation of the modulation amplitude A_{mod} with time. By squaring this equation we get

$$A^2_{mod}(t) = 2A^2 (1 + \cos 2\omega_{mod}t).$$

Thus $A^2_{mod}(t)$ oscillates about its average value at twice the modulation frequency *i.e.* at the beat frequecy $(\omega_1 - \omega_2)$.

15. *Two tuning forks when sounded together give 4 beats per second. One is unison with a length of 96 cm of a sonometer wire under certain tension and the other with 96.8 cm of that wire under the same tension. Find the frequencies of these forks.*

Solution

From the equation of transverse vibration of a sonometer wire (see Chapter 8), we have

$$n_1 = \frac{1}{2\times 96}\sqrt{T/\mu},$$

$$n_2 = \frac{1}{2\times 96.8}\sqrt{T/\mu}$$

where T is the tension of the wire and μ is the mass per unit length of the wire.

Thus, $$\frac{n_1}{n_2} = \frac{96.8}{96}$$

and $$n_1 - n_2 = 4$$

Solving we get $n_1 = 484$ Hz and $n_2 = 480$ Hz.

16. *Two identical simple pendulums a and b with the same string length l and the same bob mass m are coupled by a massless spring of spring constant k which is attached to the bobs. The spring is unstretched when both the strings of the pendulums are vertical.*

(*i*) *Derive the differential equations of motion (for small oscillations) of the bobs.*

(*ii*) *Find the normal mode frequencies, the configurations of the system in these modes and the normal coordinates.*

(*iii*) *Find a superposition of the two modes which correspond to the initial conditions (at time t = 0) that both the pendulums have zero velocity, that bob a has amplitude 2A, and that bob b has zero amplitude. Let E_a be the total energy of bob a at time t = 0. Find the expressions for the energy of bob a and that of bob b at any time.*

(Assume weak coupling between the oscillators).

Solution

For solutions of (*i*) and (*ii*) see problem 13 of Chapter 2.

(*iii*) Suppose X and Y are the normal coordinates satisfying the uncoupled differential equations

$$\ddot{X} = -\omega_1^2 X, \quad ...(2.58)$$

$$\ddot{Y} = -\omega_2^2 Y \quad ...(2.59)$$

with $$X = \frac{1}{2}(x_1 + x_2),$$

$$Y = \frac{1}{2}(x_1 - x_2)$$

where x_1 and x_2 are the displacements of the bobs a and b from their respective equilibrium positions.

The solutions of Eqns (2.58) and (2.59) are

$$X = A_1 \cos (\omega_1 t + \phi_1),$$
$$Y = A_2 \cos (\omega_2 t + \phi_2)$$

so that the displacements of the bobs are

$$x_1 = A_1 \cos (\omega_1 t + \phi_1) + A_2 \cos (\omega_2 t + \phi_2),$$
$$x_2 = A_1 \cos (\omega_1 t + \phi_1) - A_2 \cos (\omega_2 t + \phi_2).$$

The initial conditions give four equations

$$x_1(0) = A_1 \cos \phi_1 + A_2 \cos \phi_2 = 2A, \quad ...(2.60)$$

$$x_2(0) = A_1 \cos \phi_1 - A_2 \cos \phi_2 = 0, \quad ...(2.61)$$

$$\dot{x}_2(0) = -\omega_1 A_1 \sin \phi_1 - \omega_2 A_2 \sin \phi_2 = 0, \quad ...(2.62)$$

$$\dot{x}_2(0) = -\omega_1 A_1 \sin \phi_1 + \omega_2 A_2 \sin \phi_2 = 0. \quad ...(2.63)$$

From Eqns. (2.62) and (2.63), we find

$$\sin \phi_1 = \sin \phi_2 = 0.$$

We make the choice of the phase constants ϕ_1 and ϕ_2 consistent with Eqns. (2.60) and (2.61). We take $\phi_1 = \phi_2 = 0$. We have from Eqns. (2.60) and (2.61) $A_1 = A_2 = A$ so that the displacements of the bobs are given by

$$x_1(t) = A \cos \omega_1 t + A \cos \omega_2 t.$$

$$x_2(t) = A \cos \omega_1 t - A \cos \omega_2 t.$$

The displacement of the bobs are obtained by the superposition of two normal modes of frequencies ω_1 and ω_2 which differ by a small amount when the coupling is weak *i.e.* k is small. We shall get the beat effect when the two modes X and Y are present with equal amplitudes $A_1 = A_2 = A$. The pendulum a is under the influence of two harmonic modes of equal amplitudes with slightly different frequencies. Thus the motion a exhibits beats. The pendulum b also exhibits beats.

Initially the bob b is at zero and the bob a has displacement $2A$. Both of them are released from rest at time $t = 0$. Gradually the oscillation amplitude of pendulum a decreases and that of pendulum b increases, until pendulum a comes to rest and pendulum b oscillates with the amplitude and energy that pendulum a started out with. (Here we neglect the frictional forces). The vibration energy is transferred completely from one pendulum to the other and the process continues. The vibration energy slowly flows back and forth between a and b. One complete round trip for the energy from a to b and back to a is a beat. The beat period is the time for the round trip.

We put

$$\omega_1 = \omega_{av} - \omega_{mod},$$
$$\omega_2 = \omega_{av} + \omega_{mod}$$

and we get

$$x_1(t) = A_{mod}(t) \cos \omega_{av} t.$$
$$x_2(t) = B_{mod}(t) \sin \omega_{av} t$$

where

$$A_{mod}(t) = 2A \cos \omega_{mod} t,$$

$$B_{mod}(t) = 2A \sin \omega_{mod} t$$

Thus we get almost harmonic oscillations for x_1 and x_2. The beat frequency is

$$\omega_{beat} = \omega_2 - \omega_1 = \left[\frac{g}{l} + \frac{2k}{m}\right]^{1/2} - \sqrt{\frac{g}{l}} \approx \frac{k}{m}\sqrt{\frac{l}{g}}$$

when k is small.

Thus $x_2 = 0$ when $|x_1|$ is maximum and $x_1 = 0$ when $|x_2|$ is maximum.

Let us find an expression for the total energy (kinetic + potential) of each pendulum. For the pendulum a the oscillation amplitude $A_{mod}(t)$ is almost constant over one cycle of the fast oscillation. If the spring is weak it does not have significant amount of stored energy. Thus during one fast oscillation we think of pendulum a as a harmonic oscillator of frequency ω_{av} with constant amplitude A_{mod}.

Average KE of the pendulum a (averaged over one fast cycle) is $\frac{1}{4} m\omega^2_{av} A^2_{mod}$. The total energy is twice the average value of the KE:

$$E_a = \frac{1}{2} m\omega^2_{av} A^2_{mod} = 2mA^2\omega^2_{av} \cos^2 \omega_{mod} t.$$

Similarly, we have

$$E_b = \frac{1}{2} m\omega^2_{av} B^2_{mod} = 2mA^2\omega^2_{av} \sin^2 \omega_{mod} t.$$

and

$$E_a + E_b = 2mA^2\omega^2_{av} = E$$

where E is the total energy of the system. The energy difference between the two pendulums is

$$E_a - E_b = E \cos 2\omega_{mod} t = E \cos (\omega_2 - \omega_1)t.$$

Hence

$$E_a = \frac{1}{2} E\, [1 + \cos (\omega_2 - \omega_1)t],$$

$$E_b = \frac{1}{2} E\, [1 - \cos (\omega_2 - \omega_1)t].$$

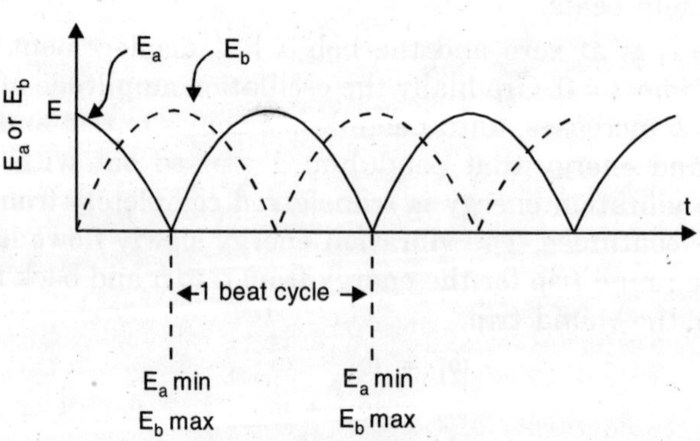

Fig. 2.19

The energy E_a and E_b are plotted against t in Fig. 2.19. When the pendulum a has the maximum energy, the pendulum b has zero energy, and vice versa. The total energy of the system E is constant and it flows back and forth between the two pendulums at the beat frequency.

17. *Two radio stations broadcast their programmes at the same amplitude A, and a slightly different frequencies ν_1 and ν_2 respectively, where $\nu_2 - \nu_1 = 10^3$ Hz. A detector receives the signals from the two stations simultaneously. It can only detect signals of intensity $\geqslant 2A^2$.*

(*a*) *Find the time interval between successive maxima of the intensity of the signal received by the detector.*

(*b*) *Find the time for which the detector remains idle in each cycle of the intensity of the signal.* (I.I.T. 1993)

Solution

(*a*) The expression for the superposition of two waves having the same amplitude is

$$\psi = A \cos \omega_1 t + A \cos \omega_2 t$$

$$= \left[2A \cos \frac{\omega_2 - \omega_1}{2} t\right] \cos \frac{\omega_1 + \omega_2}{2} t,$$

The resultant wave has the slowly varying amplitude

$$A' = 2A \cos (\omega_2 - \omega_1)\, t/2.$$

It is maximum when $(\omega_2 - \omega_1)\, t = 0, 2\pi, 4\pi, ...$

Hence the time interval between successive maxima is

$$\Delta t = \frac{2\pi}{\omega_2 - \omega_1} = \frac{2\pi}{2\pi(\nu_2 - \nu_1)} = 10^{-3} \text{ s}$$

(*b*) For the intensity $2A^2$, the amplitude would be $\pm\sqrt{2}\, A$.

Thus, $$2A \cos (\omega_2 - \omega_1)t/2 = \pm\sqrt{2}\, A$$

or $$\cos (\omega_2 - \omega_1)t/2 = \pm\, 1/\sqrt{2}$$

or $$(\omega_2 - \omega_1)t_1/2 = \frac{\pi}{4},\ (\omega_2 - \omega_1)t_2/2 = \frac{3\pi}{4}.$$

In the interval $\pi/4$ to $3\pi/4$ the cosine function decreases.

$$t_1 = \frac{\pi}{2(\omega_2 - \omega_1)} = \frac{1}{4(\nu_2 - \nu_1)}$$

$$t_2 = \frac{3\pi}{2(\omega_2 - \omega_1)} = \frac{3}{4(\nu_2 - \nu_1)}$$

Thus, the time interval during which the amplitude becomes less than $\sqrt{2}\, A$ is

$$\Delta t = t_2 - t_1 = \frac{1}{2(\nu_2 - \nu_1)} = 5 \times 10^{-4} \text{ s}.$$

During this time interval the detector remains idle for each cycle of the intensity of the signal.

18. *Three simple harmonic motions in the same direction having the same amplitude a and same period are superimposed. If each differs in phase from the next by 45°, then*

(*a*) *The resultant amplitude is $(1+\sqrt{2})a$.*

(*b*) *The phase of the resultant motion relative to the first is 90°.*

(*c*) *The energy associated with the resulting motion is $(3+2\sqrt{2})$ times the energy associated with any single motion.*

(*d*) *The resulting motion is not simple harmonic.* (I.I.T. 1999)

Solution

From problem 2, we know that the resultant motion is given by

$$X = R \sin (\omega t + \theta)$$

with

$$R = a \frac{\sin (N \phi / 2)}{\sin (\phi / 2)}$$

$$\theta = (N - 1) \phi/2$$

Here, $N = 3$ and $\phi = 45°$.

$$R = a \frac{\sin 3\psi}{\sin \psi} = a (2 \cos^2 \psi + \cos 2\psi) = a (1 + 2 \cos 2\psi)$$

where

$$\psi = 45°/2$$

Thus, $R = a (1+\sqrt{2})$

The resultant motion is simple harmonic with amplitude $(1+\sqrt{2})a$.

$$\theta = \phi = 45°.$$

Total energy of SHM is

$$E = \frac{1}{2} m\omega^2 A^2$$

$$= \frac{1}{2} m\omega^2 ((1+\sqrt{2}))^2 a^2$$

$$= \frac{1}{2} m\omega^2 a^2 (3+2\sqrt{2})$$

Correct Choice: (*a*), (*c*).

19. *The function*

$$x = A \sin^2 \omega t + B \cos^2 \omega t + C \sin \omega t \cos \omega t$$

represents SHM

(*a*) *for every value of A, B and C (except C = 0)*

(*b*) *if* $A = -B$, $C = 2B$, *amplitude* = $|B\sqrt{2}|$

(*c*) *if* $A = B$, $C = 0$

(*d*) *if* $A = B$, $C = 2B$, *amplitude* = $|B|$ (I.I.T. 2006)

Solution

$$x = \frac{A}{2}(1 - \cos 2\omega t) + \frac{B}{2}(1 + \cos 2\omega t) + \frac{C}{2} \sin 2\omega t$$

$$= \frac{A+B}{2} + \frac{B-A}{2} \cos 2\omega t + \frac{C}{2} \sin 2\omega t$$

When $A = B$ and $C = 0$, the motion is not simple harmonic. The motion is simple harmonic for any value of A, B and C (except $C = 0$).

If

$$A = -B \text{ and } C = 2B,$$

$$x = B \cos 2\omega t + B \sin 2\omega t$$

which is SHM with amplitude = $\sqrt{B^2 + B^2} = |B\sqrt{2}|$.

If $$A = B \text{ and } C = 2B,$$

$$x = B + B \sin 2\omega t$$

which is SHM with amplitude $= |B|$

Correct Choice: (*a*), (*b*), (*d*).

20. *Two masses m_1 and m_2 connected by a light spring of natural length l is compressed completely and tied by a light thread. This system while moving with velocity v along the positive x-axis passes through the origin at time $t = 0$. At this position the thread snaps. Find the positions of two masses as functions of time for $t \geq 0$.*

Solution

Suppose x_1 and x_2 are the positions of two masses m_1 and m_2 at time t. The differential equations satisfied by x_1 and x_2 are

$$m_1 \ddot{x}_1 = k(x_2 - x_1 - l)$$

$$m_2 \ddot{x}_2 = -k(x_2 - x_1 - l)$$

where k is the spring constant of the spring.

From these two equations, we get

$$\ddot{x}_2 - \ddot{x}_1 = -\frac{k}{m_1}(x_2 - x_1 - l) - \frac{k}{m_2}(x_2 - x_1 - l)$$

or $$\frac{d^2}{dt^2}(x_2 - x_1 - l) + \left(\frac{k}{m_1} + \frac{k}{m_2}\right)(x_2 - x_1 - l) = 0$$

Thus, $$x_2 - x_1 - l = A \cos(\omega t + \phi)$$

where $$\omega^2 = \frac{k}{m_1} + \frac{k}{m_2}$$

At $t = 0$, $x_2 = x_1 = 0$,

$$-l = A \cos \phi.$$

Again, at $t = 0$, $\dot{x}_2 - \dot{x}_1 = 0$.

$$0 = -A\omega \sin \phi.$$

Thus, $\phi = 0$, $A = -l$.

and $$x_2 - x_1 - l = -l \cos \omega t$$

or $$x_2 - x_1 - l = l(1 - \cos \omega t)$$

We know that the centre of mass moves with velocity V.

$$x_{cm} = \frac{m_1 x_1 + m_2 x_2}{m_1 + m_2} = Vt.$$

Solving these two equations, we get

$$x_1 = Vt - \frac{m_2}{m_1 + m_2} l(1 - \cos \omega t)$$

$$x_2 = Vt + \frac{m_1}{m_1 + m_2} l(1 - \cos \omega t).$$

21. *The coordinates of a particle moving in a plane are given by*

$$x(t) = a \cos pt,$$
$$y(t) = b \sin pt$$

where a, b (< a) and p are positive constants of appropriate dimensions. Then

(a) *the path of the particle is an ellipse.*

(b) *the velocity and acceleration of the particle are normal to each other at* $t = \pi/(2p)$.

(c) *the acceleration of the particle is always directed towards a focus.*

(d) *the distance travelled by the particle in time interval* $t = 0$ *to* $t = \pi/2p$ *is a.*

(I.I.T. 1999)

Solution

We have

$$x^2/a^2 + y^2/b^2 = 1$$

which is the equation of an ellipse with semimajor axis = a and semiminor axis = b.

Let $\vec{r}(t) = a \cos pt\ \hat{i} + b \sin pt\ \hat{j}$

Velocity $\vec{V}(t) = -ap \sin pt\ \hat{i} + bp \cos pt\ \hat{j}$

Acceleration $\vec{f}(t) = -ap^2 \cos pt\ \hat{i} - bp^2 \sin pt\ \hat{j}$

At $t = \dfrac{\pi}{2p}$, $\vec{V} = -ap\,\hat{i}$ and $= -\vec{f} - bp^2\ \hat{j}$.

Thus, at $t = \dfrac{\pi}{2p}$, $\vec{V} \cdot \vec{f} = 0$, the velocity and acceleration of the particle are normal to each other.

At $t = \dfrac{\pi}{2p}$, $\vec{f} = -\ bp^2\ \hat{j}$ which is directed along the y-axis. The acceleration is not always directed towards a focus.

At $t = 0$, $r = a$ and $t = \dfrac{\pi}{2p}$, $r = b$. Thus, the distance travelled by the particle in the time interval $t = 0$ to $t = \pi/2p$ is not a.

Correct Choice : (*a*), (*b*).

SUPPLEMENTARY PROBLEMS

1. Two simple harmonic motions

$$x_1 = a \sin(\omega t + \phi),$$
$$x_2 = b \cos(\omega t + \phi)$$

act on a particle along the same straight line. Find the amplitude of the resultant motion.

2. A point moves in the xy plane according to the law

$$x = A \sin \omega t,$$
$$y = B \cos \omega t.$$

Find (a) the trajectory equation $y(x)$ of the point and the direction of its motion along the trajectory, (b) the acceleration $\vec{a}$ of the point as a function of its radius vector $\vec{r}$ relative to the origin of the co-ordinates.

3. Find the trajectory $y(x)$ of a point if it moves according to the following equations
(a) $x = a \sin \omega t,\ y = a \sin 2\omega t$
(b) $x = a \sin \omega t,\ y = a \cos 2\omega t$
Plot these trajectories.

4. Discuss the formation of Lissajous' figures due to superposition of two vibrations

$$x = a \sin (2\omega t + \delta),$$
$$y = b \sin \omega t.$$

in a plane acting at right angles to each other.

5. In problem 4 consider the cases when (a) $\delta = 0$, (b) $\delta = \frac{\pi}{4}$, (c) $\delta = \frac{\pi}{2}$, (d) $\delta = \frac{3\pi}{4}$.

6. A particle of mass m moves in two dimensions under the following potential energy function:

$$V(\vec{r}) = \frac{1}{2}k(x^2 + 4y^2).$$

Find the resulting motion, given the initial conditions at $t = 0 : x = a,\ y = 0,\ \dot{x} = 0,\ \dot{y} = V_0$.

[**Hints:** The force function is

$$\vec{F} = -\vec{\nabla}V = -kx\hat{i} - 4ky\,\hat{j}$$

and the differential equations are

$$m\ddot{x} + kx = 0\ ;\ m\ddot{y} + 4ky = 0]$$

7. Two tuning forks A and B are used to produce Lissajous' figures. It is found that the figure completes its cycle in 10 seconds. If the frequency of A (slightly greater than that of B) is 256 Hz, calculate the frequency of B.

8. Lissajous' figures are produced with two tuning forks whose frequencies are approximately in the ratio 2 : 1. It takes 5s to go through a cycle of changes. On loading slightly the fork of higher frequency, the period of cycle is raised to 10s. If the frequency of the lower fork is 256 Hz, what is the frequency of the other fork, before and after loading?

9. Two tuning forks A and B of nearly equal frequencies are employed to obtain Lissajous' figures and it is observed that the figures go through a cycle of changes in 10 s. On slightly loading A with wax, the figures go through cycle in 5 s. If the frequency of B is 256 Hz, what is the frequency of A before and after loading?

10. Two bodies of masses m and $2m$ are attached to each other and to two fixed points by three identical light springs along a straight line. The springs are stretched so the tension in each spring is T and its length L (much greater than its unstretched

length). Show that the angular frequencies of the normal modes for longitudinal oscillations of small amplitude are given by

$$\omega^2 = \frac{3 \pm \sqrt{3}}{2} \frac{T}{mL}.$$

[**Hints:** Since L is much greater than the unstretched length of the springs, $T = kL$, where k is the spring constant.]

11. Two bodies of masses m and $3m$ are attached to each other to two fixed points by three identical light springs of relaxed length a_0 along the x-axis. At equilibrium each of the springs has length a and exerts a tension, $T = k\ (a - a_0)$ where k is the spring constant. Show that the angular frequencies of the normal modes for transverse oscillations of small amplitude are given by

$$\omega^2 = \frac{4 \pm \sqrt{7}}{3} \frac{T}{ma}.$$

12. Two masses m_1 and m_2, initially at rest on a frictionless surface are connected by a spring of force constant k and natural length a. The spring is compressed to $\frac{3}{4}$ of its natural length and released from rest. Determine the subsequent displacements of both masses.

[**Hints:**
$$m_1 \ddot{x}_1 = k(x_2 - x_1 - a),$$
$$m_2 \ddot{x}_2 = -k(x_2 - x_1 - a)$$

where x_1 is the displacement of m_1 from its original position and x_2 is the displacement of m_2 from the original position of m_1. Thus, $x_1\ (0) = 0$ and $x_2\ (0) = \frac{3}{4}a$. We have

$$x_2 - x_1 - a = A \cos(\omega t + \phi)$$

with
$$\omega^2 = \frac{k(m_1 + m_2)}{m_1 m_2},\ A = -\frac{1}{4}a \text{ and } \phi = 0\Bigg].$$

13. Two masses m_1 and m_2 are connected by springs of spring constants k_1 and k_2 and natural lengths L_1 and L_2 (Fig. 2.20). The point P is fixed and O_1 and O_2 mark the equilibrium positions of the springs. Suppose at $t = 0$, m_1 is displaced a distance a_1 from O_1 and m_2 is displaced a distance a_2 from O_2 and then both masses are released from rest. Find the position $x_1\ (t)$ of m_1 and position $x_2\ (t)$ of m_2 for all $t > 0$. Assume no friction or external forces. Take $m_1 = m_2 = 1$, $k_1 = 3$, $k_2 = 2$, $a_1 = -1$ and $a_2 = 2$.

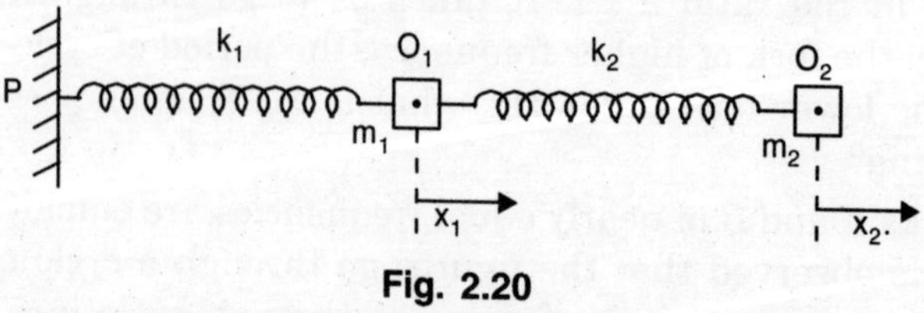

Fig. 2.20

14. A uniform bar AB of length L and mass m is supported at its end by identical springs with spring constant k. Motion is set by depressing the end B by a small distance a and releasing it from rest (Fig. 2.21). Solve the problem of motion, identifying normal modes and frequencies.

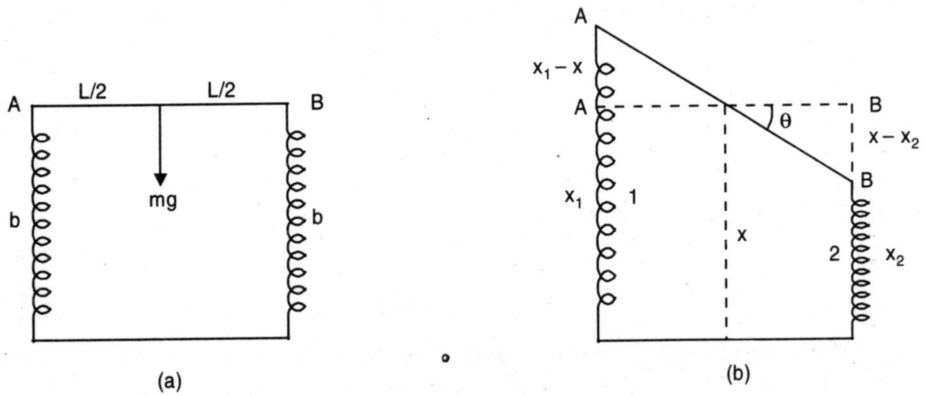

Fig. 2.21

[**Hints:** At equilibrium position, length of each spring = $b = l - \frac{mg}{2k}$, where l = relaxed length of each spring. At the general configuration, we have

$$x_1 + x_2 = 2x,$$

$$m\frac{d^2}{dt^2}(x - b) = -2k(x - b),$$

$$I\ddot{\theta} = -\frac{kL^2}{2}\theta$$

where x_1, x_2 and x are the lengths of spring 1, spring 2 and height of the centre of mass of the rod in Fig. 2.21 (*b*).

$$I = \text{Moment of inertia of the rod} = \frac{mL^2}{12}\Bigg].$$

15. Solve problem 11 (page 72) of longitudinal oscillations of two coupled masses in which $m_1 = m_2 = m$ and the springs 1 and 3 have spring constant k_1 and the spring 2 (middle spring) has spring constant k_2.

16. Consider the motion of a three-particle system in which the particles all lie in a straight line. The two end particles, each of mass *m*, are bound to the central particle of mass *M* through the springs of stiffness *k* as shown in Fig. 2.22. x_1, x_2 and x_3 are the displacements of three masses from their respective equilibrium positions. (*i*) Find the differential equations of motion for the particles. (*ii*) Show that the normal mode frequencies are $\omega_1 = 0$.

$$\omega_2 = \sqrt{k/m} \text{ and } \omega_3 = \left(\frac{k}{m} + \frac{2k}{M}\right)^{\frac{1}{2}}$$

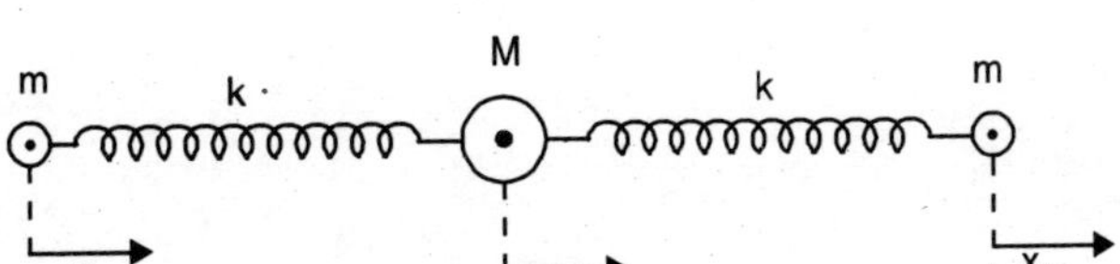

Fig. 2.22

(*iii*) Describe the motion of the three particles in these modes.

(*iv*) Assuming such a model for carbon dioxide molecule find the ratio ω_3/ω_2.

17. A particle of mass *m* moves in the plane $z = 0$ under the attractive force $2\pi^2 mr$ towards the origin, *r* being the distance from the origin. In addition there is a force of magnitude $m\pi v$ in the direction $\vec{v} \times \hat{k}$ where $\vec{v}$ is the velocity of the particle and $\hat{k}$ is a unit vector perpendicular to $z = 0$. Show that

$$\pi\dot{y} - 2\pi^2 x = \ddot{x},$$

$$-\pi\dot{x} - 2\pi^2 y = \ddot{y}.$$

Solve these equations for *x* and *y* and show that the motion in the *x* and *y* directions is the sum of two simple harmonic oscillations of periods 1 and 2.

[**Hints:** We have

$$m(\ddot{x}\hat{i} + \ddot{y}\hat{j}) = -2\pi^2 m(x\hat{i} + y\hat{j}) + \frac{m\pi v}{v}\begin{vmatrix} \hat{i} & \hat{j} & \hat{k} \\ \dot{x} & \dot{y} & 0 \\ 0 & 0 & 1 \end{vmatrix}.$$

18. If two tuning forks *A* and *B* are sounded together they produce 5 beats per second. When *A* is slightly loaded with wax they produce 3 beats per second when sounded together. Find the original frequency of *A* if the frequency of *B* is 380 Hz.

19. You are given four tuning forks. The fork with the lowest frequency vibrates at 500 Hz. By using two tuning forks at a time, the following beat frequencies are heard: 1, 2, 3, 5, 7 and 8 Hz. What are the possible frequencies of the other tuning forks?

20. You are given five tuning forks, each of which has a different frequency. By trying every pair of tuning forks find (*a*) maximum number of different beat frequencies (*b*) minimum number of different beat frequencies.

21. Two identical piano wires have a fundamental frequency of 400 Hz when kept under the same tension. What fractional increase in the tension of one wire will lead to the occurrence of 4 beats/s when both wires vibrate simultaneously?

3 The Damped Harmonic Oscillator

3.1 DAMPED HARMONIC MOTION

Various frictional forces may act on a harmonic oscillator, tending to reduce its successive amplitudes. Such a motion is called damped harmonic motion. Suppose a particle of mass m is subject to a restoring force proportional to the distance from a fixed point on the x-axis and a damping force proportional to the velocity. Then the equation of motion becomes

$$m\ddot{x} = -kx - \beta\dot{x} \qquad \text{...(3.1)}$$

x being the displacement of the particle from the fixed point at any instant t, k and β are positive constants. The damping force is $-\beta\dot{x}$ where β is the damping coefficient. Equation (3.1) can be written as

$$\ddot{x} + 2b\dot{x} + \omega^2 x = 0 \qquad \text{...(3.2)}$$

where $2b = \beta/m$ and $\omega = \sqrt{k/m}$ is the natural frequency of the oscillator. The relaxation time τ is defined by

$$\tau = \frac{1}{2b} = \frac{m}{\beta}. \qquad \text{...(3.3)}$$

3.2 DAMPED LC OSCILLATIONS (LCR CIRCUIT)

If resistance R is present in an LC circuit, the total energy

$$U = \frac{1}{2}Li^2 + \frac{q^2}{2C} \qquad \text{...(3.4)}$$

is no longer constant, but decreases with time as it is transformed steadily to thermal energy in the resistor:

$$\frac{dU}{dt} = -i^2R$$

Hence,

$$Li\frac{di}{dt} + \frac{q}{C}\frac{dq}{dt} = -i^2R$$

Substituting $i = \dfrac{dq}{dt}$ and $\dfrac{di}{dt} = \dfrac{d^2q}{dt^2}$ and dividing by i, we get

$$L\frac{d^2q}{dt^2} + R\frac{dq}{dt} + \frac{1}{C}q = 0 \qquad \text{...(3.5)}$$

or $$\ddot{q} + 2b\,\dot{q} + \omega^2 q = 0 \qquad ...(3.6)$$

which is the differential equation that describes the damped oscillations. This is equivalent to Eqn. (3.2).

The general solution of Eqn. (3.6) can be written as (see Eqn. (3.10)) $[b < \omega]$

$$q = Q\, e^{-bt} \cos(\omega' t - \theta). \qquad ...(3.7)$$

Here, $$b = \frac{R}{2L},\quad \omega = \frac{1}{\sqrt{Lc}},\quad \omega' = \sqrt{\omega^2 - b^2}.$$

SOLVED PROBLEMS

1. *Obtain an expression for the displacement of the damped harmonic oscillator where the damping force is proportional to the velocity. Discuss the effect of the damping on the displacement and frequency of the oscillator.*

Solution

The differential equation of the damped harmonic motion is given by Eqn. (3.2):

$$\ddot{x} + 2\dot{x} + \omega^2 x = 0$$

where $(-2mb\dot{x}) = (-\beta\dot{x})$ is the damping force acting on the particle of mass m and ω is the natural frequency of the oscillator.

Let $x = \exp(\alpha t)$ be the trial solution of above equation. Then, we have

$$\alpha^2 + 2b\alpha + \omega^2 = 0.$$

The two roots of α are

$$\alpha_{1,2} = \frac{-2b \pm \sqrt{4b^2 - 4\omega^2}}{2} = -b \pm \sqrt{b^2 - \omega^2}$$

So, the general solution of Eqn. (3.2) is

$$x = e^{-bt}\left[A_1 \exp(\sqrt{b^2 - \omega^2}\,t) + A_2 \exp(-\sqrt{b^2 - \omega^2}\,t)\right] \qquad ...(3.8)$$

where A_1 and A_2 are two constants whose values can be determined from the initial conditions.

Case I: $b > \omega$ (the damping force is large)

The expression (3.8) for x represents a damped dead beat motion, the displacement x decreasing exponentially to zero (Fig. 3.1).

Case II: $b < \omega$ (the damping is small)

In this case $\sqrt{b^2 - \omega^2} = i\sqrt{\omega^2 - b^2} = i\omega'$ and Eqn. (3.8) becomes

$$x = e^{-bt}\,[B_1 \cos \omega' t + B_2 \sin \omega' t] \qquad ...(3.9)$$

where $\omega' = \sqrt{\omega^2 - b^2}$, $B_1 = A_1 + A_2$ and $B_2 = i\,(A_1 - A_2)$. If we write $B_1 = R \cos\theta$ and $B_2 = R \sin\theta$, Eqn. (3.9) reduces to

$$x = Re^{-bt} \cos(\omega' t - \theta) \qquad ...(3.10)$$

where $R = \sqrt{B_1^2 + B_2^2}$ and $\theta = \tan^{-1}(B_2/B_1)$

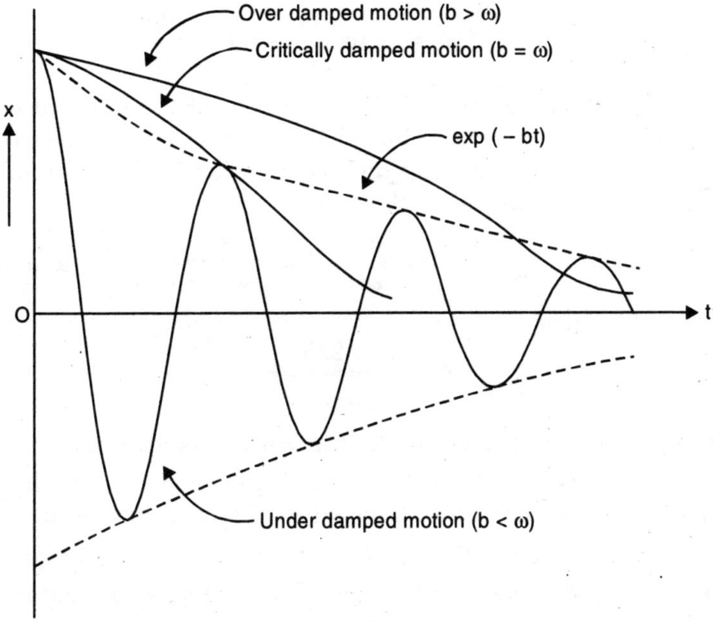

Fig. 3.1

Equation (3.10) gives a damped oscillatory motion (Fig. 3.1). Its amplitude R exp $(-bt)$ decreases exponentially with time. The time period of damped oscillation is

$$T = \frac{2\pi}{\omega'} = \frac{2\pi}{\sqrt{\omega^2 - b^2}} \qquad ...(3.11)$$

whereas the undamped time period is

$$T_0 = \frac{2\pi}{\omega}.$$

Thus the time period of damped oscillation is slightly greater than the undamped natural time period when $b \ll \omega$. In other words, the frequency of the damped oscillation $\omega' = \sqrt{\omega^2 - b^2}$ is less than the undamped natural frequency ω.

Let us consider the simple case in which $\theta = 0$ in Eqn. (3.10). Then $\cos \omega' t = \pm 1$ when $t = 0, t_1 = \frac{T}{2}, t_2 = T, t_3 = \frac{3T}{2}$ etc. Suppose that the values of x in both directions corresponding to these times are x_0, x_1, x_2, x_3 etc. so that

$$\begin{aligned} x_0 &= R \\ x_1 &= -Re^{-bT/2}, \\ x_2 &= Re^{-bT}, \\ x_3 &= -Re^{-3bT/2}, \\ &\vdots \end{aligned}$$

Considering the absolute values of the displacements, we get

$$\left|\frac{x_0}{x_1}\right| = \left|\frac{x_1}{x_2}\right| = \left|\frac{x_2}{x_3}\right| = \ldots = e^{bT/2}$$

The quantity $$\delta = 2 \ln \left| \frac{x_n}{x_{n+1}} \right| = bT = \frac{b}{\nu} \qquad ...(3.12)$$

is called logarithmic decrement. Here ν is the frequency of the damped oscillatory motion. The logarithmic decrement is the logarithm of the ratio of two successive maxima in one direction = ln (x_n/x_{n+2}). Thus the damping coefficient b can be found from an experimental measurement of consecutive amplitudes. Since

$$b = \frac{\beta}{2m},\ T = \frac{2\pi}{\sqrt{\omega^2 - b^2}} = \frac{1}{\nu} \text{ and } \omega^2 = k/m,$$

We have from Eqn. (3.12)

$$\delta = \frac{2\pi\beta}{\sqrt{4mk - \beta^2}}. \qquad ...(3.13)$$

The energy equation of the damped harmonic oscillator:

We can regard equation (3.10) as a cosine function whose amplitude R exp $(-\ bt)$ gradually decreases with time. For an undamped oscillator of amplitude R, the mechanical energy is constant and is given by $E = \frac{1}{2}kR^2$. If the oscillator is damped, the mechanical energy is not constant but decreases with time. For a damped oscillator the amplitude is R exp $(-\ bt)$ and the mechanical energy is

where $$E\ (0h) = \frac{1}{2}\ kR^2 \text{ is the initial mechanical energy.}$$

$$E\ (t) = \frac{1}{2}\ kR^2 \exp\ (-2bt)$$

Like the amplitude the mechanical energy decreases exponentially with time.

Case III: $b = \omega$ *(critically damped motion)*

When $b = \omega$, we get only one root $\alpha = -\ b$. One solution of Eqn. (3.2) is

$$x_1 = A_1 \exp\ (-bt)$$

and the other solution is

$$x_2 = A_2 t \exp\ (-bt).$$

So the general solution for critically damped motion (Fig. 3.1) is

$$x = e^{-bt}(A_1 + A_2 t) \qquad ...(3.14)$$

The motion is non-oscillatory and the particle approaches origin slowly.

2. *A particle of mass 3 moves along the x-axis attracted toward origin by a force whose magnitude is numerically equal to 12x. The particle is also subjected to a damping force whose magnitude is numerically equal to 12 times the instantaneous speed. If it is initially at rest at x = 10, find the position and the velocity of the particle at any time.*

Solution

The equation of motion of the particle is

$$3\ddot{x}\hat{i} = -12\dot{x}\hat{i} - 12x\hat{i}$$

or $$\ddot{x} + 4\ddot{x} + 4x = 0.$$

The solution of this equation is

$$x = e^{-2t}(A_1 + A_2 t).$$

[See Eqn. (3.14)]. The motion is critically damped ($b = \omega = 2$).

When $t = 0$, $x = 10$ and $\dot{x} = 0$; therefore $A_1 = 10$ and $A_2 = 20$. The position of the particle at any time t is

$$x = 10e^{-2t}(1 + 2t).$$

The velocity is given by

$$\vec{v} = \dot{x}\hat{i} = -40t\, e^{-2t}\hat{i}\,.$$

Initially $\vec{v} = 0$ and the velocity becomes very small after a long time. The magnitude of the velocity is maximum when $t = \frac{1}{2}$.

3. *A particle of mass 1 g moves along the x-axis under the influence of two forces: (i) a force of attraction toward origin which is numerically equal to 4x dynes, and (ii) a damping force whose magnitude in dynes is numerically equal to twice the instantaneous speed. Assuming that the particle starts from rest at a distance 10 cm from the origin, (a) set up the differential equation of motion of the particle, (b) find the position of the particle at any time, (c) determine the amplitude, period and frequency of the damped oscillation, and (d) find the logarithmic decrement of the problem.*

Solution

(*a*) The differential equation for the motion of the particle is

$$\ddot{x}\hat{i} = -4x\hat{i} - 2\dot{x}\hat{i}$$

or

$$\ddot{x} + 2\dot{x} + 4x = 0.$$

This is an example where the motion is damped oscillation ($b = 1$, $\omega = 2$ and $b < \omega$).

(*b*) The solution of this equation is

$$x = R\, e^{-t} \cos\,(\sqrt{3}\,t - \theta)$$

[see Eqn. (3.10)]. Since $x = 10$ cm at $t = 0$, we find that $R \cos \theta = 10$.

Since $\dot{x} = 0$ at $t = 0$, we have

$$-R\cos\theta + \sqrt{3}R\sin\theta = 0$$

Hence $$\theta = \frac{\pi}{6} \text{ and } R = \frac{20}{\sqrt{3}} \text{ cm}.$$

Thus, we obtain $$x = \frac{20}{\sqrt{3}} e^{-t} \cos\left(\sqrt{3}\,t - \frac{\pi}{6}\right)$$

(*c*) Amplitude $= \frac{20}{\sqrt{3}} e^{-t}$ cm. The amplitudes of oscillation decrease towards zero as t increases.

$$\text{Period} = \frac{2\pi}{\sqrt{3}}\text{s and frequency} = \frac{\sqrt{3}}{2\pi}\text{Hz}.$$

(*d*) From Eqn. (3.12), the logarithmic decrement δ is given by

$$\delta = bT = \frac{2\pi}{\sqrt{3}}.$$

4. *A system of unit mass whose natural angular frequency in the absence of damping is 4 rad s^{-1} is subject to a damping force which is proportional to the velocity of the system, the constant of proportionality being 10 s^{-1}. Show that the system is over damped and that the general solution for the displacement is*

$$x = A \exp(-2t) + B \exp(-8t).$$

The mass is initially at $x = 0.5$ m and given an initial velocity v towards the equilibrium position. Find the smallest value of v that will produce negative displacement.

Solution

The equation of motion is

$$m\ddot{x} + \beta\dot{x} + kx = 0$$

In the present problem, $m = 1$, $\beta = 10$ and $\omega = \sqrt{k/m} = 4$ rad s^{-1}. Thus $2b = \beta/m = 10$ and $k = 16$. Since $b > \omega$, the damping force is large and the motion is over damped. The general solution for x is [see Eqn. (3.8)]

$$\begin{aligned} x &= \exp(-5t)\,[A \exp(3t) + B \exp(-3t)] \\ &= A \exp(-2t) + B \exp(-8t). \end{aligned}$$

Now at $t = 0$, $x = 0.5$ m and $\dot{x} = -v$, where v is a positive number. Thus, we have

$$\begin{aligned} A + B &= 0.5 \\ -2A - 8B &= -v \end{aligned}$$

which give

$$x = \frac{4-v}{6}\exp(-2t) + \frac{v-1}{6}\exp(-8t).$$

In order to make x negative it is necessary that $\frac{4-v}{6}$ should be negative or, $v > 4$. Thus $v = 4$ ms^{-1} is the minimum value of v required to give a negative displacement to the system.

5. *In a damped oscillatory motion an object oscillates with a frequency of 1 Hz and its amplitude of vibration is halved in 5 s. Find the differential equation for the oscillation. Find also the logarithmic decrement of the problem.*

Solution

The variation of x with t correspond to an underdamped decaying oscillation with differential equation

$$\ddot{x} + 2b\dot{x} + \omega^2 x = 0$$

where $b < \omega$. The solution of this equation is

$$x = R\,e^{-bt}\cos[\sqrt{\omega^2 - b^2}\,t - \theta].$$

It is given that $\sqrt{\omega^2 - b^2} = 2\pi.1 = 2\pi$

and

$$Re^{-5b} = R/2 \text{ or, } b = \frac{1}{5}\ln 2 = 0.139$$

Thus,

$$\omega^2 = b^2 + 4\pi^2 = 39.50.$$

Hence, the required differential equation is

$$\ddot{x} + 0.278\,\dot{x} + 39.50x = 0.$$

The logarithmic decrement $\delta = b/v = 0.139$.

6. *The energy of recoil of a rocket launcher of mass m = 4500 kg is absorbed in a recoil spring. At the end of the recoil, a damping dashpot is arranged in such a way that the launcher returns to the firing position without any oscillation (critical damping). The launcher recoils 3 m with an initial speed of 10 ms^{-1}. Find (a) the recoil's spring constant and (b) the dashpot's coefficient of critical damping.*

Solution

(*a*) We use the principle of conservation of energy for the rocket launcher and the recoil spring:

$$\text{K.E.} + (\text{P.E})_{\text{elastic}} = \text{constant.}$$

Thus,

$$\frac{1}{2}mv_0^2 + 0 = 0 + \frac{1}{2}kx_{\max}^2$$

giving

$$k = \frac{mv_0^2}{x_{\max}^2} = \frac{(4500)\,(10)^2}{3^2}\text{s} = 50 \text{ kN m}^{-1}$$

(*b*) The coefficient of critical damping is given by

$$\beta = 2mb = 2m\omega = 2\sqrt{mk} = 2\sqrt{(4500)\,(50000)} = 30 \text{ kN sm}^{-1}.$$

[Since $\beta\dot{x}$ is force, β has the unit of Nsm^{-1}].

7. *Solve the problem of simple pendulum if a damping force proportional to the instantaneous tangential velocity is taken into account.*

Solution

The instantaneous tangential velocity of a simple pendulum of length l and mass m is $l\frac{d\psi}{dt}$ [see problem 24 of Chapter 1]. The equation of motion of the damped simple pendulum is given by

$$ml\,\frac{d^2\psi}{dt^2} = -\,mg\sin\psi - \beta l\frac{d\psi}{dt}$$

Replacing $\sin\psi$ by ψ for small oscillations we have the equation

$$\frac{d^2\psi}{dt^2} + 2b\,\frac{d\psi}{dt} + \omega^2\psi = 0$$

where

$$b = \frac{\beta}{2m} \text{ and } \omega = \sqrt{g/l}$$

is the undamped frequency of the simple pendulum. Now we may discuss in the light of Eqn. (3.2). Three cases arise.

(1) $\frac{\beta^2}{4m^2} > \frac{g}{l}$ (over damped motion),

$$\psi = \exp\frac{-\beta t}{2m}\,[A_1\exp(\lambda t) + A_2\exp(-\lambda t)]$$

where

$$\lambda = \left[\frac{\beta^2}{4m^2} - \frac{g}{l}\right]^{1/2}.$$

(2) $\dfrac{\beta^2}{4m^2} = \dfrac{g}{l}$ (critically damped motion),

$$\psi = \exp\left(-\frac{\beta t}{2m}\right)[A_1 + A_2 t].$$

(3) $\dfrac{\beta^2}{4m^2} < \dfrac{g}{l}$ (damped oscillations or under damped motion)

$$\psi = \exp\left(-\frac{\beta t}{2m}\right)[B_1 \cos \omega' t + B_2 \sin \omega' t]$$

where
$$\omega' = \left[\frac{g}{l} - \frac{\beta^2}{4m^2}\right]^{1/2}.$$

In cases (1) and (2) the pendulum bob gradually returns to the equilibrium position without oscillation. In case (3), the pendulum oscillates with frequency ω' and its amplitude decreases exponentially with time.

8. *(a) A steady force of 50 N is required to lift a mass of 2 kg vertically through water at a constant velocity of 2.5 m/s. Assuming that the effect of viscosity can be described by a force proportional to velocity, determine the constant of proportionality. (The effect of buoyancy is neglected.)*

(b) The same mass is then suspended in water by a spring with spring constant $k = 120\ Nm^{-1}$. Determine the equilibrium extension of the spring. The mass is further pulled through a small distance below its equilibrium position and released from rest at time $t = 0$. Show that it will vibrate about the equilibrium position according to an equation of the form

$$\ddot{x} + 2b\dot{x} + \omega^2 x = 0$$

and determine b and ω for this system. Show that the motion is under damped and find its period of oscillation. Find the time in which the amplitude of oscillation falls by a factor of e.

Solution

(*a*) When the mass is lifted vertically it moves with a constant velocity.

Thus the net force acting on it is zero:

$$50 - mg - \beta v = 0$$

or
$$\beta = \frac{50 - mg}{v} = \frac{50 - 2 \times 9.8}{2.5} = 12.16 \text{ N sm}^{-1} (\equiv \text{kg s}^{-1})$$

where β is the constant of proportionality.

(*b*) When the mass is suspended by a spring of spring constant *k*, it will be stationary at equilibrium and no viscous force will act on it. If *X* is the extension of the spring, we have

$$mg = kX$$

or
$$X = \frac{2 \times 9.8}{120} = 0.163 \text{ m}.$$

When the mass is pulled through a small distance below its equilibrium position and then released, it will vibrate to and fro about the equilibrium position. Now a resistive force of amount $\beta\dot{x}$ will act on the body. Therefore the equation of motion is given by

$$m\ddot{x} = -\beta\dot{x} - kx$$

or $$\ddot{x} + 2b\dot{x} + \omega^2 x = 0$$

where $$b = \frac{\beta}{2m} = \frac{12.16}{2 \times 2} = 3.04 \text{ s}^{-1}.$$

and $$\omega = \sqrt{k/m} = \sqrt{120/2} = 7.746 \text{ rad.s}^{-1}$$

Since $b < \omega$, the body will execute damped simple harmonic motion:

$$x = Re^{-bt} \cos[\sqrt{\omega^2 - b^2}\, t - \theta].$$

Its period of oscillation is

$$T = \frac{2\pi}{\sqrt{\omega^2 - b^2}} = 0.88 \text{ s}.$$

If t is the time in which the amplitude falls by a factor of e, then

$$Re^{-bt} = R/e \text{ or, } t = \frac{1}{b} = 0.33 \text{ s}.$$

9. *A point performs damped oscillations with frequency ω and amplitude $a_0 \exp(-bt)$. Find the velocity amplitude of the point as a function of time t if at the initial moment $t = 0$ (i) its displacement is equal to a_0, (ii) its displacement is zero.*

Solution

(*i*) $$x = a_0 e^{-bt} \cos \omega t.$$

$$\dot{x} = -ba_0 e^{-bt} \cos \omega t - a_0 \omega e^{-bt} \sin \omega t$$

$$= Re^{-bt} \cos(\omega t - \theta)$$

where $$R \cos\theta = -ba_0 \text{ and } R \sin\theta = -a_0\omega.$$

Thus, $$R = a_0 \sqrt{\omega^2 + b^2}$$

and $$\text{the velocity amplitude} = a_0 \sqrt{\omega^2 - b^2}\, e^{-bt}.$$

(*ii*) $$x = a_0 e^{-bt} \sin \omega t,$$

$$\dot{x} = -ba_0 e^{-bt} \sin \omega t + a_0 \omega e^{-bt} \cos \omega t$$

$$= Re^{-bt} \cos(\omega t + \theta),$$

where $$R \cos\theta = a_0\omega \text{ and } R \sin\theta = ba_0.$$

Thus $$R = a_0 \sqrt{\omega^2 + b^2}$$

and $$\text{the velocity amplitude} = a_0 \sqrt{\omega^2 + b^2}\, e^{-bt}.$$

10. *A spring supports a mass of 5 g which performs damped oscillations. It is found to have successive maxima of 2.1 cm and 1.3 cm. If the damping coefficient β is 10 g s^{-1}, find the stiffness of the spring.*

Solution

$$\text{log decrement } \delta = \ln\left(\frac{2.1}{1.3}\right) = b.\frac{2\pi}{\omega'},$$

where $$b = \frac{\beta}{2m} = \frac{10}{2 \times 5} \text{ s}^{-1} = 1 \text{ s}^{-1}.$$

Thus, $$\omega' = \frac{2\pi b}{\delta} = 13.1016$$

and $$\omega^2 = \omega'^2 + b^2 = 172.652 = k/m$$

Hence, $$k = m\omega^2 = 0.005 \times 172.652 = 0.863 \text{ N/m}.$$

11. *An automobile suspension system is critically damped and its period of oscillation with no damping is one second. If the system is initially displaced by an amount x_0, and released with zero initial velocity, find the displacement at t = 2s.*

Solution

The displacement for critically damped motion is given by

$$x = e^{-bt}(A_1 + A_2 t)$$

with $$b = \omega = 2\pi.$$

Now, $$\dot{x} = -be^{-bt}(A_1 + A_2 t) + A_2 e^{-bt}.$$

At $t = 0$, $x_0 = A_1$ and $\dot{x} = 0 = -bA_1 + A_2$.

which give $A_1 = x_0$ and $A_2 = 2\pi x_0$.

Thus, $$x = x_0(1 + 2\pi t)\, e^{-2\pi t}$$

At $$t = 2s,\ x = 0.000047\, x_0.$$

12. *For a damped harmonic oscillator with equation*

$$m\ddot{x} + \beta\dot{x} + kx = 0$$

show that the work done against the damping force in an infinitesimal time dt is equal to the loss of energy of the mass m during the same time interval dt.

Solution

In an infinitesimal time dt during which the mass m traverses a distance dx, the work done δw against the damping force $\beta\dot{x}$ is given by

$$\delta w = \beta\left(\frac{dx}{dt}\right)dx = \beta\left(\frac{dx}{dt}\right)\left(\frac{dx}{dt}\right)dt = \beta\left(\frac{dx}{dt}\right)^2 dt.$$

Total energy of the system is given by

$$E = \frac{1}{2}m\left(\frac{dx}{dt}\right)^2 + \frac{1}{2}kx^2.$$

Thus the loss of energy of the mass in time dt is

$$-\frac{dE}{dt}dt = \left[-m\frac{dx}{dt}\frac{d^2x}{dt^2} - kx\frac{dx}{dt}\right]dt$$

$$= \left[-\frac{dx}{dt}\left\{-\beta\frac{dx}{dt} - kx\right\} - kx\frac{dx}{dt}\right]dt$$

$$= \beta\left(\frac{dx}{dt}\right)^2 dt$$

$$= \delta w.$$

13. *If the quality factor Q in a damped harmonic motion is defined as*

$$Q = \frac{2\pi \times \textit{average energy stored per cycle}}{\textit{Average energy dissipated per cycle}}$$

then show that

$$Q = \frac{\omega}{2b}.$$

Solution

The displacement x of the mass m at any time is given by

$$x = R\, e^{-bt} \cos(\omega' t - \theta).$$

Total energy at any instant is given by

$$E = \frac{1}{2} m\dot{x}^2 + \frac{1}{2} kx^2$$

$$= \frac{1}{2} m\, [b^2R^2e^{-2bt} \cos^2(\omega' t - \theta) + \omega'^2R^2e^{-2bt} \cdot \sin^2(\omega' t - \theta) + b\omega' R^2e^{-2bt} \sin^2(\omega' t - \theta)]$$

$$+ \frac{1}{2} kR^2e^{-2bt} \cos^2(\omega' t - \theta).$$

Loss of energy of the mass in time dt is

$$-\frac{dE}{dt} dt = \beta\dot{x}^2\, dt.$$

Assuming that e^{-2bt} remains almost constant over a cycle, the average energy over a complete cycle at that time is

$$E_{av} = \frac{1}{4} mR^2e^{-2bt}(b^2 + \omega'^2) + \frac{1}{4} kR^2e^{-2bt}.$$

$$= \frac{1}{4} R^2e^{-2bt}(m\omega^2 + k) = \frac{1}{2} m\omega^2R^2e^{-2bt}$$

since the average value of $\cos^2(\omega' t - \theta)$ and $\sin^2(\omega' t - \theta)$ are each 1/2 and that of $\sin 2(\omega' t - \theta)$ is zero.

Average energy dissipated per cycle at that time is

$$\beta(\dot{x}^2)_{av} T = \beta\left(\frac{1}{2}\omega^2R^2e^{-2bt}\right)T.$$

Hence

$$Q = \frac{2\pi \times \frac{1}{2} m\omega^2R^2e^{-2bt}}{\frac{1}{2}\beta T\omega^2R^2e^{-2bt}} = \frac{2\pi m}{\beta T} = \frac{\omega}{2b}.$$

14. *Find the quality factor Q for the damped oscillations of Problem 5.*

Solution

$$Q = \frac{\omega}{2b} = \frac{6.285}{2 \times 0.139} = 22.6.$$

15. *An LCR circuit has L = 10 mH, C = 1.0 μF, and R = 1 Ω. (a) After what time t will the amplitude of the charge oscillations drop to one-half of its initial value? (b) To how many periods of oscillation does this correspond?*

Solution

Amplitude of charge oscillation = qe^{-bt}

(*a*) We have the condition $e^{-bt} = \frac{1}{2}$

or $$t = \frac{\ln 2}{b} = \frac{(\ln 2)\,2L}{R} = \frac{(\ln 2)\times 2\times 10\times 10^{-3}}{1}$$

$$= 13.86 \text{ ms.}$$

(*b*) $$\omega = \frac{1}{\sqrt{LC}} = \frac{1}{\sqrt{10\times 10^{-3}\times 1\times 10^{-6}}} 10^4 \text{ rad/s.}$$

$$b = \frac{R}{2L} = 50$$

Since $\omega >> b$, $\omega' \approx \omega$.

The elapsed time, expressed in terms of the period of oscillation, is

$$\frac{t}{T} = \frac{\omega t}{2\pi} \approx 22.$$

Thus the amplitude drops to one-half after about 22 cycles of oscillation.

16. *In a damped LCR circuit show that the fraction of the energy lost per cycle of oscillation, $\Delta U/U$, is given to a close approximation by $\frac{2\pi R}{\omega L}$. [Assume that R is small]*

Solution

We assume that initially the current $i = 0$ and $\omega' \approx \omega$ since $b = \frac{R}{2L}$ is small.

Initially the energy of the capacitor = $U = \frac{q^2}{2C}$ and the energy of the capacitor after a time T is

$$U' = \frac{q^2}{2C} e^{-2bT}.$$

Thus, $$\frac{U'}{U} = e^{-2bT} = e^{\frac{-RT}{L}}$$

or $$1 - \frac{U'}{U} = \frac{\Delta U}{U} = 1 - e^{-\frac{R}{L}\frac{2\pi}{\omega}}$$

or $$\frac{\Delta U}{U} \approx 1 - \left(1 - \frac{2\pi R}{\omega L}\right) = \frac{2\pi R}{\omega L}$$

The quantity $\frac{\omega}{2b} = \frac{\omega L}{R}$ is called the Q of the circuit (for 'quality'). A high-Q circuit has low resistance and a low fractional energy loss per cycle $\left(= \frac{2\pi}{Q}\right)$.

17. *An object of mass 0.1 kg moves along the x-axis under the influence of two forces: (i) a force of attraction towards origin which is numerically equal to 85 x N and (ii) a damping force whose magnitude is 0.07 dx/dt N. (a) What is the period to the motion? (b) How long does it take for the amplitude of the damped oscillations at drop to half its initial value? (c) How long does it take for the mechanical energy to drop to one-half its initial value?*

Solution

(*a*) Here k = 85 N/m and β = 0.07 kg/s

$$\omega = \sqrt{k/m} = \sqrt{850}\,\text{s}^{-1} = 29.15\,\text{s}^{-1}$$

$$b = \beta/2m = 0.35$$

$b < \omega$ and the motion is damped simple harmonic.

$$T = \frac{2\pi}{\sqrt{\omega^2 - b^2}} = 0.216 \text{ s}$$

(*b*) If t is the time in which the amplitude falls by a factor of 2, then

$$R\,e^{-bt} = R/2$$

or
$$t = \frac{\ln 2}{b} = 1.98 \text{ s}$$

(*c*)
$$E(t) = E(\text{o})\,e^{-2bt} = E(\text{o})/2$$

or
$$t = \frac{\ln 2}{2b} = 0.99 \text{ s.}$$

18. *An object moves on the x-axis in such a way that its velocity and displacement from the origin satisfy the relation v = – kx, where k is a positive constant. Show that the object does not change its direction and the kinetic energy of the object keeps on decreasing.*

Solution

We have $\dfrac{dx}{dt} + kx = 0,\ k > 0$

or
$$x = A\,e^{-kt},\ A = \text{constant.}$$

and
$$\dot{x} = -Ak\,e^{-kt} = -kx$$

(*a*) Let A be positive. At $t = 0$, $x = A$ and as t increases x decreases continuously and after a long time x becomes zero.

(*b*) Let A be negative. At $t = 0$, $x = -|A|$ and as t increases x increases continuously from the negative value and after a long time it becomes zero.

So the object does not change its direction.

The kinetic energy of the object is

$$\frac{1}{2}m\dot{x}^2 = \frac{1}{2}m\,A^2k^2\,e^{-2kt}$$

which decreases with time.

19. *A particle of mass 10^{-2} kg is moving along the positive x-axis under the influence of a force*

$$F(x) = -\frac{K}{2x^2}$$

where $K = 10^{-2}$ *N* m^2. *At* $t = 0$ *it is at* $x = 1.0$ *m and its velocity* $v = 0$.

(*a*) *Find its velocity when it reaches* $x = 0.5$ *m*

(*b*) *Find the time at which it reaches* $x = 0.25$ *m.* (I.I.T. 1998)

Solution

The force experienced by the body is

$$F(x) = m\frac{dv}{dt} = m\frac{dv}{dt}\frac{dx}{dt} = mv\frac{dv}{dx} = -\frac{K}{2x^2}$$

or
$$mv\,dv = -\frac{K}{2x^2}\,dx.$$

At $t = 0$, $v = 0$ and $x = 1$ m.

Integrating from $v = 0$ to $v = v$ and $x = 1$ to $x = x$, we have

$$\frac{1}{2}mv^2 = \frac{K}{2x} + C$$

At $x = 1$, $v = 0$ and $C = -\frac{K}{2}$

Thus,
$$v = \pm\sqrt{\frac{K}{m}}\left(\frac{1-x}{x}\right)^{\frac{1}{2}}$$

when
$$x = 0.5 \text{ m},\ v = \pm\sqrt{\frac{K}{m}} = \pm 1 \text{ ms}^{-1}.$$

At time $t = 0$, $v = 0$ and the particle starts moving opposite to the direction of increasing x since the force is opposite to the direction of increasing x.

Thus, we have to choose the –ve sign.

(*a*) $v = -1$ ms^{-1} when $x = 0.5$ m

(*b*) Since $\frac{K}{m} = 1$, $v = \frac{dx}{dt} = -\sqrt{\frac{1-x}{x}}$.

or
$$dx\sqrt{\frac{x}{1-x}} = -\,dt.$$

Integrating from $x = 1$ to $x = 0.25$ and $t = 0$ to $t = t$, we get

$$-t = \int_1^{0.25}\sqrt{\frac{x}{1-x}}\,dx$$

We put
$$x = \sin^2\theta$$

or
$$dx = 2\sin\theta\cos\theta\,d\theta.$$

$$-t = \int_{\pi/2}^{\pi/6} 2\sin^2\theta\,d\theta = -\frac{\pi}{3} - \frac{\sqrt{3}}{4}$$

$$t = \left[\frac{\pi}{3} + \frac{\sqrt{3}}{4}\right] \text{s.}$$

SUPPLEMENTARY PROBLEMS

1. A particle of 2 g moves along the x-axis under the influence of two forces: (*i*) force of attraction toward origin which is numerically equal to four times the instantaneous distance from the origin, and (*ii*) a damping force proportional to the instantaneous speed. For what range of values of the damping constant β will the motion be (*a*) over damped, (*b*) under damped or damped oscillatory, (*c*) critically damped?

2. (*a*) Solve the differential equation

$$5\ddot{x} + 10\dot{x} + 25x = 0$$

subject to the conditions $x = 2, \dot{x} = -1$ at $t = 0$.

(*b*) Give the physical interpretation of the result.

3. Show that the time period of damped harmonic oscillator with equation is given by

$$m\ddot{x} + \beta\dot{x} + kx = 0$$

is given by $$T = \frac{4\pi m}{\sqrt{4km - \beta^2}}.$$

Find the time in which the amplitude of oscillation falls by a factor of e.

4. Find the frequency of oscillation of an object satisfying the differential equation

$$\ddot{x} + 0.693\dot{x} + 9.99x = 0.$$

Find the time in which its amplitude of vibration is halved. Find also the logarithmic decrement of the problem.

5. In a damped oscillatory motion an object oscillates with a frequency of 2 Hz and its amplitude of vibration is halved in 2 s. Find the differential equation for the oscillation. Find also the logarithmic decrement of the problem.

6. A 1.5 kg weight hung on a vertical spring stretches it 0.4 m. The weight is then pulled down 1 m and released. (*a*) Find the differential equation of motion of the body with boundary conditions if a damping force numerically equal to 15 times the instantaneous speed is acting on it. (*b*) Is the motion damped, over damped or critically damped? Find the position of the body at any time.

7. The natural frequency of a mass vibrating on a spring is 20 Hz, while its frequency with damping is 16 Hz. Find the logarithmic decrement.

8. A point performs damped oscillations according to the law

$$x = a_0 e^{-bt} \sin \omega t.$$

Find (*a*) oscillation amplitude and the velocity of the point at the moment $t = 0$, (*b*) the moments of time at which the point reaches the extreme positions.

9. Show that for damping which is less than 10% of the critical value, the undamped natural frequency and the damped frequency agree to within 0.5%.

10. A body of mass 10 g is suspended by a spring of stiffness 0.25 N/m and subject to damping which is 1% of the critical value. After approximately how many oscillations will the amplitude of the system be halved?

11. A block is suspended by a spring and a dashpot with a strong damping action. Show that if the block is displaced downwards and given a downward velocity, it will never pass through its equilibrium position again.

12. The angular frequency of a harmonic oscillator is 16 rad/s. With weak damping imposed it is found that the amplitudes of two consecutive oscillations in the same direction are 5 cm and 0.25 cm. Find the new period of the system.

13. The frequency of a damped oscillator is one-half the frequency of the same oscillator with no damping. Find the ratio of the maxima of successive oscillations.

14. A system of unit mass whose natural angular frequency in the absence of damping is 6 rad s^{-1} is subject to a damping force of magnitude $20\dot{x}$ where x is the displacement from the equilibrium position. Show that the general solution for x is

$$x = Ae^{-2t} + Be^{-18t}$$

The mass is initially at $x = 1$ m and given an initial velocity of magnitude 34 ms^{-1} towards the equilibrium position. Find the time when the displacement becomes greatest in the negative x-direction and the value of the negative displacement.

15. For a damped oscillator let us assume that m = 250 g, k = 85 N/m, and β = 70 g/s. (*a*) How long does it take for the amplitude of the damped oscillator to drop to half its initial value? (*b*) How long does it take for the mechanical energy to drop one-half of its initial value? (*c*) What is the ratio of the amplitude of the damped oscillations after 20 full cycles have elapsed to the initial amplitude?

16. What resistance R should be connected to an inductor L = 100 mH and capacitor C = 10 μF in series in order that the maximum charge on the capacitor decays to 90% of its initial value in 50 cycles?

17. A single loop circuit consists of a 7.2Ω resistor, a 12 H inductor, and a 3.2 μF capacitor. Initially the capacitor has a charge of 6.2 μC and the current is zero. Calculate the charge on the capacitor after 10 and 100 complete cycles of oscillations.

18. The equation of motion for the angle of twist θ of the moving coil galvanometer is given by

$$I\frac{d^2\theta}{dt^2} = -\beta\frac{d\theta}{dt} - C\theta$$

where I is the moment of inertia of the moving system about the axis of rotation, $-\beta\frac{d\theta}{dt}$ is the damping force (due to (*i*) mechanical damping proportional to the angular velocity $\frac{d\theta}{dt}$ and (*ii*) electromagnetic damping proportional to $\frac{d\theta}{dt}$ — Lenz' law) which opposes the motion of the galvanometer and C is the restoring couple per unit twist of the suspension wire.

(*a*) Find the condition of the ballistic motion of the galvanometer. (*b*) If the deflection on the scale of the lamp and scale arrangement of the galvanometer is 25 cm and 20 cm in one direction on the first and tenth oscillation, what is the value of the logarithmic decrement?

19. A bell rings at a frequency of 100 Hz. Its amplitude of vibration is halved in 10s. Find the quality factor of the bell.

4

Forced Vibrations and Resonance

4.1 FORCED VIBRATIONS

Suppose that the particle of mass m is under the influence of an external force $F(t)\hat{i}$ in addition to the restoring force $-kx\hat{i}$ and damping force $-\beta\frac{dx}{dt}\hat{i}$.

Then the equation of motion becomes

$$m\ddot{x} = -kx - \beta\dot{x} + F(t). \quad ...(4.1)$$

If the external force is periodic, $F(t) = F \sin pt$, we can write Eqn. (4.1) in the following form

$$\ddot{x} + 2b\dot{x} + \omega^2 x = f \sin pt \quad ...(4.2)$$

where $$b = \frac{\beta}{2m},\ \omega^2 = \frac{k}{m},\ f = \frac{F}{m}.$$

The general solution of Eqn. (4.2) is

$$x = x_1 + x_2 \quad ...(4.3)$$

where x_1 is the general solution of the homogeneous equation (see problem 3–1):

$$\ddot{x}_1 + 2b\dot{x}_1 + \omega^2 x_1 = 0 \quad ...(4.4)$$

and x_2 is any particular integral of Eqn. (4.2). A particular solution of Eqn. (4.2) is given by (see problem 1)

$$x_2 = \frac{f}{\sqrt{(\omega^2 - p^2)^2 + 4b^2p^2}} \sin(pt - \alpha) \quad ...(4.5)$$

where $$\tan\alpha = \frac{2bp}{\omega^2 - p^2},\ 0 \le \alpha \le \pi. \quad ...(4.6)$$

We have seen in problem 3–1 that x_1 becomes negligible within a short time and so we call this solution the *transient solution*. After a long time when x_1 becomes negligible the motion of the mass m is given by Eqn. (4.5) which is called the *steady-state solution*. In the steady state x_2 has a frequency p which is equal to the frequency of the impressed force but lags behind by a phase angle α. The vibrations or oscillations represented by x_2 are called *forced vibrations* or *forced oscillations*.

4.2 RESONANCE

The amplitude of the steady-state oscillation [Eqn. (4.5)] is given by

$$A = \frac{f}{\sqrt{(\omega^2 - p^2)^2 + 4b^2p^2}} \qquad ...(4.7)$$

The condition for maximum amplitude is

$$p = \omega\left(1 - \frac{2b^2}{\omega^2}\right)^{1/2} \qquad ...(4.8)$$

where we have assumed $b^2 < \frac{1}{2}\omega^2$. Near the frequency p of the impressed force given by equation (4.8), very large oscillations may set in and the phenomenon is called *amplitude resonance* and the frequency is called the *frequency of amplitude resonance* or *amplitude resonant frequency*. If the damping is very small, $b^2 << \frac{1}{2}\omega^2$, the resonance frequency $p \approx \omega - \frac{b^2}{\omega}$ which is very close to the natural frequency ω of the undamped oscillator. However, the velocity amplitude of the system becomes maximum for $p = \omega$ for any given value of b (see problem 4–2). This phenomenon is known as *velocity resonance* or simply *resonance*.

4.3 QUALITY FACTOR Q

The quality factor Q is the ratio

$$Q = \frac{\sqrt{AC}}{B} \qquad ...(4.9)$$

where A, B, C are coefficients of the differential Eqn..

$$A\ddot{x} + B\dot{x} + Cx = F \qquad ...(4.10)$$

describing the motion of a resonant system with F as the external force and x as displacement. Thus from Eqn. (4.1), we have

$$Q = \frac{\sqrt{mk}}{\beta} = \frac{\omega}{2b}. \qquad ...(4.11)$$

4.4 HELMHOLTZ RESONATOR

Helmholtz resonator is used to determine the frequency of a vibrating body with the help of the phenomenon of resonance. The resonator consists of either a spherical or a cylindrical air cavity with a small neck (Fig. 4.1). The dimension of the cavity is small in comparison with the wavelength of sound to be detected. In case of spherical cavity the volume of the cavity is fixed whereas the volume is variable in case of cylindrical cavity.

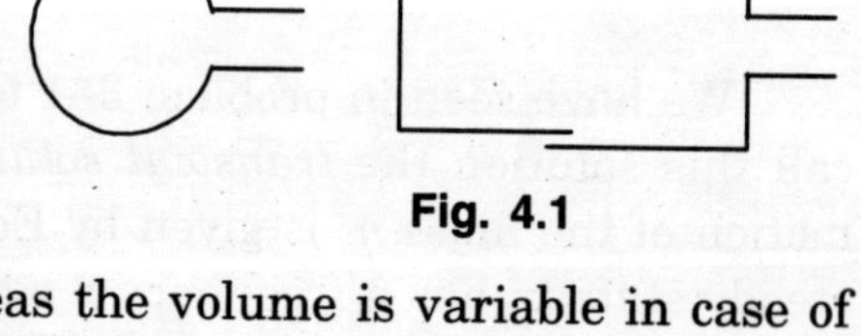

Fig. 4.1

The air contained at the neck of the resonator acts like a piston alternately compressing and rarefying the air within the cavity of the resonator. The natural frequency of vibration of Helmholtz resonator is given by

$$\nu = \frac{v}{2\pi}\sqrt{\frac{S}{lV}} \qquad ...(4.12)$$

where v = Velocity of propagation of sound in air
l = Length of the neck of the resonator
S = Area of cross-section of the neck
V = Volume of the resonator.

The natural frequency of the resonator can be changed by changing the volume V of the resonator. When the sound wave of frequency resonant with the natural frequency of the resonator is incident on it, the resonator will produce sharp response. The frequency of the vibrating body is then equal to the natural frequency of the resonator given by Eqn. (4.12).

SOLVED PROBLEMS

1. *A particle of mass m moves under the influence of external periodic force $F \sin pt\,\hat{i}$ in addition to the restoring force $-kx\,\hat{i}$ and damping force $-\beta\dot{x}\hat{i}$. Set up the differential equation of motion and find the steady-state solution.*

Solution

The differential Eqn. of motion is given by equation (4.1):

$$m\ddot{x} = -kx - \beta\dot{x} + F\sin pt$$

which can be rewritten as Eqn. (4.2):

$$\ddot{x} + 2b\dot{x} + \omega^2 x = f\sin pt$$

where $b = \dfrac{\beta}{2m}$, $\omega^2 = \dfrac{k}{m}$ and $f = \dfrac{F}{m}$.

The general solution for x is

$$x = x_1 + x_2$$

where x_1 is the general solution of the homogeneous equation (see problem 1 of Chapter 3):

$$\ddot{x}_1 + 2b\dot{x}_1 + \omega^2 x_1 = 0$$

and x_2 is any particular integral of Eqn. (4.2). The solution x_1 is the displacement of the damped harmonic oscillator.

$$x_1 = e^{-bt}\left[A_1 \exp(\sqrt{b^2-\omega^2}\,t) + A_2 \exp(-\sqrt{b^2-\omega^2}\,t)\right]$$

x_1 is the same as the damped oscillation ($b < \omega$) or dead beat motion ($b > \omega$) or critically damped motion ($b = \omega$).

To find x_2 let us take the solution

$$x_2 = A\sin(pt - \alpha). \qquad ...(4.13)$$

This supposition is justifiable on the ground that the system will ultimately vibrate with the same frequency p as that of the impressed sustained harmonic force.

Now, $$\dot{x}_2 = A\,p\cos(pt - \alpha)$$

$$\ddot{x}_2 = -Ap^2\sin(pt - \alpha).$$

Substituting in Eqn. (4.2), we get

$$A(\omega^2 - p^2)\sin(pt - \alpha) + 2Abp\cos(pt - \alpha) = f\sin\{(pt - \alpha) + \alpha\}$$

$$= f\sin(pt - \alpha)\cos\alpha + f\cos(pt - \alpha)\sin\alpha. \quad \text{...(4.14)}$$

Since Eqn. (4.14) is true for all values of t we can equate the coefficients of $\sin(pt - \alpha)$ and $\cos(pt - \alpha)$ from both sides:

$$A(\omega^2 - p^2) = f\cos\alpha \quad \text{...(4.15)}$$

$$2Abp = f\sin\alpha \quad \text{...(4.16)}$$

Hence, we get

$$A = \frac{f}{\sqrt{(\omega^2 - p^2)^2 + 4b^2p^2}} \quad \text{...(4.17)}$$

and $$\tan\alpha = \frac{2bp}{\omega^2 - p^2} \quad \text{...(4.18)}$$

Eqns. (4.15) and (4.16) give

$$\sin\alpha = \frac{2bp}{\sqrt{(\omega^2 - p^2)^2 + 4b^2p^2}} \quad \text{...(4.19)}$$

$$\cos\alpha = \frac{\omega^2 - p^2}{\sqrt{(\omega^2 - p^2)^2 + 4b^2p^2}} \quad \text{...(4.20)}$$

Since $\sin\alpha$ is never negative, the range of α is $0 \le \alpha \le \pi$.

The complete solution of Eqn. (4.2) is

$$x = x_1 + \frac{f}{\sqrt{(\omega^2 - p^2)^2 + 4b^2p^2}}\sin(pt - \alpha). \quad \text{...(4.21)}$$

When $b < \omega$ the first part x_1 represents natural vibrations of damped harmonic oscillator. These vibrations become negligible very soon as the amplitude diminishes exponentially with time. If the damping is very small the natural vibrations will persist for a longer time. After a long time when x_1 becomes negligible we can write

$$x = \frac{f}{\sqrt{(\omega^2 - p^2)^2 + 4b^2p^2}}\sin(pt - \alpha). \quad \text{...(4.22)}$$

which is called the *steady-state solution*. Eqn. (4.22) represents the *sustained forced vibrations*. If the frequency of vibrations of x_1, that is, $\sqrt{\omega^2 - b^2}$ and that of x_2 i.e. p are nearly equal, at the initial stage, beats are produced. These beats are transient, as the natural vibrations become small after a short interval of time.

2. *Obtain the expression for the velocity of the mass m when it is in the steady state forced vibration of problem 1. Show that the velocity amplitude is maximum when the resonant frequency is p = ω and the velocity amplitude at resonance is*

$$v_{ar} = \frac{f}{2b}.$$

Solution

For the steady-state forced vibration the displacement x of the particle is given by

$$x = A \sin (pt - \alpha)$$

The velocity of the particle at any instant is

$$\dot{x} = Ap \cos (pt - \alpha) = v_a \cos (pt - \alpha)$$

where the velocity amplitude v_a is given by

$$v_a = Ap = \frac{fp}{\sqrt{(\omega^2 - p^2)^2 + 4b^2p^2}}$$

$$= \frac{f}{\sqrt{\left[\frac{\omega^2}{p} - p\right]^2 + 4b^2}}.$$

When $p = \omega$, v_a is maximum for any given value of b. This phenomenon is known as velocity resonance. The velocity amplitude at resonance is

$$v_{ar} = \frac{f}{2b}.$$

3. *In the steady state forced vibration of problem 1 show that (a) the amplitude of the driven system is maximum when* $p = \sqrt{\omega^2 - 2b^2}$ *and*

(b) the value of the maximum amplitude is $\dfrac{f}{2b\sqrt{\omega^2 - b^2}}$.

Solution

(*a*) In the steady state forced vibration the amplitude of vibration of the driven system is given by

$$A = \frac{f}{\sqrt{(\omega^2 - p^2)^2 + 4b^2p^2}}.$$

It is maximum when the denominator (or the square of the denominator) is a minimum.

Let $$u = (\omega^2 - p^2)^2 + 4b^2p^2.$$

The function u has a minimum or maximum when

$$\frac{du}{dp} = -2(\omega^2 - p^2)\, 2p + 8b^2p = 0$$

or $$p(p^2 - \omega^2 + 2b^2) = 0$$

i.e. $$p = 0 \text{ or, } p = \sqrt{\omega^2 - 2b^2} \text{ where } \omega^2 > 2b^2.$$

For $\quad p = 0, \quad \frac{d^2u}{dp^2} = -4\ (\omega^2 - 3p^2 - 2b^2) = -4(\omega^2 - 2b^2) < 0.$

For $\quad p = \sqrt{\omega^2 - 2b^2}, \quad \frac{d^2u}{dp^2} = -\ 4[\omega^2 - 2b^2 - 3(\omega^2 - 2b^2)]$

$$= 8\ (\omega^2 - 2b^2) > 0.$$

Thus $p = \sqrt{\omega^2 - 2b^2}$ gives the minimum of u.

Note that the angular frequency p of the periodic impressed force at amplitude resonance is slightly smaller than that at velocity resonance.

(*b*) The maximum amplitude at resonance is

$$A_{max} = \frac{f}{\sqrt{4b^4 + 4b^2(\omega^2 - 2b^2)}} = \frac{f}{2b\sqrt{\omega^2 - b^2}}$$

4. *In the steady state forced vibration describe how the phase of the driven system changes with the frequency of the driving system.*

Solution

The phase angle α is given by Eqns. (4.18–4.20). Suppose that the angular frequency of the impressed force is increased gradually from 0 to ∞.

(*i*) When $p = 0$, $\alpha = 0$. There is no difference of phase between the driven system and the impressed force.

(*ii*) When $p < \omega$, $\tan\alpha$ = +ve, $\cos\alpha$ = +ve and $\sin\alpha$ = +ve. Thus α has a value intermediate between 0 and $\frac{\pi}{2}$.

(*iii*) When $p \to \omega$, $\tan\alpha \to \infty$, $\sin\alpha \to 1$ and $\cos\alpha \to 0$. So, $\alpha \to \frac{\pi}{2}$. Thus at velocity resonance the driven system lags behind the driver by an angle $\frac{\pi}{2}$.

(*iv*) When $p > \omega$, $\tan\alpha$ = –ve, $\sin\alpha$ = +ve and $\cos\alpha$ = –ve. Here, $\frac{\pi}{2} < \alpha < \pi$.

(*v*) When $p \to \infty$, $\tan\alpha \to 0$, $\sin\alpha \to 0$ and $\cos\alpha \to -1$. Hence $\alpha \to \pi$.

The variation of α with p is shown in Fig. 4.2.

We know that

$$\alpha = \tan^{-1}\frac{2bp}{\omega^2 - p^2}.$$

Thus the rate of change of α with p is

$$\frac{d\alpha}{dp} = \frac{2b(p^2 + \omega^2)}{(\omega^2 - p^2)^2 + 4b^2p^2}.$$

At resonance $(p = \omega)$, $\frac{d\alpha}{dp} = \frac{1}{b}$.

Hence smaller the value of b, the greater is the rate of change of phase angle near the resonance frequency.

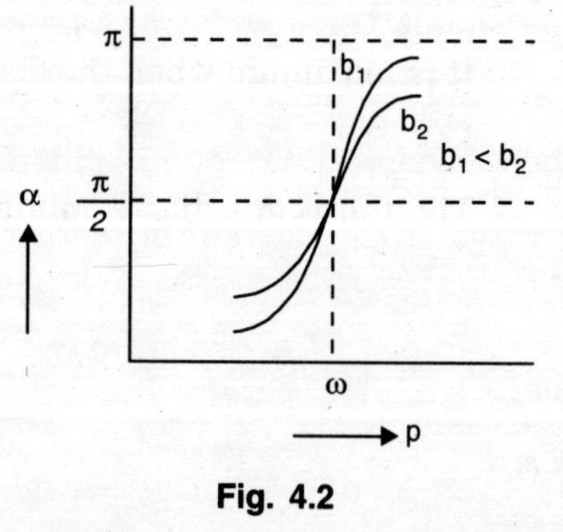

Fig. 4.2

5. *Show that in the steady state forced vibration the rate of dissipation of energy due to frictional force is equal to the rate of supply of energy by the driving force in each cycle.*

Solution

In the steady state forced vibration the displacement of the particle is given by

$$x = A \sin (pt - \alpha).$$

Suppose at any instant the force $F \sin pt$ moves through a distance dx in time dt. Then the work done by the force = $F \sin pt\, dx$. The rate of work done averaged over a cycle is

$$\frac{1}{T}\int_0^T F \sin pt \left(\frac{dx}{dt}\right) dt = \frac{1}{T}\int_0^T F \sin pt\, pA \cos (pt - \alpha)\, dt.$$

$$= \int_0^T [\sin pt \cos pt \cos \alpha + \sin^2 pt \sin \alpha]$$

$$= \frac{1}{2} FpA \sin \alpha \qquad ...(4.23)$$

where we have put $T = \dfrac{2\pi}{p}$.

Work done against the frictional force $\beta\dot{x}$ for the displacement dx is $\beta\dot{x}\, dx$. Rate of work done against the frictional force averaged over a cycle is

$$\frac{1}{T}\int_0^T \beta\left(\frac{dx}{dt}\right)^2 dt = \frac{\beta}{T} A^2p^2 \int_0^T \cos^2 (pt - \alpha)\, dt$$

$$= \frac{1}{2}\beta A^2 p^2. \qquad ...(4.24)$$

Now we have to show that the expression (4.23) is equal to expression (4.24).

Now, $\sin \alpha = 2bp\dfrac{A}{f} = \dfrac{\beta}{m} p \dfrac{A}{f}$

and $F = fm$

Thus $\dfrac{1}{2} FpA \sin \alpha = \dfrac{1}{2}\, fmp\, A \dfrac{\beta pA}{mf} = \dfrac{1}{2}\beta A^2p^2.$

Since energy is dissipated in each cycle due to frictional force, this loss is made up by the energy supplied by the driving force to maintain the steady forced vibration.

In the steady state forced vibration the displacement of the particle is sinusoidal and the mechanical energy remains fixed at the steady value $\frac{1}{2} kA^2$.

6. (*a*) *Show that in the steady state forced vibration the power supplied by the driving force averaged over a cycle is given by*

$$P = \frac{mb\, p^2 f^2}{(\omega^2 - p^2)^2 + 4b^2p^2}$$

and the power is maximum when $p = \omega$

(*b*) *Find the two values of p, namely p_1 and p_2 at which the power P is half of that at resonance and show that $p_1 \cdot p_2 = \omega^2$.*

(*c*) *Show that the full width of the power resonance curve at half maximum = 2b. If sharpness of resonance s is defined by*

$$\textit{Sharpness of resonance} = \frac{\textit{Frequency at resonance}}{\textit{Full width at half maximum power}} \textit{ then show that } s = \frac{\omega}{2b}.$$

Solution

(*a*) From problem 5, we have

$$P = \frac{1}{2}\beta A^2 p^2 = \frac{mb\, p^2 f^2}{(\omega^2 - p^2)^2 + 4b^2 p^2}$$

$$= \frac{mb\, f^2}{\omega^2\left(\frac{\omega}{p} - \frac{p}{\omega}\right)^2 + 4b^2}.$$

Thus, the power P is maximum when $p = \omega$ and the maximum value is

$$P_{\max} = \frac{mb f^2}{4b^2} = \frac{mf^2}{4b}.$$

(*b*)
$$\frac{mb\, f^2}{\omega^2\left(\frac{\omega}{p} - \frac{p}{\omega}\right)^2 + 4b^2} = \frac{1}{2} P_{\max}$$

or
$$p^4 - (2\omega^2 + 4b^2)\, p^2 + \omega^4 = 0$$

The two values of p^2 are

$$p_1^2 = \omega^2 + 2b^2 + \sqrt{(\omega^2 + 2b^2)^2 - \omega^4}$$

$$p_2^2 = \omega^2 + 2b^2 - \sqrt{(\omega^2 + 2b^2)^2 - \omega^4}$$

The power resonance curve is shown in Fig. 4.3

We find that

$$p_1^2, p_2^2 = \omega^2 + 2b^2 \pm 2b\sqrt{\omega^2 + b^2}$$

$$= (\sqrt{\omega^2 + b^2} \pm b)^2$$

Thus,
$$p_1, p_2 = \sqrt{\omega^2 + b^2} \pm b$$

and
$$p_1 \cdot p_2 = \omega^2.$$

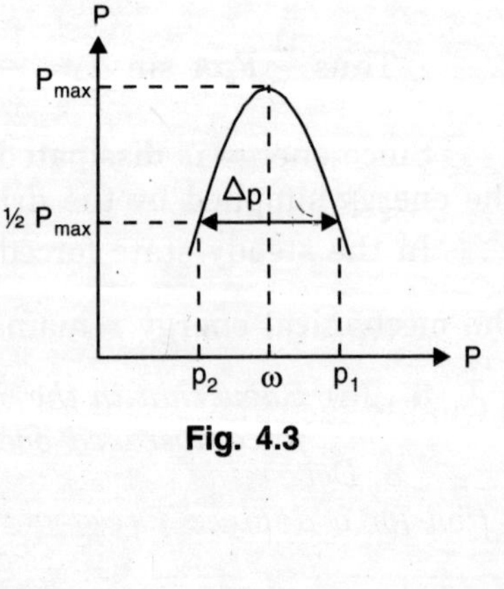

Fig. 4.3

(*c*) $\Delta p = p_1 - p_2$ = Full width of the power resonance curve at half maximum = 2 b. Hence,

$$\frac{\text{Frequency at resonance}}{\text{Full width at half maximum power}} = \frac{\omega}{\Delta p} = \frac{\omega}{2b}$$

= sharpness of resonance.

7. *If the quality factor Q in the steady state forced vibration is defined as*

$$Q = \frac{2\pi \times \textit{Average energy stored per cycle}}{\textit{Average energy dissipated per cycle}}$$

then show that

$$Q = \frac{p}{4b}\left(1+\frac{\omega^2}{p^2}\right).$$

Show that the quality factor is minimum at resonance $p = \omega$ *and its minimum value is*

$$Q_{min} = \frac{\omega}{2b}.$$

Solution

Total energy at any instant of time in the steady state is

$$E = \frac{1}{2} m\dot{x}^2 + \frac{1}{2} kx^2$$

$$= \frac{1}{2} mA^2p^2 \cos^2 (pt - \alpha) + \frac{1}{2} kA^2 \sin^2 (pt - \alpha)$$

The average value of $\cos^2 (pt - \alpha)$ and $\sin^2 (pt - \alpha)$ are each $\frac{1}{2}$ per cycle.

Thus, $$E_{av} = \frac{1}{4} mA^2 (p^2 + \omega^2).$$

From problem 5 we know that the average power dissipated $= \frac{1}{2} \beta A^2p^2$.

Average power dissipated per cycle $= T \times \frac{1}{2} \beta A^2p^2$,

where $$T = \frac{2\pi}{p}.$$

Thus, $$Q = \frac{2\pi \times \frac{1}{4} mA^2(p^2+\omega^2)}{T \cdot \frac{1}{2} \beta A^2 p^2} = \frac{\pi}{2b} \frac{p^2+\omega^2}{Tp^2}$$

$$= \frac{p}{4b}\left(1+\frac{\omega^2}{p^2}\right)$$

Q is minimum when $\frac{dQ}{dp} = 0$ or, $p = \omega$ and $\frac{d^2Q}{dp^2}\Big|_{p=\omega} = \frac{1}{2b\omega} > 0.$

Thus the minimum of Q occurs at resonance when $p = \omega$ and the minimum value is $\frac{\omega}{2b}$.

Quality factor at resonance = sharpness of resonance.

8. *Determine the root-mean-square (rms) values of displacement, velocity and acceleration for a damped forced harmonic oscillator operating at steady state.*

Solution

The defining expression for the rms value is

$$g_{rms} = \left[\frac{\int_0^T g^2 dt}{\int_0^T dt}\right]^{1/2} = \left[\frac{1}{T}\int_0^T g^2 dt\right]^{1/2}$$

where g is an arbitrary periodic function of time with period T.

The expression for the displacement in the steady state forced vibration is

$$x = A \sin (pt - \alpha) \text{ with } T = \frac{2\pi}{p}.$$

Thus,
$$x_{rms} = A\left[\frac{1}{T}\int_0^T \sin^2(pt-\alpha)\right]^{1/2} = \frac{A}{\sqrt{2}}.$$

The rms values of velocity and acceleration are

$$v_{rms} = \frac{pA}{\sqrt{2}} \text{ and } a_{rms} = \frac{p^2 A}{\sqrt{2}}.$$

9. *A machine of total mass 90 kg is supported by a spring resting on the floor and its motion is constrained to be in the vertical direction only. The system is lightly damped with a damping constant 900 Ns/m. The machine contains an eccentrically mounted shaft which, when rotating at an angular frequency p, produces a vertical force on the system of $Fp^2 \sin pt$ where F is a constant. It is found that resonance occurs at 1200 r.p.m. (revolutions per minute) and the amplitude of vibration in the steady state is then 1 cm. Find the amplitude of vibration in the steady state when the driving frequency is (a) 2400 r.p.m. (b) 3000 r.p.m. (c) very large. Find also the quality factor Q at resonance. Assume that the gravity has a negligible effect on the motion.*

Solution

At resonance,
$$\omega = p = \frac{2\pi \times 1200}{60} = 40\pi \text{ s}^{-1}.$$

$$b = \frac{\beta}{2m} = \frac{900}{2 \times 90} = 5 \text{ s}^{-1}$$

Here the periodic force = $Fp^2 \sin pt$.

Thus the amplitude at resonance $A = \dfrac{fp^2}{2bp} = 0.01$ m

Hence,
$$f = \frac{2 \times 5 \times 0.01}{40\pi} = \frac{0.01}{4\pi}.$$

(*a*) At 2400 r.p.m., $p = 80\pi \text{ s}^{-1}$

$$\text{Amplitude} = \frac{fp^2}{\sqrt{(\omega^2 - p^2)^2 + 4b^2 p^2}}$$

$$= \frac{\frac{0.01}{4\pi} \times (80\pi)^2}{[(1600\pi^2 - 6400\pi^2)^2 + 100 \times 6400\pi^2]^{1/2}}.$$

≈ 0.11 cm.

(*b*) At 3000 r.p.m., $p = 100\pi\ s^{-1}$

Amplitude ≈ 0.09 cm.

(*c*) As $p \to \infty$, Amplitude $\to \frac{fp^2}{p^2} = f = 0.08$ cm.

Quality factor at resonance

$$Q = \frac{\omega}{2b} = \frac{40\pi}{2 \times 5} = 12.57.$$

10. *Two bodies of masses m_1 and m_2 connected by a spring of spring constant k, can move along a horizontal line (axis of the spring). A periodic force F cos ωt is exerted on the body of mass m_1 along the line. Find expression for the displacements of the two masses and indicate by a sketch graph the dependence of the amplitude of motion of m_1 on frequency ω.*

Solution

Let x_1 and x_2 be the respective displacements of the masses m_1 and m_2 from their equilibrium positions. The extension of the spring is $x_2 - x_1$. Thus the Eqns. of motion of m_1 and m_2 are

$$m_1\ddot{x}_1 = k\,(x_2 - x_1) + F\exp(i\omega t) \qquad ...(4.25)$$

$$m_2\ddot{x}_2 = -k\,(x_2 - x_1). \qquad ...(4.26)$$

[We use complex exponential motion to simplify the calculation]

Since we are forcing the bodies at frequency ω, let us try solutions

$$x_1 = A\exp(i\omega t) \qquad ...(4.27)$$

$$x_2 = B\exp(i\omega t) \qquad ...(4.28)$$

Substituting Eqns. (4.27) and (4.28) into Eqns. (4.25) and (4.26), we get

$$kB - kA + F = -m_1A\omega^2$$

$$-kB + kA = -m_2B\omega^2$$

which give

$$B = \frac{kA}{k - m_2\omega^2} \qquad ...(4.29)$$

and

$$A = \frac{F(k - m_2\omega^2)}{\omega^2[m_1m_2\omega^2 - k(m_1 + m_2)]} \qquad ...(4.30)$$

Thus,

$$x_1 = \frac{F(k - m_2\omega^2)\cos\omega t}{\omega^2[m_1m_2\omega^2 - k(m_1 + m_2)]} \qquad ...(4.31)$$

and

$$x_2 = \frac{Fk\cos\omega t}{\omega^2[m_1m_2\omega^2 - k(m_1 + m_2)]} \qquad ...(4.32)$$

The amplitude of the motion of m_1 is A [Eqn. (4.30)]. It is zero when $\omega^2 = k/m_2$ and infinite (resonance) when $\omega^2 = \dfrac{k(m_1 + m_2)}{m_1 m_2}$. The amplitude tends to zero as ω tends to infinity. The amplitude will also be infinite when $\omega = 0$ (this corresponds to a steady force accelerating the whole system).

Fig. 4.4 shows a sketch of A as a function of ω.

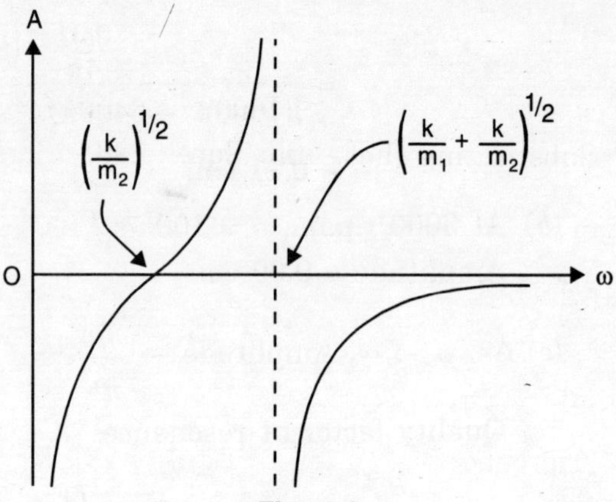

Fig. 4.4

11. *A periodic external force acts on a 3 kg mass suspended from the lower end of a vertical spring having spring constant 75 N/m. The damping force is proportional to the instantaneous speed of the mass and is 20 N when the speed is 1 m/s. Find the frequency at which amplitude resonance occurs.*

Solution

Natural angular frequency of the spring = $\omega = \sqrt{75/3} = 5$ rad s^{-1}.

Damping force = $\beta\dot{x}$ = 20N when $\dot{x}$ = 1 m/s.

Thus, $$\beta = 20 \text{ Nsm}^{-1} \text{ and } b = \frac{\beta}{2m} = \frac{10}{3} \text{ s}^{-1}.$$

For amplitude resonance, angular frequency = $\sqrt{\omega^2 - 2b^2} = \dfrac{5}{3}$ rad s^{-1} and resonance frequency = $\nu = \dfrac{5}{6\pi}$ Hz.

12. *The mass on a vertical spring undergoes forced vibrations according to the Eqn.*

$$\ddot{x} + \omega^2 x = f \sin \omega t.$$

where there is no damping and the impressed frequency is equal to the natural frequency of oscillation. (a) Obtain the solution of the above differential equation (b) Give a physical interpretation.

Solution

(*a*) The general solution of the equation

$$\ddot{x} + \omega^2 x = f \sin \omega t \qquad ...(4.33)$$

is

$$x = x_1 + x_2 \qquad ...(4.34)$$

where x_1 is the general solution of the homogeneous equation

$$\ddot{x}_1 + \omega^2 x_1 = 0 \qquad ...(4.35)$$

and x_2 is the particular integral of Eqn. (4.33). Now the general solution of Eqn. (4.35) is

$$x_1 = A \cos \omega t + B \sin \omega t. \qquad ...(4.36)$$

The particular solution of Eqn. (4.33) has the form

$$x = t[c_1 \cos \omega t + c_2 \sin \omega t]$$

which gives

$$\dot{x} = (c_1 \cos \omega t + c_2 \sin \omega t) + t(-\omega c_1 \sin \omega t + \omega c_2 \cos \omega t)$$

$$\ddot{x} = 2(-\omega c_1 \sin \omega t + \omega c_2 \cos \omega t) - t\omega^2(c_1 \cos \omega t + c_2 \sin \omega t).$$

Substituting these into Eqn. (4.33) and simplifying, we obtain

$$2c_2\omega \cos \omega t - 2c_1\omega \sin \omega t = f \sin \omega t$$

from which $c_2 = 0$ and $c_1 = -\dfrac{f}{2\omega}$. Thus the particular integral is

$$x_2 = -\frac{ft}{2\omega} \cos \omega t \qquad ...(4.37)$$

The general solution of Eqn. (4.33) is therefore

$$x = A \cos \omega t + B \sin \omega t - \frac{ft}{2\omega} \cos \omega t \qquad ...(4.38)$$

(*b*) The first two terms of Eqn. (4.38) are oscillatory with constant amplitude. The last term involving t increases with time to such an extent that the spring breaks finally. A graph of the last term is shown in Fig. 4.5. This example illustrates the phenomenon of resonance. Here the natural frequency of the spring equals the frequency of the impressed force.

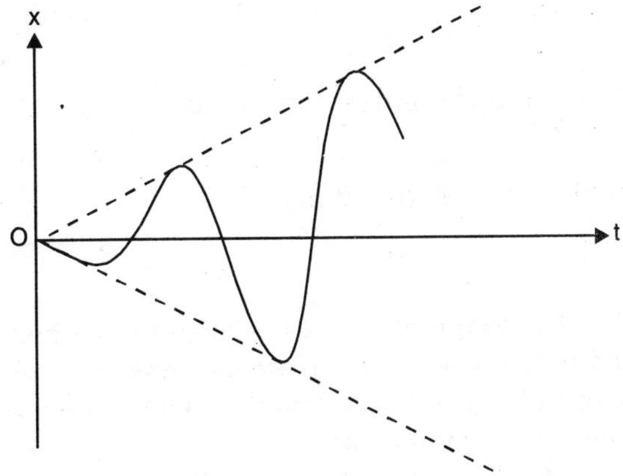

Fig. 4.5

13. *A vertical spring has a spring constant 50 N/m. At t = 0 a force given in newtons by F (t) = 48 cos 7t, t ≥ 0 is applied to a 20 N weight which hangs in equilibrium at the end of the spring. Neglecting damping find the position of the weight at any later time t.*

Solution

We have the equation of motion

$$\frac{20}{g}\ddot{x} = -50x + 48 \cos 7t$$

or

$$\ddot{x} + 25x = 24 \cos 7t \qquad ...(4.39)$$

where we put $g = 10$ m/s^2.

The complementary function of Eqn. (4.39) is

$$x = A \cos 5t + B \sin 5t \qquad ...(4.40)$$

and the particular integral is given by

$$x = c_1 \cos 7t + c_2 \sin 7t \qquad ...(4.41)$$

Substituting Eqn. (4.41) into Eqn. (4.39), we get $c_1 = -1$ and $c_2 = 0$. Thus the general solution is

$$x = A \cos 5t + B \sin 5t - \cos 7t \qquad ...(4.42)$$

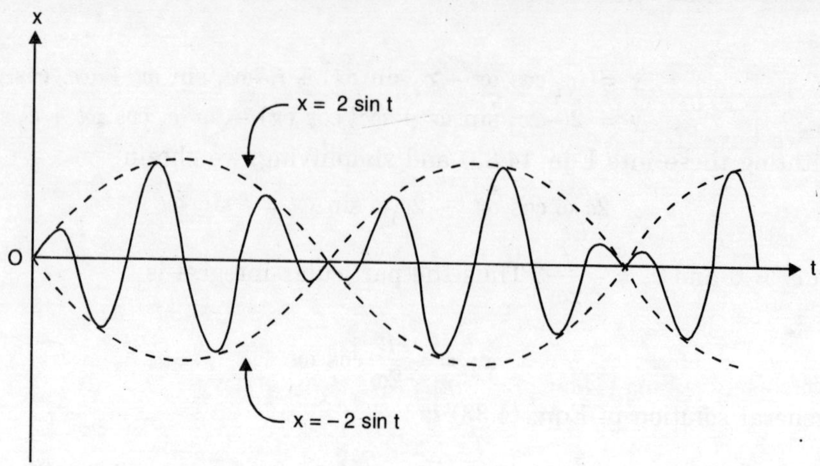

Fig. 4.6

Using the initial conditions

$$x = 0, \ \dot{x} = 0 \text{ at } t = 0$$

we find, $A = 1$, $B = 0$ and thus

$$x = \cos 5t - \cos 7t = 2 \sin t \sin 6t. \quad ...(4.43)$$

The graph of x vs. t is shown by the heavy curve of Fig. 4.6. The dashed curves are the curves of $x = \pm 2 \sin t$. If we consider that $2 \sin t$ is the amplitude of $\sin 6t$, we see that the amplitude varies sinusoidally. The phenomenon is known as amplitude modulation (see problem 14 chapter 2).

14. *Show that the natural frequency of vibration of Helmholtz resonator is given by*

$$\nu = \frac{v}{2\pi}\sqrt{\frac{S}{lV}}$$

where v = *Velocity of propagation of sound in air*
l = *Length of the neck of the resonator*
V = *Volume of the resonator*
S = *Area of cross-section of the neck.*

Solution

We assume that the air in the neck of Helmholtz resonator acts as a piston alternately compressing and rarefying the air within the cavity of the resonator.

Let x = Displacement towards the cavity of the piston of sectional area S at any instant t and δP be the increase in the pressure in the cavity.

Total force acting on the piston is

$$Sl\rho \frac{d^2x}{dt^2} = S\,\delta P$$

where ρ = Density of air and $Sl\rho$ = Mass of air in the neck.

Since the pressure change in the cavity is adiabatic

$$PV^{\gamma} = \text{Constant}$$

or

$$\delta P V^{\gamma} + P\gamma V^{\gamma-1}\delta V = 0$$

or $$\delta P = -\gamma P \frac{\delta V}{V}$$

Thus, we have

$$\frac{d^2x}{dt^2} + \omega^2 x = 0$$

where $$\omega = \sqrt{\frac{\gamma PS}{\rho l V}}.$$

The velocity of propagation of sound in a gas is given by

$$v = \sqrt{\frac{\gamma P}{\rho}}$$

[See Chapter 5]

The frequency of vibration is thus given by

$$\nu = \frac{\omega}{2\pi} = \frac{v}{2\pi}\sqrt{\frac{S}{lV}}.$$

Since the damping is small, a Helmholtz resonator is highly selective and the response is very sharp.

15. *A Helmholtz resonator has a cylindrical neck of cross-section 2 cm² and length 1 cm. What must be the volume of the resonator in order to have resonance at frequency of 500 Hz?*

[Velocity of sound in air = 340 m/s]

Solution

We know

$$\nu = \frac{v}{2\pi}\sqrt{\frac{S}{lV}}$$

or $$V = \frac{S}{l}\left(\frac{v}{2\pi\nu}\right)^2 = 2 \times \left(\frac{340 \times 100}{2\pi \times 500}\right)^2 \text{ cm}^3$$

$$= 234.25 \text{ cm}^3.$$

16. *If an alternating emf $E = E_0 \sin \omega t$ is applied to a series LCR circuit (Fig. 4.7) the resulting alternating current in the circuit is given by (steady-state)*

$$i = I \sin(\omega t - \phi)$$

(a) Find the current amplitude I and the phase constant ϕ.

(b) Show that the current amplitude I has the maximum value (resonance) when $\omega = \omega_0$, where $\omega_0 = \frac{1}{\sqrt{LC}}$ is the natural frequency.

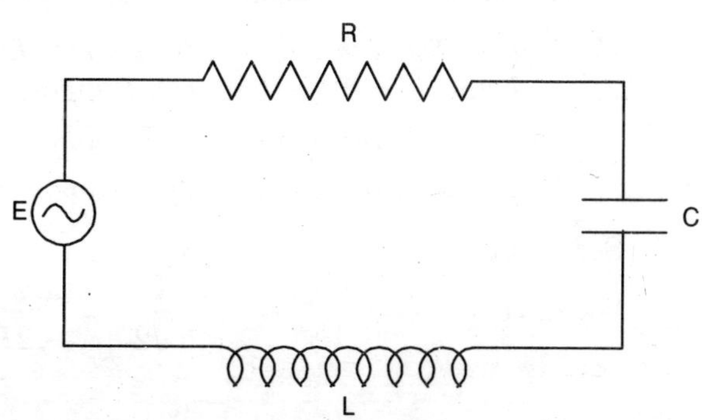

Fig. 4.7

(c) Show that the value of I and the phase angle ϕ at resonance are E_0/R and zero respectively.

Solution

(*a*) The equation for the current i in the LCR circuit can be written as

$$L\frac{di}{dt} + Ri + \frac{q}{C} = E_0 \sin \omega t \qquad ...(4.44)$$

where the charge q on the capacitor is given by

$$q = \int i dt. \qquad ...(4.45)$$

We consider a steady-state solution of Eqn. (4.44) in the form (after the alternating emf has been applied for some time)

$$i = I \sin (\omega t - \phi) \qquad ...(4.46)$$

Substituting Eqn. (4.46) into Eqn. (4.44) and equating the coefficients of sin ωt and cos ωt from both sides, we get

$$I\left(L\omega - \frac{1}{\omega C}\right) \sin \phi + RI \cos \phi = E_0 \qquad ...(4.47)$$

$$I\left(L\omega - \frac{1}{\omega C}\right) \cos \phi - RI \sin \phi = 0 \qquad ...(4.48)$$

which give

$$I = \frac{E_0}{\left[R^2 + \left(L\omega - \frac{1}{\omega C}\right)^2\right]^{1/2}} \qquad ...(4.49)$$

$$\tan \phi = \frac{L\omega - \frac{1}{\omega C}}{R}. \qquad ...(4.50)$$

(*b*) From Eqn. (4.49), we find that the maximum value of I occurs when

$$L\omega = \frac{1}{\omega C} \text{ or, } \omega = \frac{1}{\sqrt{LC}} = \omega_0 \text{ (resonance).}$$

(*c*) The value of I at resonance is $I_0 = E_0/R$ and the phase angle ϕ is zero at resonance.

17. *A curve between I and ω in a series LCR circuit connected to an emf E_0 sin ωt has a peak ($I = I_0$) at $\omega = \omega_0$ (resonance). Suppose ω_1 and ω_2 are two values of ω on both sides of ω_0 at which the value of I is $I_0/\sqrt{2}$. Show that (a) ω_1 and ω_2 are half-power points and $\Delta\omega = \omega_2 - \omega_1 = R/L$, (b) the Quality factor of the circuit is $Q = \omega_0 L/R$.*

Solution

We have
$$\frac{I_0}{\sqrt{2}} = \frac{E_0}{\sqrt{2}R} = \frac{E_0}{\left[R^2 + \left(L\omega - \frac{1}{\omega C}\right)^2\right]^{1/2}}.$$

or
$$L\omega - \frac{1}{\omega C} = \pm R \qquad ...(4.51)$$

At resonance, $\omega = \omega_0$ and $L\omega_0 - \dfrac{1}{\omega_0 C} = 0$

$$\text{At } \omega = \omega_2 > \omega_0, \quad L\omega_2 - \frac{1}{\omega_2 C} = +R \qquad ...(4.52)$$

$$\text{and at } \omega = \omega_1 < \omega_0, \quad L\omega_1 - \frac{1}{\omega_1 C} = -R \qquad ...(4.53)$$

Multiplying Eqn. (4.52) by $\dfrac{1}{\omega_1}$ and Eqn. (4.53) by $\dfrac{1}{\omega_2}$ and then subtracting, we get

$$\frac{L\omega_2}{\omega_1} - \frac{L\omega_1}{\omega_2} = \frac{R}{\omega_1} + \frac{R}{\omega_2}$$

or
$$L(\omega_2 - \omega_1) = R$$

Thus,
$$\Delta\omega = \frac{R}{L}.$$

Since power $\propto i^2$, ω_1 and ω_2 correspond to frequencies at which $P = P_0/2$, where P_0 is the maximum power at resonance.

(*b*) Quality factor $= \dfrac{\text{Frequency at resonance}}{\text{Full width at half maximum power}}$

$$= \frac{\omega_0 L}{R}.$$

18. *A wind-powered generator converts wind energy into electrical energy. Assume that the generator converts a fixed fraction of the wind energy intercepted by its blade into electrical energy. For wind speed V, the electrical power output will be proportional to*

(*a*) V (*b*) V^2 (*c*) V^3 (*d*) V^4. (I.I.T. 2000)

Solution

$$\text{Power} = \vec{F}\cdot\frac{d\vec{s}}{dt} = \vec{F}\cdot\vec{V}$$

$$\text{Force } F \propto \frac{d}{dt}(mV) \text{ or, } F \propto \frac{dm}{dt}V$$

where mass per unit time = Area of cross-section × velocity × density.

Thus, $F \propto (A\, V\, \rho).\ V \propto V^2$.

Power delivered $\propto V^3$.

Correct Choice : (*c*)

SUPPLEMENTARY PROBLEMS

1. In steady state forced vibration of problem 1 show that (*a*) at low frequencies p, the phase α of the driven system is zero and the amplitude A is independent of p (*b*) at high frequencies, $\alpha = \pi$ and A depends on p.

2. Show that for steady state forced vibration

(*a*) the power P is given by

$$P = \frac{mbf^2}{\omega^2\Delta^2 + 4b^2}$$

where $\Delta = \frac{\omega}{p} - \frac{p}{\omega}$.

(*b*) $\frac{P_{max}}{P} = 1 + \frac{\omega^2\Delta^2}{4b^2}$.

3. The sharpness of resonance s may be defined as the reciprocal of Δ at which the power is half of that at resonance. Show that the sharpness of resonance is given by $s = \frac{\omega}{2b}$. Illustrate the concept of sharpness of resonance by plotting P/P_{max} against Δ for different values of b.

4. Show that the power in the steady state forced vibration is the same whether the angular frequency p of the impressed periodic force is q times ω or $\frac{1}{q}$ of ω, where ω is the undamped natural frequency of the oscillator.

5. (*a*) Show that the steady state complex amplitude of a damped oscillator driven by an external force F exp (ipt) is given by the expression

$$A = \frac{F}{m(\omega^2 - p^2) + i\beta p}.$$

where m = mass of the system, ω = natural frequency of the oscillator in the absence of damping, and β = damping constant.

(*b*) Using the above result show that the amplitude of vibration may be written as

$$|A| = \frac{F}{\left[m^2\left(\omega^2 - p^2\right)^2 + \beta^2 p^2\right]^{1/2}}.$$

[Assume that $x = A$ exp (ipt)].

6. A machine of total mass 100 kg is supported by a spring resting on the floor and its motion is constrained to be in the vertical direction only. The system is lightly damped with a damping constant 20 Nsm^{-1}. The machine contains an eccentrically mounted shaft which, when rotating at an angular frequency p, produces a vertical force on the system of F sin pt where F is a constant. It is found that resonance occurs at 1200 r.p.m. and the amplitude of vibration in the steady state is then 1 cm. Find the amplitude of vibration in the steady state when the driving frequency is (*a*) 3000 r.p.m. (*b*) very large. Find also the quality factor Q at resonance. Neglect the effect of gravity on the motion.

7. In a resonance experiment the frequency of a sinusoidal driving force is increased gradually. If the amplitude of the forced vibration increases from 0.01 mm at very low frequencies to a maximum value of 5 mm when the frequency is 250 Hz. Calculate the Q-value of the system and the full width at half maximum power.

[**Hints:** At low frequencies, $A = f/\omega^2$ and at resonance, $A_r = f/(2b\omega)$].

8. (*a*) In the steady state forced vibration of problem 1 [page 107] show that no amplitude resonance occurs if $b \geq \omega/\sqrt{2}$. In the limiting case $b = \omega/\sqrt{2}$, show that the amplitude is given by $A = f/\sqrt{\omega^4 + p^4}$.

   (*b*) Prove that the velocity amplitude function exhibits a maximum at $p = \omega$ for any value of the damping factor.

9. The position of a particle moving along the x-axis is determined by the equation

$$\ddot{x} + 4\dot{x} + 8x = 20 \sin 2t.$$

If the particle starts from rest at $x = 0$, find (*a*) x as a function of t. (*b*) the amplitude, period and frequency of the oscillation after a long time has elapsed.

[**Hints:** $x = Re^{-2t}\cos(2t - \theta) + \sqrt{5}\sin(2t - \alpha)$ with $\sin\alpha = 2/\sqrt{5}$, $\cos\alpha = 1/\sqrt{5}$, $x(0) = \dot{x}(0) = 0$.]

10. Find an expression for the acceleration amplitude of a damped mechanical oscillator driven by a force $F \sin \omega t$ and hence calculate the frequency at which it will become maximum.

11. A Helmholtz resonator of volume 289 cm^3 has a cylindrical neck of cross-section 1 cm^2 and length 1 cm. Find the natural frequency of vibration of the resonator. Velocity of sound in air = 340 m/s.

12. In problem 16 let $R = 160\ \Omega$, $C = 15\ \mu F$, $L = 230$ mH, $\nu = 60$ Hz and $E_0 = 36$ V. Find the current amplitude I and the phase constant ϕ. Find the frequency at which the circuit will resonate.

13. For an LCR circuit connected to an alternating emf = $E_0 \sin \omega t$, at what angular frequency ω_0 will the current have its maximum value (resonance)? What is this maximum value? At what angular frequencies ω_1 and ω_2 will the current amplitude have one-half of this maximum value? What is the fractional half-width $\left[= \dfrac{\Delta\omega}{\omega_0} = \dfrac{\omega_2 - \omega_1}{\omega_0}\right]$ of the resonance curve?

14. An LCR circuit is acted on by an alternating electromotive force $E_0 \sin \omega t$. Show that the frequency at which the voltage across the condenser becomes maximum is given by

$$\omega = \omega_0 \left[1 - \frac{1}{2Q^2}\right]^{1/2}$$

where $\omega_0 = \dfrac{1}{\sqrt{LC}}$ and $Q = \dfrac{\omega_C L}{R}$

5 Waves

5.1 WAVES

A wave is a disturbance that moves through a medium without giving the medium, as a whole, any permanent displacement. The general name for these waves is progressive wave. If the disturbance takes place perpendicular to the direction of propagation of the wave, the wave is called transverse. If the disturbance is along the direction of propagation of the wave, it is called longitudinal. At any point, the disturbance is a function of time and at any instant, the disturbance is a function of the position of the point. In a sound wave, the disturbance is pressure-variation in a medium. In the transmission of light in a medium or vacuum, the disturbance is the variation of the strengths of the electric and magnetic fields. In a progressive wave motion, it is the disturbance that moves and not the particles of the medium.

5.2 WAVES IN ONE DIMENSION

Suppose a wave moves along the x-axis with constant velocity v and without any change of shape (*i.e.* with no dispersion) and the disturbance takes place parallel to the y-axis, then

$$y(x, t) = f(x - vt) \qquad ...(5.1)$$

defines a one-dimensional wave along the positive direction of the x-axis (forward wave). A wave which is the same in all respect but moving in the opposite direction (*i.e.* along the direction of x decreasing) is given by Eqn. (5.1) with the sign of v changed:

$$y(x, t) = f(x + vt) \qquad ...(5.2)$$

This is known as backward wave. Eqns. (5.1) and (5.2) satisfy the second-order partial differential equation:

$$\frac{\partial^2 y}{\partial x^2} = \frac{1}{v^2}\frac{\partial^2 y}{\partial t^2} \qquad ...(5.3)$$

Eqn. (5.3) is known as the non-dispersive wave equation.

A wave whose profile is that of a sine or cosine function is called a *harmonic wave*. We can express such a wave as

$$y = f(x - vt) = A \sin k(x - vt) \qquad ...(5.4)$$

where A is the amplitude of the wave and k is called the circular wave number. For a particular point $x = x_1$, we may write Eqn. (5.4) as

$$y = -A \sin k(vt - x_1) \qquad ...(5.5a)$$

$$= A \cos k\left(vt - x_1 + \frac{\pi}{2k}\right). \qquad ...(5.5b)$$

Sine and cosine functions have exactly the same form, the only difference between them being the point at which the origin is chosen. Since the choice of origin is always completely arbitrary, the first minus sign in Eqn. (5.5*a*) can be removed by a new choice of origin.

We know that a point executing simple harmonic motion has the equation of motion

$$y = A \sin(\omega t - \alpha) \qquad ...(5.6)$$

Comparing Eqns. (5.5*a*) and (5.6), we have

$$\omega = k\nu \text{ or, } k = \frac{\omega}{v} = \frac{2\pi\nu}{v}, \qquad ...(5.7)$$

where ν is the frequency of oscillations caused by the wave. Since $T = 1/\nu$, we can identify the period of the wave as

$$T = \frac{2\pi}{kv}. \qquad ...(5.8)$$

Since the sine function is periodic, the wave profile repeats itself after fixed interval of x. The repeat distance is known as the *wavelength* and is designated by λ. Since

$$y = A \sin k(x - vt) = A \sin k\,[(x + \lambda) - vt)]$$

we have

$$k\lambda = 2\pi \text{ or, } k = \frac{2\pi}{\lambda} = \frac{2\pi\nu}{v} \qquad ...(5.9)$$

and

$$v = \nu\lambda = \frac{\omega}{k}. \qquad ...(5.10)$$

v is called the 'phase velocity' of the travelling wave.

We can write Eqn. (5.4) in a number of equivalent forms:

$$y = A \sin(kx - \omega t) \qquad ...(5.11)$$

$$y = A \sin\frac{2\pi}{\lambda}(x - vt) \qquad ...(5.12)$$

$$y = A \sin 2\pi\left(\frac{x}{\lambda} - \frac{t}{T}\right). \qquad ...(5.13)$$

By a different choice of the origin, we could equally well arrive at the expression

$$y = A \sin(\omega t - kx) \qquad ...(5.14)$$

Again,

$$y = A \exp[i(\omega t - kx)] \qquad ...(5.15)$$

is the exponential representation of a harmonic wave. When a sine wave is expressed in the form of Eqn. (5.15), it is the imaginary part of the expression that has physical meaning.

If we compare two similar waves

$$y_1 = A \sin\frac{2\pi}{\lambda}(x - vt),$$

$$y_2 = A \sin\left[\frac{2\pi}{\lambda}(x - vt) + \delta\right] = A \sin\frac{2\pi}{\lambda}\left[\left(x + \frac{\lambda}{2\pi}\delta\right) - vt\right]$$

we see that y_2 is the same as y_1 except that it is displaced by a distance $d = \frac{\lambda}{2\pi}\delta$; δ is called the phase of y_2 relative to y_1 and d the path difference:

$$\text{Path difference} = \frac{\lambda}{2\pi} \times \text{phase difference.} \qquad ...(5.16)$$

If $\delta = 2\pi, 4\pi, ...,$ then $d = \lambda, 2\lambda, ...,$ and we say that the waves are in *phase,* and $y_1 = y_2$. If $\delta = \pi, 3\pi, ...,$ then the two waves are exactly out of phase and $y_1 = -y_2$.

5.3 THREE DIMENSIONAL WAVE EQUATION

The three-dimensional wave equation is given by

$$\nabla^2\phi = \frac{1}{v^2}\frac{\partial^2\phi}{\partial t^2} \qquad ...(5.17)$$

where ∇^2 is the Laplacian operator. The vector representation of a harmonic plane wave in three dimensions is given by

$$\phi = A \sin (\vec{k}\cdot\vec{r} - \omega t) \qquad ...(5.18)$$

where $\vec{k}$ is the vector along the direction of propagation of the wave, known as propagation vector and $|\vec{k}| = 2\pi/\lambda$.

5.4 TRANSVERSE WAVES ON A STRETCHED STRING

The speed v of transverse waves on an infinitely long stretched elastic string of mass per unit length μ and tension T is

$$v = \sqrt{T/\mu} \qquad ...(5.19)$$

5.5 STROBOSCOPE OR STROBE

It is an instrument used to make a rotating, oscillating or vibrating body appear to be *stationary* or *slow-moving.* In a simple instrument a rotating disc with evenly spaced holes is placed in the line of sight between the observer and the cyclically moving body. The frequency of the rotational disc is adjusted so that it becomes perfectly *synchronised* with the cyclically moving object which appear to be completely stationary to the observer. This illusion caused by the synchronised motion of two bodies is known as *stroboscopic effect.* This method is used to find the frequency of the periodically moving object. Very short flashes of light are used to produce still photographs of first moving objects. In medicine stroboscopes are used to view the motion of vocal chords.

Determination of the frequency of a tuning fork by stroboscopic method: Two thin light aluminium plates are attached to the inner sides of the prongs of the tuning fork. When the prongs during vibrations are wide apart, a rectangular slit is formed once per vibration of the prongs at the time of maximum displacement of the prongs away from each other. A circular disc with equispaced radial stripes is rotated and viewed through the slit when the fork is vibrating. The angular velocity of the disc is gradually increased. At a certain minimum speed of the disc the stripes appear stationary. This will occur when one stripe exactly succeeds the preceding one in the time between two successive openings of the slit. If N = number of stripes in the disc, p = number of revolutions of the disc per second, then the frequency of the tuning fork is given by

$$n = p\,N.$$

The disc can be run by a motor at any desired speed. If the speed of the disc is increased gradually the pattern appears stationary to produce the stroboscopic effect at the angular velocity of the disc which is the multiple of the minimum required speed.

SOLVED PROBLEMS

1. *A wave displacement is given by*

$$y = 0.1 \sin (0.1x - 0.1t) \text{ m}.$$

Find (a) the amplitude of the wave, (b) the magnitude of the propagation vector, (c) the wavelength, (d) the time period, and (e) the wave velocity.

Solution

The various parameters of the given harmonic wave can be found by comparing it with the standard form

$$y = A \sin (kx - \omega t)$$

for a wave propagating in the positive x-direction.

(*a*) The amplitude $A = 0.1$ m

(*b*) The propagation vector $k = 0.1 \text{ m}^{-1}$

(*c*) The wavelength $\lambda = 2\pi/k = 20\pi$ m

(*d*) The angular frequency $\omega = 0.1 \text{ s}^{-1}$,
time period $T = 2\pi/\omega = 20\pi$ s

(*e*) The wave velocity $v = \omega/k = 1$ m/s.

2. *Show that*

$$y(x, t) = f(x - vt)$$

represents a one-dimensional travelling wave (or progressive wave) moving with constant velocity v and without any change of shape along the positive direction of x.

Solution

Consider a disturbance y which propagates along the x-axis with velocity v (Fig. 5.1). The disturbance y may refer to the elevation of water wave or the magnitude of the y-displacement of a string. Since the disturbance is moving, y will depend on x and t. When $t = 0$, y will be some function of x which we may call $f(x)$. We assume that the wave propagates without change of shape. At a later time t, the wave profile will be identical with that at $t = 0$, except that the wave profile has moved a distance (vt) in the positive x-direction. If we take a new origin O' at the point $x = vt$, and denote distances measured from O' by X, then $x = X + vt$, and the equation of the wave profile, referred to this new origin, is

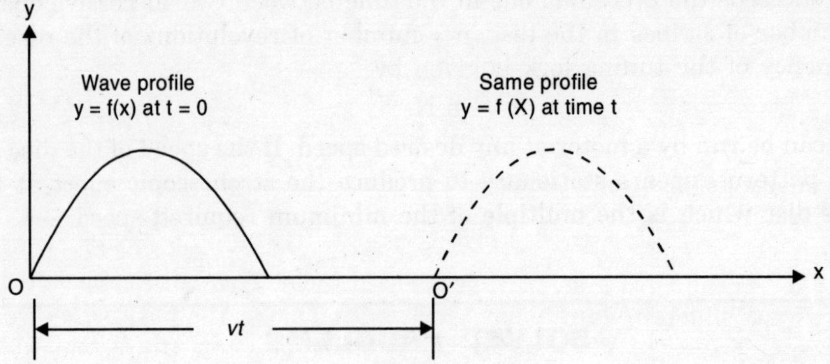

Fig. 5.1

$$y = f(X),$$

or $$y(x,t) = f(x - vt).$$

This equation is the most general expression of a wave moving with constant velocity v without change of shape along the positive direction of x.

3. *Show that the most general solution of one dimensional wave equation*

$$\frac{\partial^2 y}{\partial x^2} = \frac{1}{v^2}\frac{\partial^2 y}{\partial t^2}$$

is $y = f(x - vt) + g(x + vt)$ where f and g are arbitrary functions of $x - vt$ and $x + vt$ respectively.

Solution

Let $u = x - vt$ and $w = x + vt$.

Thus, $$\frac{\partial}{\partial x} = \frac{\partial u}{\partial x}\frac{\partial}{\partial u} + \frac{\partial w}{\partial x}\frac{\partial}{\partial w} = \frac{\partial}{\partial u} + \frac{\partial}{\partial w}$$

$$\frac{\partial^2}{\partial x^2} = \frac{\partial}{\partial x}\frac{\partial}{\partial x} = \left(\frac{\partial}{\partial u} + \frac{\partial}{\partial w}\right)\left(\frac{\partial}{\partial u} + \frac{\partial}{\partial w}\right)$$

$$= \frac{\partial^2}{\partial u^2} + \frac{2\partial^2}{\partial u \partial w} + \frac{\partial^2}{\partial w^2}.$$

$$\frac{\partial}{\partial t} = \frac{\partial u}{\partial t}\frac{\partial}{\partial u} + \frac{\partial w}{\partial t}\frac{\partial}{\partial w} = -v\frac{\partial}{\partial u} + v\frac{\partial}{\partial w}$$

$$\frac{\partial^2}{\partial t^2} = \left(-v\frac{\partial}{\partial u} + v\frac{\partial}{\partial w}\right)\left(-v\frac{\partial}{\partial u} + v\frac{\partial}{\partial w}\right)$$

$$= v^2\left[\frac{\partial^2}{\partial u^2} - \frac{2\partial^2}{\partial u \partial w} + \frac{\partial^2}{\partial w^2}\right].$$

Using the wave equation, we get

$$\frac{\partial^2 y}{\partial u \partial \omega} = 0$$

or $$\frac{\partial y}{\partial u} = F(u)$$

where F is an arbitrary function of u.

Integrating we get

$$y = \int F(u)du + g(w)$$

$$= f\ (u) + g\ (w)$$

Thus, the general solution of the wave equation is

$$y = f\ (x - vt) + g\ (x + vt).$$

The waves $f\ (x - vt)$ and $g\ (x + vt)$ travel with same velocity v, but in opposite directions. The wave $f\ (x - vt)$ is called forward wave which moves along the positive x-direction and the wave $g\ (x + vt)$ is backward wave which moves in the negative x-direction. The method described here for solving the partial differential equation is known as D' Alembert's method.

4. (*i*) *Show that the equation of a plane perpendicular to the unit vector* $\hat{s}$ *is*

$$\hat{s} \cdot \vec{r} = \text{Constant}.$$

(*ii*) *Find a plane wave solution of the three dimensional wave equation*

$$\nabla^2\phi = \frac{1}{v^2}\frac{\partial^2\phi}{\partial t^2}$$

Solution

(*i*) Let $\hat{s}$ be a unit vector in a fixed direction. We consider a plane perpendicular to this fixed direction so that the distance of this plane from the origin is ON = Constant (Fig. 5.2). Let $\vec{r}$ (x, y, z) be the position vector of a point P on this plane. Then,

$$\vec{r} \cdot \hat{s} = ON = \text{Constant}.$$

Thus, the equation of the plane perpendicular to the unit vector $\hat{s}$ is

$$\hat{s} \cdot \vec{r} = \text{Constant}$$

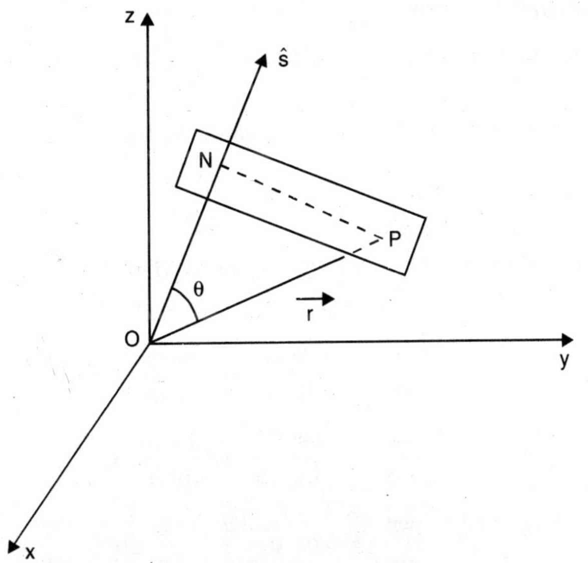

Fig. 5.2

(*ii*) A plane wave is one in which the disturbance is constant for all points of a plane drawn perpendicular to the direction of propagation. Such a plane is called a *plane wave front,* and the wave front moves perpendicular to itself with the velocity of propagation v. Let $\hat{s}\,(s_x, s_y, s_z)$ be a unit vector in the direction of propagation of the wave. A solution of the wave equation of the form

$$\phi = \phi\,(\vec{r}\cdot\hat{s}, t)$$

is said to represent a plane wave, since at each instant of time, ϕ is a constant over each of the planes $\vec{r}\cdot\hat{s}$ = constant, which is perpendicular to the unit vector $\hat{s}$.

It will be convenient to choose a new set of Cartesian axes $O\xi$, $O\eta$ and $O\tau$ with $O\tau$ in the direction of s. Then,

$$\vec{r}\cdot\hat{s} = \tau$$

or

$$xs_x + ys_y + zs_z = \tau.$$

Thus,

$$\frac{\partial}{\partial x} = \frac{\partial \tau}{\partial x}\frac{\partial}{\partial \tau} = s_x\frac{\partial}{\partial \tau},\ \frac{\partial}{\partial y} = s_y\frac{\partial}{\partial \tau} \text{ and } \frac{\partial}{\partial z} = s_z\frac{\partial}{\partial \tau},$$

so that

$$\nabla^2\phi = (s_x^2 + s_y^2 + s_z^2)\frac{\partial^2\phi}{\partial \tau^2} = \frac{\partial^2\phi}{\partial \tau^2},$$

and the wave equation becomes

$$\frac{\partial^2\phi}{\partial \tau^2} = \frac{1}{v^2}\frac{\partial^2\phi}{\partial t^2}.$$

If we put $u = \tau - vt$ and $w = \tau + vt$, we can solve this equation by applying D' Alembert's method (See problem 3):

$$\phi = f(\vec{r}\cdot\hat{s} - vt) + g(\vec{r}\cdot\hat{s} + vt)$$

where f and g are arbitrary functions.

5. *Deduce the two-dimensional wave equation in polar coordinates (r, ϕ).*

Solution

The two-dimensional wave equation in Cartesian coordinates is given by

$$\frac{\partial^2\psi}{\partial x^2} + \frac{\partial^2\psi}{\partial y^2} = \frac{1}{v^2}\frac{\partial^2\psi}{\partial t^2}$$

where ψ is the wave disturbance and x and y are related to r and ϕ via the equations

$$x = r\cos\phi,\ y = r\sin\phi.$$

Now,

$$\frac{\partial\psi}{\partial r} = \frac{\partial\psi}{\partial x}\frac{\partial x}{\partial r} + \frac{\partial\psi}{\partial y}\frac{\partial y}{\partial r} = \cos\phi\frac{\partial\psi}{\partial x} + \sin\phi\frac{\partial\psi}{\partial y}$$

$$\frac{\partial\psi}{\partial \phi} = \frac{\partial\psi}{\partial x}\frac{\partial x}{\partial \phi} + \frac{\partial\psi}{\partial y}\frac{\partial y}{\partial \phi} = -r\sin\phi\frac{\partial\psi}{\partial x} + r\cos\phi\frac{\partial\psi}{\partial y}$$

which gives

$$\frac{\partial\psi}{\partial x} = \cos\phi\,\frac{\partial\psi}{\partial r} - \frac{1}{r}\sin\phi\,\frac{\partial\psi}{\partial \phi}$$

$$\frac{\partial\psi}{\partial y} = \sin\phi\,\frac{\partial\psi}{\partial r} + \frac{1}{r}\cos\phi\,\frac{\partial\psi}{\partial \phi}$$

It follows then that

$$\frac{\partial^2\psi}{\partial x^2} = \left(\cos\phi\frac{\partial}{\partial r}-\frac{1}{r}\sin\phi\frac{\partial}{\partial\phi}\right)\left(\cos\phi\frac{\partial\psi}{\partial r}-\frac{1}{r}\sin\phi\frac{\partial\psi}{\partial\phi}\right)$$

$$= \cos^2\phi\frac{\partial^2\psi}{\partial r^2}+\frac{2}{r^2}\sin\phi\cos\phi\frac{\partial\psi}{\partial\phi}-\frac{2}{r}\sin\phi\cos\phi\frac{\partial^2\psi}{\partial r\partial\phi}$$

$$+\frac{1}{r}\sin^2\phi\frac{\partial\psi}{\partial r}+\frac{1}{r^2}\sin^2\phi\frac{\partial^2\psi}{\partial\phi^2}.$$

Similarly

$$\frac{\partial^2\psi}{\partial y^2} = \left(\sin\phi\frac{\partial}{\partial r}+\frac{1}{r}\cos\phi\frac{\partial}{\partial\phi}\right)\left(\sin\phi\frac{\partial\psi}{\partial r}+\frac{1}{r}\cos\phi\frac{\partial\psi}{\partial\phi}\right)$$

$$= \sin^2\phi\frac{\partial^2\psi}{\partial r^2}-\frac{2}{r^2}\sin\phi\cos\phi\frac{\partial\psi}{\partial\phi}+\frac{2}{r}\sin\phi\cos\phi\frac{\partial^2\psi}{\partial r\partial\phi}$$

$$+\frac{1}{r}\cos^2\phi\frac{\partial\psi}{\partial r}+\frac{1}{r^2}\cos^2\phi\frac{\partial^2\psi}{\partial\phi^2}.$$

Adding these, we get

$$\frac{\partial^2\psi}{\partial x^2}+\frac{\partial^2\psi}{\partial y^2} = \frac{\partial^2\psi}{\partial r^2}+\frac{1}{r}\frac{\partial\psi}{\partial r}+\frac{1}{r^2}\frac{\partial^2\psi}{\partial\phi^2}.$$

The two-dimensional wave equation in the polar co-ordinates (r, ϕ) is

$$\frac{\partial^2\psi}{\partial r^2}+\frac{1}{r}\frac{\partial\psi}{\partial r}+\frac{1}{r^2}\frac{\partial^2\psi}{\partial\phi^2}=\frac{1}{v^2}\frac{\partial^2\psi}{\partial t^2}.$$

6. *Consider three dimensional wave equation*

$$\nabla^2\psi = \frac{1}{v^2}\frac{\partial^2\psi}{\partial t^2}$$

where ψ is a function of r and t. Find solutions representing spherical waves i.e. solutions of the form $\psi = \psi(r, t)$.

Solution

We know, $r = |\vec{r}| = \sqrt{x^2+y^2+z^2}$ and

$$\frac{\partial}{\partial x} = \frac{\partial r}{\partial x}\frac{\partial}{\partial r} = \frac{x}{r}\frac{\partial}{\partial r}$$

$$\frac{\partial^2}{\partial x^2}=\frac{x}{r}\frac{\partial}{\partial r}\left(\frac{x}{r}\frac{\partial}{\partial r}\right) = \frac{x^2}{r^2}\frac{\partial^2}{\partial r^2} - \frac{x^2}{r^3}\frac{\partial}{\partial r} + \frac{x}{r^2}\frac{\partial x}{\partial r}\frac{\partial}{\partial r}$$

$$= \frac{x^2}{r^2}\frac{\partial^2}{\partial r^2} - \frac{x^2}{r^3}\frac{\partial}{\partial r} + \frac{1}{r}\frac{\partial}{\partial r}$$

Similarly

$$\frac{\partial^2}{\partial y^2} = \frac{y^2}{r^2}\frac{\partial^2}{\partial r^2} - \frac{y^2}{r^3}\frac{\partial}{\partial r} + \frac{1}{r}\frac{\partial}{\partial r}$$

and
$$\frac{\partial^2}{\partial z^2} = \frac{z^2}{r^2}\frac{\partial^2}{\partial r^2} - \frac{z^2}{r^3}\frac{\partial}{\partial r} + \frac{1}{r}\frac{\partial}{\partial r}$$

Thus,
$$\nabla^2 = \frac{\partial^2}{\partial r^2} + \frac{2}{r}\frac{\partial}{\partial r}$$

and
$$\nabla^2\psi = \frac{1}{r}\frac{\partial^2}{\partial r^2}(r\psi).$$

The wave equation now becomes

$$\frac{1}{r}\frac{\partial^2}{\partial r^2}(r\psi) - \frac{1}{v^2}\frac{\partial^2\psi}{\partial t^2} = 0$$

or
$$\frac{\partial^2}{\partial r^2}(r\psi) - \frac{1}{v^2}\frac{\partial^2}{\partial t^2}(r\psi) = 0$$

The general solution of this equation is

$$r\psi = f(r - vt) + g\,(r + vt)$$

where f and g are arbitrary functions. Hence we get

$$\psi = \frac{1}{r}\,f\,(r - vt) + \frac{1}{r}g\,(r + vt).$$

The first term represents a spherical wave diverging from the origin, the second a spherical wave converging towards the origin. The velocity of propagation in both cases is v.

7. *A compressional wave of frequency 300 Hz is set up in an iron rod and passes from the iron rod into air. The speed of the wave is 4800 m/s in iron and 330 m/s in air. Find the wavelength in each material.*

Solution

The frequency of a wave remains unchanged as it passes from one medium to another.

In iron,
$$\lambda = \frac{v}{\nu} = \frac{4800}{300} = 16 \text{ m}.$$

In air,
$$\lambda = \frac{v}{\nu} = \frac{330}{300} = 1.1. \text{ m}.$$

8. *Verify that the wave function*

$$y(x, t) = Ae^{-B(x - vt)^2}$$

satisfies the one-dimensional wave equation.

Solution

Let $f\,(x, t) = Ae^{-B(x - vt)^2}$

Now,
$$\frac{\partial f}{\partial t} = 2ABv(x - vt)e^{-B(x - vt)^2}$$

$$\frac{\partial^2 f}{\partial t^2} = v^2[-2AB + 4AB^2(x - vt)^2]\,e^{-B(x - vt)^2}$$

$$\frac{\partial f}{\partial x} = -2AB(x - vt)e^{-B(x - vt)^2}$$

$$\frac{\partial^2 f}{\partial x^2} = [-2AB + 4AB^2(x - vt)^2]e^{-B(x - vt)^2}$$

Thus, we see that

$$\frac{\partial^2 f}{\partial x^2} = \frac{1}{v^2}\frac{\partial^2 f}{\partial t^2}$$

9. *A sinusoidal wave travelling in the positive x-direction on a stretched string has amplitude 2.0 cm, wavelength 1.0 m and velocity 5.0 ms^{-1}. The initial conditions are: $y = 0$ and $\frac{\partial y}{\partial t} < 0$ at $x = 0$ and $t = 0$. Find the wave function $y = f(x, t)$.*

Solution

The general form of a wave travelling in the positive x direction is

$$y(x, t) = A \cos\left[2\pi\left(\frac{x}{\lambda} - \frac{t}{T}\right) + \delta\right].$$

Here, $A = 0.02$ m, $\lambda = 1.0$ m and $v = \frac{\lambda}{T} = 5$ ms^{-1}. So, $T = 0.2$ s. Thus,

$$y = 0.02 \cos[2\pi(x - 5t) + \delta].$$

Putting $t = 0$, $x = 0$ and $y = 0$, we obtain $0.02 \cos \delta = 0$ or, $\cos \delta = 0$.

$$\text{Now, } \frac{\partial y}{\partial t} = 0.02 \times 10\pi \sin[2\pi(x - 5t) + \delta].$$

It is given that $\frac{\partial y}{\partial t} < 0$ at $x = 0$, $t = 0$, i.e. $\sin \delta < 0$. Hence we may conclude that

$$\delta = \left(-\frac{\pi}{2}\right) + 2n\pi,$$

where n is an integer. Thus, the wavefunction is

$$y = 0.02 \cos\left[2\pi(x - 5t) - \frac{\pi}{2}\right]$$

$$= 0.02 \sin 2\pi(x - 5t).$$

10. *Show that the speed v of transverse waves on an infinitely long stretched elastic string of mass per unit length μ and tension T is*

$$v = \sqrt{T/\mu}.$$

Solution

We consider a very small segment of the continuous string. At equilibrium the segment occupies a small length Δx centered at x (Fig. 5.3). Let the mass of this segment be Δm. Now mass density $\mu = \frac{\Delta m}{\Delta x}$. The mass density is assumed to be uniform along the string. The string tension at equilibrium, denoted by T, is also assumed to be uniform.

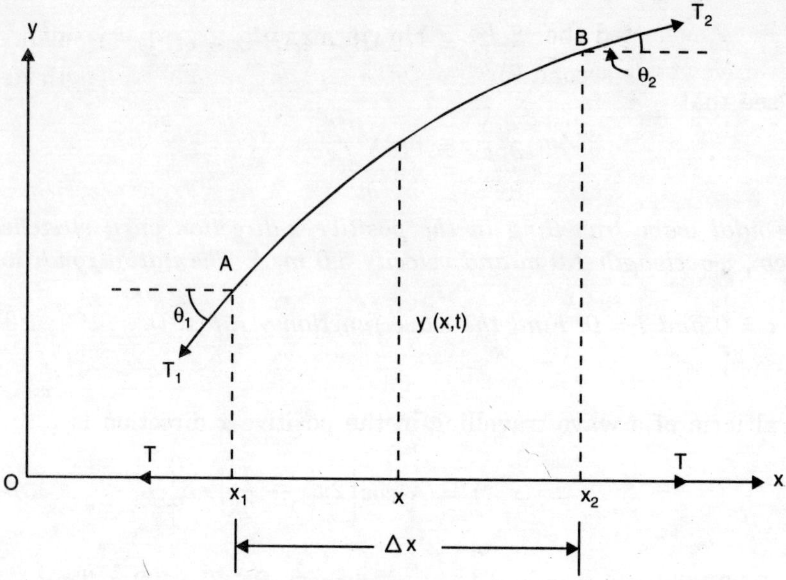

Fig. 5.3

For a general non-equilibrium situation the segment has a transverse displacement $y(x, t)$ averaged over the segment AB. The segment AB is no longer exactly straight. It has (generally) a slight curvature. We draw tangents at the points A and B to the displaced segment AB. Tensions T_1 and T_2 to the segment AB act along the tangents. The tension in the segment is no longer T, since the segment is longer than its equilibrium length Δx. The tangents at the points A and B make angles θ_1 and θ_2 with the horizontal line. Let us find the net upward force F_y on the segment. At its left end the segment is pulled downward with a force $T_1 \sin \theta_1$. At its right end it is pulled upward with a force $T_2 \sin \theta_2$. Thus, the net upward force on the segment Δx of the string is

$$F_y(t) = T_2 \sin \theta_2 - T_1 \sin \theta_1$$

In the small oscillation approximation, we may neglect the increase in length of the segment, and the angles θ_1 and θ_2 are very small. So, $\sin \theta_2 \approx \tan \theta_2$ and $\sin \theta_1 \approx \tan \theta_1$, $T_2 \approx T$ and $T_1 \approx T$. We now have

$$F_y(t) = T \tan \theta_2 - T \tan \theta_1.$$

$$= T \left[\left(\frac{\partial y}{\partial x} \right)_{x_2} - \left(\frac{\partial y}{\partial x} \right)_{x_1} \right]$$

$$= T \left[\left(\frac{\partial y}{\partial x} \right)_{x_1 + \Delta x} - \left(\frac{\partial y}{\partial x} \right)_{x_1} \right] \qquad ...(5.20)$$

Let us write $f(x) = \dfrac{\partial y}{\partial x}$. Thus,

$$f(x_1 + \Delta x) - f(x_1) \approx \Delta x f'(x).$$

We have from Eqn. (5.20), (when $\Delta x \to 0$),

$$F_y(t) = T \Delta x \left(\frac{\partial^2 y}{\partial x^2} \right)_x$$

Here, we have neglected the higher order terms since Δx is very small.

According to Newton's second law the force on the segment of length Δx is equal to

$$\Delta m \frac{\partial^2 y}{\partial t^2} = \mu \, \Delta x \frac{\partial^2 y}{\partial t^2}.$$

Hence, we have

$$\mu \, \Delta x \frac{\partial^2 y}{\partial t^2} = T \, \Delta x \frac{\partial^2 y}{\partial x^2}$$

or,

$$\frac{\partial^2 y}{\partial t^2} = \frac{T}{\mu} \frac{\partial^2 y}{\partial x^2}.$$

This has the form of classical wave equation

$$\frac{\partial^2 y}{\partial x^2} = \frac{1}{v^2} \frac{\partial^2 y}{\partial t^2}$$

where $v = \sqrt{T/\mu}$ = velocity of propagation of the wave.

11. *(a) An elastic string fixed at the top end has mass 1 g and natural length 0.1 m. A mass 1.1 kg is attached to its lower end, and the spring is stretched by 0.022 m. Calculate the speed of propagation of transverse waves along the string. If the mass makes small vertical oscillations, find its time period of oscillation.*

(b) A transverse disturbance travels down the string starting from the upper end and is reflected at lower end. The reflected wave travels up the string and is reflected again from the upper point. If this process continues, how many times will this disturbance pass the middle point of the string in one period of a small vertical oscillation of the mass?

[Take the tension in the string to be uniform]

Solution

(*a*)

$$v_{tr} = \sqrt{\frac{T}{\mu}} = \sqrt{\frac{1.1 \times 9.8}{10^{-3}/0.1}}$$

$$= 32.83 \text{ m/s}.$$

Time period of vertical oscillation of the mass is

$$2\pi \sqrt{\frac{0.022}{9.8}} = 0.298 \text{ s}.$$

(*b*) Distance traversed by the transverse wave in time 0.298 s is 32.83 × 0.298 = 9.78 m. The transverse wave passes the middle point first time when it traverses a distance = 0.1/2 = 0.05 m. In traversing the remaining distance (9.78 – 0.05 = 9.78 m), it passes the middle point 9.73/0.1 = 97.3 times. Thus, the total number of times = 1 + 97 = 98.

12. *A wire of mass m and length l is fixed at the top end, and hangs freely under its own weight. Deduce the time for a transverse wave to travel the length of the wire.*

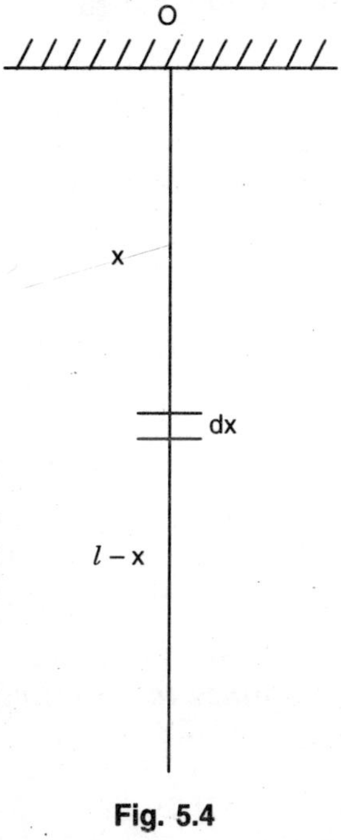

Fig. 5.4

Solution

Mass per unit length of the wire = $\mu = \frac{m}{l}$.

Consider a small portion of the string of length dx at a distance x from the point of suspension O (Fig. 5.4). The tension at this point is $(l-x)\mu g$ and the velocity of propagation of transverse wave at this point is

$$\sqrt{\frac{(l-x)\mu g}{\mu}} = \sqrt{(l-x)\,g}.$$

Time taken by the transverse wave to traverse a distance dx at this point is

$$\frac{dx}{\sqrt{(l-x)\,g}}.$$

Total time to travel the length of the wire is

$$\int_0^l \frac{dx}{\sqrt{(l-x)\,g}} = 2\sqrt{\frac{l}{g}}.$$

13. *A uniform inextensible string of length l and total mass M is fixed at one end and hangs freely under its own weight. It is tapped at the top end so that a transverse wave runs down it. At the same moment a small stone is released from rest and falls freely from the top of the string.*

How far from the top does the stone pass the wave?

Solution

From the previous problem we find that the time taken by the wave to travel from $x = 0$ to $x = x$ is

$$T(x) = \int_0^x \frac{dx}{\sqrt{(l-x)\,g}} = \frac{2}{\sqrt{g}}[\sqrt{l} - \sqrt{l-x}]$$

The time $t(x)$ for the stone to fall from rest through a distance x is

$$t(x) = \sqrt{\frac{2x}{g}}.$$

Equating $t(x)$ and $T(x)$ to find the value of x at which the wave and the stone meet, we get

$$\sqrt{2x} = 2[\sqrt{l} - \sqrt{l-x}].$$

Solving this equation we obtain $x = 0$ or $8l/9$. Now, $x = 0$ corresponds to the situation when the wave and the stone just start moving. The two again meet at a distance $8l/9$ from the top of the string.

14. *Show that the velocity of propagation of longitudinal waves in a fluid (liquid or gas) contained in an infinitely long tube is given by*

$$v = \sqrt{K/\rho_0}$$

where K = *Bulk modulus of the fluid,*

ρ_0 = *Equilibrium density of the fluid.*

Solution

We consider an infinitely long tube of cross-section A (Fig. 5.5), containing a fluid (liquid or gas). Suppose that originally the fluid is at rest, its density is ρ_0 and pressure P_0. Let R be a section of the medium of area A, perpendicular to the direction of propagation of the wave. Let S be a parallel section of equal area δx apart, δx being an elementary thickness of the layer. The coordinates of R and S are x and $x + \delta x$ respectively. Originally the small cylinder of fluid RS experiences an equal pressure P_0 exerted at both ends by the surrounding fluid. Suppose the fluid is set into longitudinal agitation, for example by inserting a piston in the tube somewhere to the left of R and causing it to oscillate longitudinally. This

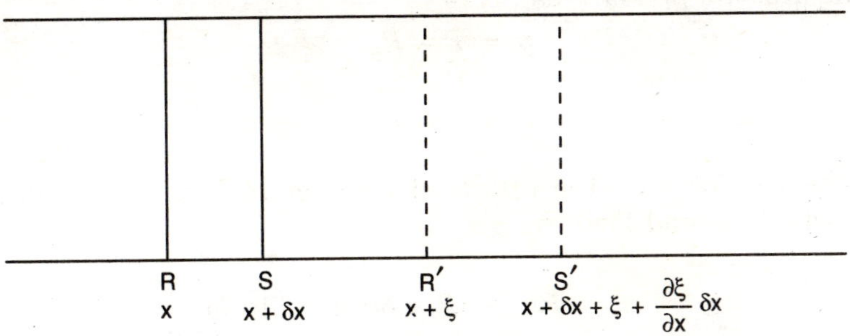

Fig. 5.5

will set the adjacent fluid into longitudinal oscillation, and this disturbance will propagate along the fluid in the form of a longitudinal wave. Suppose that at a given instant, the cylinder RS is displaced to a new position $R'S'$ such that $R'R = \xi$ and $S'S = \xi + \frac{\partial \xi}{\partial x}\delta x$. The variable ξ measures the longitudinal displacement of a point due to the passage of the wave. The thickness of the layer $R'S'$ is $\delta x + \frac{\partial \xi}{\partial x}\delta x$. Thus, the increase of thickness of the layer due to the longitudinal oscillation at that instant $= \frac{\partial \xi}{\partial x}\delta x$. Since there is no motion perpendicular to the direction of propagation of the wave, the corresponding increase in volume $= A\frac{\partial \xi}{\partial x}\delta x$.

We shall find the equation of motion of the fluid at $R'S'$. For this purpose we require to know its mass and the pressure at its two ends. Its mass is same as the mass of the undisturbed element RS, that is $A\rho_0\delta x$, where ρ_0 = normal average density or equilibrium density of the fluid. Let the pressure on the left-hand face R' be P and that on the right-hand face S' be $P + \delta P$. The bulk modulus K of a material is a measure of the pressure increase to change its volume by a given amount. It is defined as

$$K = \frac{\text{Stress}}{\text{Strain}} = \frac{\text{Extra pressure applied}}{\text{Fractional change in volume}}.$$

Proceeding to the limit of vanishing small change in volume, we obtain

$$K = -\frac{dP}{\frac{dV}{V}} = -V\frac{dP}{dV}. \qquad \text{...(5.21)}$$

The minus sign indicates that the volume decreases if the pressure increases.

$$\text{Volume strain} = \frac{dV}{V} = \frac{\text{Increase in volume}}{\text{Original volume}} = \frac{A\dfrac{\partial \xi}{\partial x}\delta x}{A\delta x}$$

$$= \frac{\partial \xi}{\partial x} \qquad ...(5.22)$$

Thus from Eqn. (5.21), we find that the extra pressure over normal undisturbed pressure

$$= -K\frac{dV}{V} = -K\frac{\partial \xi}{\partial x}.$$

If we define *acoustic pressure* as

$$p = P - P_0 \qquad ...(5.23)$$

we have

$$p = -K\frac{\partial \xi}{\partial x}. \qquad ...(5.24)$$

Equating the net force on the displaced cylinder $R'S'$ to the product of mass and acceleration (Newton's second law), we get

$$PA - (P + \delta P)A = (A\rho_0 \delta x)\frac{\partial^2 \xi}{\partial t^2}$$

or

$$-\delta p = \rho_0 \delta x \frac{\partial^2 \xi}{\partial t^2}. \qquad ...(5.25)$$

From Eqn. (5.23), we have $\delta p = \delta P$ since P_0 is constant. Thus,

$$-\delta p = -\frac{\partial p}{\partial x}\delta x = (\rho_0 \delta x)\frac{\partial^2 \xi}{\partial t^2}.$$

Using Eqn. (5.24), we obtain

$$K\frac{\partial^2 \xi}{\partial x^2} = \rho_0 \frac{\partial^2 \xi}{\partial t^2}$$

or

$$\frac{\partial^2 \xi}{\partial x^2} = \frac{1}{K/\rho_0}\frac{\partial^2 \xi}{\partial t^2} \qquad ...(5.26)$$

which is the classical wave equation having velocity of propagation

$$v = \sqrt{K/\rho_0} \qquad ...(5.27)$$

15. *In the previous problem show that*

$$\rho = \rho_0\left(1 - \frac{\partial \xi}{\partial x}\right)$$

where ρ_0 = *Equilibrium density of the fluid*

ρ = *Density of the fluid in the disturbed position.*

and $\dfrac{\partial \xi}{\partial x}$ = *volume strain.*

Solution

Volume of the fluid at $R'S' = A\ \delta x + A\frac{\partial \xi}{\partial x}\delta x$.

Hence,
$$\rho = \frac{A\rho_0\delta x}{A\delta x + A\frac{\partial \xi}{\partial x}\delta x} = \frac{\rho_0}{1+\frac{\partial \xi}{\partial x}} \approx \rho_0\left(1-\frac{\partial \xi}{\partial x}\right) \qquad ...(5.28)$$

where $\frac{\partial \xi}{\partial x} \ll 1$.

Note: If we define condensation (s) as

$$s = \frac{\rho-\rho_0}{\rho_0}, \text{ then we find } s = -\frac{\partial \xi}{\partial x}$$

and
$$\rho = \rho_0\,(1 + s)$$

16. *Show that the velocity of propagation of longitudinal wave in a fluid is given by*

$$v = \sqrt{\frac{dP}{d\rho}}$$

where P and ρ *are the pressure and density of the fluid.*

Solution

Eqn. (5.25) can be written as

$$-\frac{\partial P}{\partial x}\delta x = \rho_0\ \delta x\frac{\partial^2 \xi}{\partial t^2}$$

or
$$-\frac{dP}{d\rho}\frac{\partial \rho}{\partial x} = \rho_0\frac{\partial^2 \xi}{\partial t^2}.$$

Again, Eqn. (5.28) gives

$$\frac{\partial \rho}{\partial x} = -\ \rho_0\frac{\partial^2 \xi}{\partial x^2}$$

So,
$$\frac{\partial^2 \xi}{\partial x^2} = -\frac{1}{\rho_0}\frac{\partial \rho}{\partial x} = \frac{1}{\frac{dP}{d\rho}}\frac{\partial^2 \xi}{\partial t^2}$$

which gives the velocity of propagation as

$$v = \sqrt{\frac{dP}{d\rho}} \qquad ...(5.29)$$

17. *Assume that the changes in local conditions produced by the passage of sound wave in gases take place so slowly that the temperature remains constant an isothermal change. Show that under this assumption the velocity of sound wave through gases is given by*

$$v = \sqrt{\frac{RT}{M}}$$

where T = *Temperature of the gas in K*

and M = *Molar mass of the gas.*

Use this relation to calculate the velocity of sound through air at STP.

[Molar mass of air = 28.8 g]

Solution

For an isothermal process

$$P = \frac{\text{const.}}{V} = k\rho, \text{ where } k \text{ is a constant.}$$

Thus, $$\frac{dP}{d\rho} = k = \frac{P}{\rho} \text{ and hence } v = \sqrt{\frac{P}{\rho}}.$$

For one g mol of an ideal gas, we have

$$PV = RT \text{ and } \rho = \frac{M}{V}.$$

Thus, $\frac{P}{\rho} = \frac{RT}{M}$ and the speed of sound in a gas is

$$v = \sqrt{\frac{RT}{M}}.$$

At STP,

$$v_{\text{air}} = \left[\frac{8.31 \text{ JK}^{-1}\text{ mol}^{-1} \times 273 \text{ K}}{28.8 \times 10^{-3} \text{ kg mol}^{-1}}\right] = 280.67 \text{ m/s}.$$

Note: Newton assumed isothermal gas law for the propagation of sound wave through the gaseous medium and arrived at this result. This is off the actual value by 15%. Later on, Laplace corrected Newton's result by using adiabatic gas law instead of Boyle's law. The process takes place so rapidly that there is no flow of heat. There is not sufficient time for heat to flow from the compressions to rarefactions. (see problem 18).

18. *Using adiabatic gas law show that the velocity of sound wave through gases is given by*

$$v = \sqrt{\frac{\gamma RT}{M}}$$

where $\gamma = \frac{C_P}{C_V}$ = *Ratio of the principal heat capacities.*

Use this relation to calculate the velocity of sound through air at STP (For air, $\gamma = 1.4$*).*

Solution

When a fixed mass of ideal gas changes its state adiabatically, we have

$$PV^\gamma = \text{Constant},$$

or $$P = k\rho^\gamma, \text{ where } k \text{ is a constant.}$$

Thus, $$\frac{dP}{d\rho} = k\gamma\rho^{\gamma-1} = \frac{P}{\rho^\gamma}\gamma\rho^{\gamma-1} = \frac{\gamma P}{\rho},$$

and $$v = \sqrt{\frac{dP}{d\rho}} = \sqrt{\frac{\gamma P}{\rho}} = \sqrt{\frac{\gamma RT}{M}}.$$

At STP,

$$v_{air} = \sqrt{\frac{1.4 \times 8.31 \times 273}{28.8 \times 10^{-3}}} = 332.09 \text{ m/s}.$$

Note: For monoatomic gases (*e.g.* He, Ne, Ar), $\gamma = \frac{5}{3} = 1.667$;

and for diatomic gases (*e.g.* N_2, O_2, H_2, CO) $\gamma = \frac{7}{5} = 1.4$.

19. *Calculate the speed of sound in helium at –110°C [Molar mass of He = 4 g]*

Solution

$$v = \sqrt{\frac{1.667 \times 8.31 \times 163}{4 \times 10^{-3}}} = 751.33 \text{ m/s}.$$

20. *A gaseous mixture enclosed in a vessel of volume V consists of one gram mole of a gas A with $\gamma\ (= C_P/C_V) = 5/3$ and another gas B with $\gamma = 7/5$ at a certain temperature T. The gram molecular weights of the gases A and B are 4 and 32 respectively. The gases A and B do not react with each other and are assumed to be ideal. The gaseous mixture follows the equation $PV^{19/13}$ = Constant, in adiabatic processes.*

(*a*) *Find the number of gram moles of the gas B in the gaseous mixture.*

(*b*) *Compute the speed of sound in the gaseous mixture at T = 300 K.*

(*c*) *If T is raised 1K from 300K, find the percentage change in the speed of sound in the gaseous mixture.*

(*d*) *The mixture is compressed adiabatically to 1/5 of its initial volume V. Find the change in its adiabatic compressibility in terms of the given quantities.*

(I.I.T. 1995)

Solution

(*a*) γ for the mixture $= \frac{19}{13}$.

Since $C_V = \frac{R}{\gamma - 1}$, $(C_V)_A = \frac{R}{5/3 - 1} = \frac{3}{2}R$ and $(C_V)_B = \frac{5}{2}R$,

and $$(C_V)_{\text{mixture}} = \frac{R}{\frac{19}{13} - 1} = \frac{13}{6}R.$$

Now, for a change in temperature ΔT, the sum of energy change of individual gases must be equal to total energy change of the mixture:

$$n_A(C_V)_A\Delta T + n_B(C_V)_B\Delta T = (n_A + n_B)(C_V)_{mix}\Delta T$$

or $$1.\frac{3}{2}R + n_B\,\frac{5}{2}R = (1 + n_B)\,\frac{13}{6}R.$$

Thus $$n_B = 2$$

(*b*) Velocity of sound in the gaseous mixture is

$$v = \sqrt{\frac{\gamma RT}{M}}$$

where $$M = \frac{1 \times 4 + 2 \times 32}{3} = \frac{68}{3} \text{ g mol}^{-1}.$$

Hence, $$v = \left[\frac{\frac{19}{13}\times 8.31\times 300}{\frac{68}{3}\times 10^{-3}}\right]^{\frac{1}{2}} = 400.93 \text{ ms}^{-1}.$$

(*c*) Since $v^2 = \gamma RT/M$, $\dfrac{dv}{v} = \dfrac{dT}{2T}$.

Hence, the percentage change of the speed of sound

$$= \frac{1}{2}\frac{dT}{T}\times 100 = \frac{1}{2}\times\frac{1}{300}\times 100 = 0.167\%.$$

(*d*) From the relation pV^{γ} = Constant, we get

$$\frac{dP}{dV} = -\frac{\gamma P}{V}$$

or, $$\text{Bulk modulus} = -V\frac{dP}{dV} = \gamma P.$$

$$C = \text{Adiabatic compressibility} = \frac{1}{\text{Bulk modulus}} = \frac{1}{\gamma P}$$

$$|dC| = \frac{dP}{\gamma P^2}$$

Since, $P_1V^{19/13} = P_2\,(V/5)^{19/13}$ or, $P_2 = P_1 5^{19/13}$,

Hence, $dP = P_2 - P_1 = P_1\,(5^{19/13} - 1)$.

Let $P_1 = P$

$$|dC| = \frac{5^{19/13}-1}{\gamma P} = \frac{5^{19/13}-1}{\frac{19}{13}\frac{3RT}{V}}$$

$$= \frac{13V\left(5^{19/13}-1\right)}{57RT}$$

$$= 8.7\times 10^{-4}\ V\ \text{m}^2/\text{N}$$

[$R = 8.31$ J K^{-1} mol^{-1} and $T = 300$ K].

21. *The displacement ξ of any particle at a distance x at time t from its equilibrium position due to the passage of a plane progressive wave in a gas is given by*

$$\xi = A\sin\frac{2\pi}{\lambda}(x - vt).$$

Show that the energy density of the medium is $2\pi^2\rho A^2v^2/\lambda^2$ where ρ is the density of the gas.

Solution

The kinetic energy T of a layer of elementary thickness δx and unit cross-sectional area perpendicular to the direction of propagation of wave is

$$T = \frac{1}{2}\rho\,\delta x\left(\frac{\partial\xi}{\partial t}\right)^2$$

$$= \frac{1}{2}\ \rho\delta x \frac{4\pi^2 v^2 A^2}{\lambda^2} \cos^2 \frac{2\pi}{\lambda}(x - vt).$$

Thus, the average kinetic energy of the layer over one cycle of vibration

$$= \frac{1}{2}\ \rho\delta x \frac{4\pi^2 v^2 A^2}{\lambda^2} \times \frac{1}{2} = \frac{\pi^2 v^2 A^2 \rho\, \delta x}{\lambda^2}.$$

If P_0 is the pressure when the medium is undisturbed and $P = P_0 + p$ is the pressure when the medium is disturbed, then the average pressure may be written as

$$P_0 + \frac{1}{2}p = P_0 - \frac{1}{2}K\frac{\partial \xi}{\partial x}$$

[See Eqn. 5.24]

Since the change in volume $= \frac{\partial \xi}{\partial x}\delta x$, the work done on the gas is

$$-\left(P_0 - \frac{1}{2}K\frac{\partial \xi}{\partial x}\right)\frac{\partial \xi}{\partial x}\delta x = -P_0\frac{\partial \xi}{\partial x}\delta x + \frac{1}{2}K\left(\frac{\partial \xi}{\partial x}\right)^2 \delta x.$$

The average value of the first term is zero. Thus, the average potential energy of the layer per cycle is

$$\frac{1}{2}\ K\left(\frac{2\pi A}{\lambda}\right)^2 \cdot \frac{1}{2}\delta x = \frac{\pi^2 v^2 A^2 \rho\, \delta x}{\lambda^2}$$

where we have used the relation $v = \sqrt{K/\rho}$. Hence, average K.E. = Average P.E., and the total energy of a layer of thickness δx of unit cross-section is $2\pi^2 v^2 A^2 \rho\ \delta x/\lambda^2$.

Energy density = Energy per unit volume

$$= \frac{2\pi^2 \rho A^2 v^2}{\lambda^2} = 2\pi^2 \rho A^2 \nu^2$$

where $\nu = v/\lambda$.

Note: Intensity of energy at a point is the energy flowing per unit area perpendicular to the direction of propagation per unit time and it is equal to $2\pi^2\rho A^2\nu^2 v = \frac{2\pi^2 \rho A^2 v^3}{\lambda^2}$.

22. *A plane sinusoidal sound wave of displacement amplitude 1.0 × 10^{-3} mm and frequency 650 Hz is propagated in an ideal gas of density 1.29 kg m^{-3} and pressure 10^5 Nm^{-2}. The ratio of principal specific heat capacities is 1.41. Find the acoustic pressure amplitude of the wave.*

Solution

We have

$$\xi = 1.0\times 10^{-6} \sin\frac{2\pi}{\lambda}(x - vt) \text{ metre}$$

where

$$v = \sqrt{K/\rho} = \sqrt{\frac{\gamma P}{\rho}} = \left[\frac{1.41\times 10^5}{1.29}\right]^{1/2} = 330.61 \text{ m/s},$$

$$K = \gamma P = 1.41 \times 10^5 \text{ Nm}^{-2}$$

$$\lambda = \frac{v}{\nu} = \frac{330.61}{650} = 0.51 \text{ m.}$$

Thus, from Eqn. (5.24), we have

$$\text{Acoustic pressure} = -K\frac{\partial \xi}{\partial x} = -1.41 \times 10^5 \times 1.0 \times 10^{-6} \times \frac{2\pi}{\lambda} \cos \frac{2\pi}{\lambda}(x - vt)$$

$$\text{Acoustic pressure amplitude} = 1.41 \times 10^{-1} \times \frac{2\pi}{\lambda}$$

$$= 1.74 \text{ Nm}^{-2}.$$

23. *Show that the velocity for longitudinal waves in a rod is given by*

$$v = \sqrt{E/\rho}$$

where E = *Young's modulus of the material of the rod*

ρ = *Density of the material of the rod.*

Solution

We consider a section R of the rod of cross-sectional area A. Let S be a parallel section of equal area δx apart (see Fig. 5.5). The variable ξ measures the longitudinal displacement of a point due to passage of the wave. The increase in thickness of the layer = $(\partial\xi/\partial x)\delta x$. Relating change in length to force (F) acting, we have, from the definition of Young's modulus (E),

$$\frac{F}{A} = E\frac{\frac{\partial \xi}{\partial x}\delta x}{\delta x} = E\frac{\partial \xi}{\partial x}.$$

Thus,

$$\delta F = \frac{\partial F}{\partial x}\delta x = AE\frac{\partial^2 \xi}{\partial x^2}\delta x.$$

Applying Newton's second law of motion, we obtain

$$\delta F = (A\rho\delta x)\frac{\partial^2 \xi}{\partial t^2}$$

Hence,

$$\frac{\partial^2 \xi}{\partial x^2} = \frac{1}{E/\rho}\frac{\partial^2 \xi}{\partial t^2}$$

which is the wave equation for longitudinal wave in a rod. The velocity of longitudinal wave in a rod is thus

$$v = \sqrt{E/\rho}.$$

24. *P–V plots for two gases during adiabatic processes are shown in the Fig. 5.6. Plots 1 and 2 should correspond respectively to*

(*a*) *He and* O_2

(*b*) O_2 *and He*

(*c*) *He and Ar*

(*d*) O_2 *and* N_2 (I.I.T. 2001)

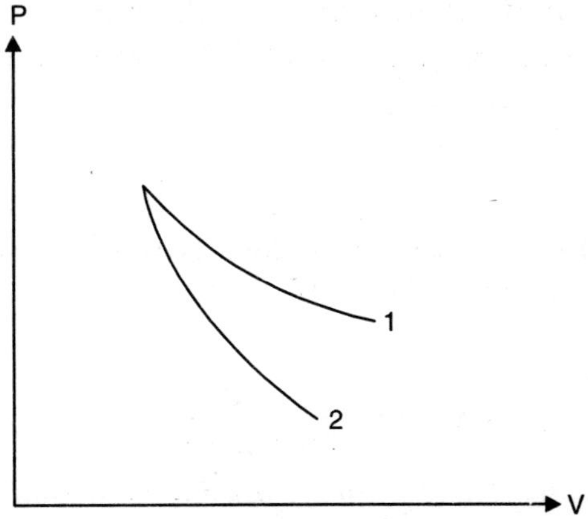

Fig. 5.6

Solution

For adiabatic process we know PV^{γ} = Constant.

Thus, $dPV^{\gamma} + PV^{\gamma-1}\, dV = 0$

or $$\gamma\frac{dP}{dV} = -\gamma\frac{P}{V}.$$

As curve 2 is steeper its γ is greater.

$\gamma = 5/3 = 1.67$ for monoatomic gases (curve 2)

$\gamma = 7/5 = 1.40$ for diatomic gases (curve 1)

Correct Choice : *b.*

25. *Starting with the same initial conditions, an ideal gas expands from volume V_1 to V_2 in three different ways. The work done by the gas is W_1 if the process is purely isothermal, W_2 is purely isobaric and W_3 is purely adiabatic. Then*

(*a*) $W_2 > W_1 > W_3$ (*b*) $W_2 > W_3 > W_1$

(*c*) $W_1 > W_2 > W_3$ (*d*) $W_1 > W_3 > W_2$

(I.I.T. 2000)

Solution

We know, PV^{η} = Constant

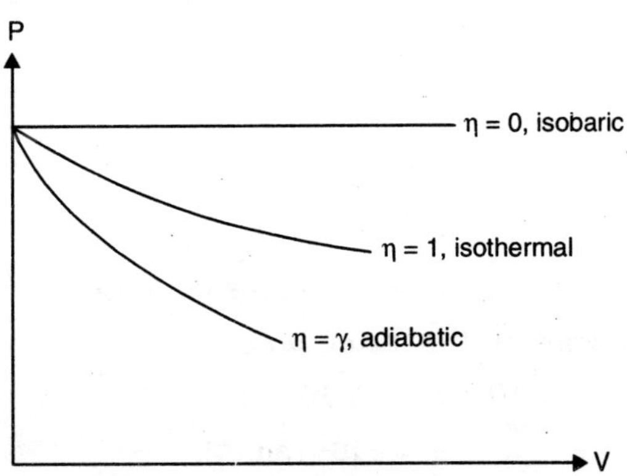

Fig. 5.7

where $\eta = 0$ for isobaric process and P = Constant

$\eta = 1$ for isothermal process and PV = Constant

$\eta = \gamma$ for adiabatic process and $\gamma > 1$.

Thus,
$$\frac{dP}{dV} = -\eta\frac{P}{V}.$$

In the P–V diagram (Fig. 5.7) the work done is the area under the P–V curve. As η increases the curve becomes steeper. Thus, $W_2 > W_1 > W_3$.

Correct Choice : *a.*

26.
$$y(x, t) = \frac{0.8}{(4x+5t)^2+5}$$

represents a moving pulse, where x and y are in meter and t in second. Then

(a) the pulse is moving in + x direction

(b) in 2s it will travel a distance of 2.5 m

(c) its maximum displacement is 0.16 m

(d) it is a symmetric pulse. (I.I.T. 1999)

Solution

$y(x, t)$ is maximum when $4x + 5t = 0$ and the maximum displacement is $0.8/5 = 0.16$ m $y(x, t)$ is not symmetric with respect to x and t.

$$y(x, t) = f(x + vt) = \frac{0.8}{16\left(x+\frac{5}{4}t\right)^2+5}$$

represents a backward wave.

Velocity of the wave $= \frac{5}{4}$ m/s.

In 2 s it travels a distance $\frac{5}{4} \times 2 = 2.5$ m.

Correct Choice : *b, c.*

27. *A transverse sinusoidal wave of amplitude a, wavelength λ and frequency f is travelling on a stretched string. The maximum speed of any point on the string is V/10, where V is the speed of propagation of the wave. If $a = 10^{-3}$ m and $V = 10$ ms^{-1}, then λ and f are given by*

(a) $\lambda = 2\pi \times 10^{-2}$ *m* *(b)* $\lambda = 10^{-3}$ *m*

(c) $f = 10^3/2\pi$ *Hz* *(d)* $f = 10^4$ *Hz.* (I.I.T. 1998)

Solution

Let
$$y = a \sin(kx - \omega t)$$

Then
$$\dot{y} = -a\omega \cos(kx - \omega t)$$

Maximum speed of a point on the string $= a\omega$.

Thus,
$$10^{-3}\,\omega = V/10 = 1$$

or
$$\omega = 10^3 \text{ rad s}^{-1}.$$
$$f = 10^3/2\pi \text{ Hz}$$

$$\lambda = \frac{V}{f} = 2\pi \times 10^{-2} \text{ m}.$$

Correct Choice : *a, c.*

28. *A string of length 0.4 m and mass 10^{-2} kg is tightly clamped at its ends. The string is under tension 16 N. The minimum value of Δt which allows constructive interference of successive pulse is*

(*a*) *0.05 s* (*b*) *0.10 s*

(*c*) *0.20 s* (*d*) *0.40 s* (I.I.T. 1998)

Solution

The velocity of the pulse down the string is

$$v = \sqrt{T/\mu} = \sqrt{\frac{16}{10^{-2}/0.4}} = 8 \text{ ms}^{-1}.$$

The pulse at A will come to A in the same direction after reflections at the end points D and C (Fig. 5.8), and meet the next pulse at A for constructive interference (see Chapter 12). Time taken by the pulse = 2 × 0.4/8 = 0.10 s.

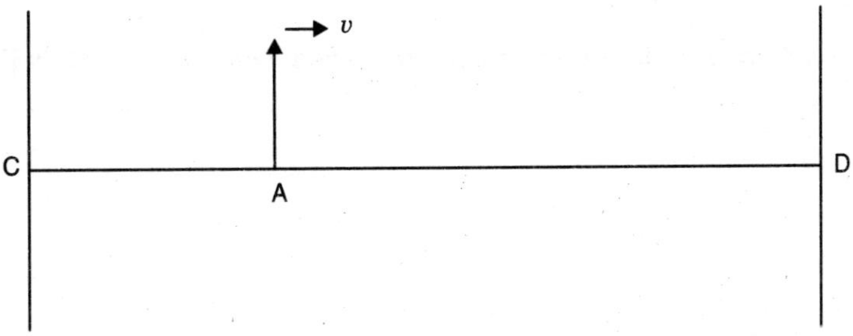

Fig. 5.8

Correct Choice : *b.*

29. *A transverse harmonic disturbance is produced in a string. The maximum transverse velocity is 3 m/s and maximum transverse acceleration is 90 m/s^2. If the wave velocity is 20 m/s then find the waveform.* (I.I.T. 2005)

Solution

Let the equation of the waveform be

$$y = A \sin (\omega t \pm kx + \phi)$$

where ϕ is the phase angle

$$\left.\frac{\partial y}{\partial t}\right|_{max} = \omega A \text{ and } \left.\frac{\partial^2 y}{\partial t^2}\right|_{max} = \omega^2 A.$$

Thus, $$\frac{\omega^2 A}{\omega A} = \omega = \frac{90}{3} = 30 \text{ rad s}^{-1}$$

$$\omega A = 3, \text{ hence } A = \frac{3}{30} = 0.1 \text{ m}.$$

The velocity of propagation of the transverse wave (wave velocity) = $v = \dfrac{\omega}{k} = 20.$

$$k = \frac{\omega}{20} = \frac{30}{20} = \frac{3}{2} \text{ m}^{-1}$$

Thus, the equation of the waveform is

$$y = 0.1 \sin (30t \pm \frac{3}{2}x + \phi).$$

30. *An ideal gas is initially at temperature T and volume V. Its volume is increased by dV due to increase in temperature by dT, the pressure remaining constant. Show that the quantity* $\delta = \frac{1}{V}\frac{dV}{dT}$ *varies with temperature as* $\frac{1}{T}$.

Solution

$$PV = nRT$$

or $$PdV = nR\,dT$$

or $$= \frac{P}{P}\frac{dV}{V} = \frac{nR\,dT}{nRT} = \frac{dT}{T}$$

or, $$\delta = \frac{1}{V}\frac{dV}{dT} = \frac{1}{T}.$$

31. *For an ideal gas at constant temperature show that the quantity* $\beta = -\frac{1}{V}\frac{dV}{dP}$ *varies with P as* $\frac{1}{P}$.

Solution

$$PV = nRT$$

or $$PdV + V\,dP = 0$$

or $$\beta = -\frac{1}{V}\frac{dV}{dP} = \frac{1}{P}.$$

[β is the isothermal compressibility and $K = \frac{1}{\beta}$ is the isothermal bulk modulus of the gas].

SUPPLEMENTARY PROBLEMS

1. A wave is represented by

$$\psi = \cos (2x + 4t)$$

where x is measured in metres and t in seconds. Find the wavelength, frequency, wave velocity and the direction of travel of the wave.

2. For the harmonic wave

$$y\,(x, t) = A \cos m\,(x - ct).$$

Show that (*i*) wavelength $\lambda = \frac{2\pi}{m}$, (*ii*) time period $T = \frac{2\pi}{mc}$.

3. A wave displacement is given by

$$y = \sin 2\pi(0.2x - 5t)\ m.$$

Find (*a*) the amplitude of the wave, (*b*) the magnitude of the propagation vector, (*c*) the wavelength, (*d*) the time period, (*e*) the wave velocity, (*f*) the frequency of the wave.

4. A harmonic plane wave is represented by

$$\phi = 0.1 \sin (0.2x - 0.3y + 0.4z - 0.5t)$$

Find (*i*) the propagation vector $\vec{k}$, (*ii*) the velocity of propagation of the wave.

5. Show that a possible solution of three-dimensional wave equation

$$\frac{\partial^2 \psi}{\partial x^2} + \frac{\partial^2 \psi}{\partial y^2} + \frac{\partial^2 \psi}{\partial z^2} = \frac{1}{v^2}\frac{\partial^2 \psi}{\partial t^2}$$

is $\psi = A \sin k_1 x \sin k_2 y \sin k_3 z \sin \omega t$. What is the wave speed?

6. Find the frequency and velocity of the progressive wave

$$y = 0.6 \sin 2\pi\left(\frac{t}{0.05} - \frac{x}{55}\right) \text{ metre.}$$

7. A wave equation which gives the displacement along the *y*-direction is given by

$$y = 10^{-4} \sin (60t + 2x)$$

where x and y are in metres and t is time in seconds. This represents a wave

(*a*) travelling with a velocity of 30 m/s in the negative *x*-direction

(*b*) of wavelength π metre

(*c*) of frequency $30/\pi$ Hz

(*d*) of amplitude 10^{-4} m travelling along the negative *x*-direction.

Tick the correct answer(s). *(I.I.T. 1982)*

8. A transverse wave is described by the equation

$$Y = Y_0 \sin 2\pi\ (ft - x/\lambda).$$

The maximum particle velocity is equal to four times the wave velocity if

(*a*) $\lambda = \pi Y_0/4$

(*b*) $\lambda = \pi Y_0/2$

(*c*) $\lambda = \pi Y_0$

(*d*) $\lambda = 2\pi Y_0$

Tick the correct answer(s). *(I.I.T. 1984)*

9. A steel cable 3.0 cm in diameter is kept under a tension of 10 kN. The density of steel is 7.8 g/cm^3. Find the speed of transverse waves along the cable.

10. A copper wire is held at the two ends by rigid supports. At 30°C the wire is just taut, with negligible tension. Find the speed of transverse waves in this wire at 10°C.

[α = Coefficient of linear expansion of copper = 1.7×10^{-5}/°C,

Y = Young's modulus of copper = 1.3×10^{11} N/m^2,

ρ = Density of copper = 9×10^3 kg/m^3]. *(I.I.T. 1979)*

[**Hints:** $\frac{dl}{l} = \frac{\text{Contraction of length}}{\text{Original length}} = (30 - 10)\alpha = 20\alpha$

$$Y = \frac{T/A}{dl/l} \text{ and } v = \sqrt{T/\mu} = \sqrt{T/(A\rho)}$$

11. A uniform flexible cable is 20 m long. It hangs vertically under its own weight and is vibrated from its upper end. Find the speed of transverse wave on the cable at its midpoint.

12. The speed of a wave on a string is 120 m/s when the tension is 120 N. To what value must the tension be increased in order to raise the wave speed to 150 m/s?

13. The linear density of a vibrating string is 1.6×10^{-2} kg/m. A transverse wave is propagating on the string and is described by the equation

$$y = 0.02 \sin(2x + 30t)\ m$$

(*a*) What is the wave speed? (*b*) What is the tension in the string?

14. A stretched string has a mass per unit length of 2.5 g/cm and a tension 9 N. A wave on this string has an amplitude of 0.12 mm and a frequency of 100 Hz and is travelling in the negative *x*-direction. Write an equation for this wave.

15. A continuous sinusoidal wave is travelling on a string having linear density 4 g/cm. The displacement of the particle of the string at $x = 10$ cm is found to vary with time according to the equation

$$y = 0.05 \sin(1 - 4t)\ m.$$

(*a*) What is the frequency of the wave?

(*b*) What is the wavelength of the wave?

(*c*) What is the velocity of the wave?

(*d*) Calculate the tension in the string.

(*e*) Write the general equation giving the transverse displacement of the particles of the string as a function of position and time.

16. Calculate the velocity of sound in a gas in which two waves of lengths 2 m and 2.02 m produce 7 beats in 4 s.

17. As the longitudinal wave passes through a fluid show that the acoustic pressure *p* varies in such a way that it satisfies the classical wave equation

$$\frac{\partial^2 p}{\partial x^2} = \frac{1}{v^2}\frac{\partial^2 p}{\partial t^2}$$

with velocity of propagation of the wave = $v = \sqrt{K/\rho_0}$.

$$\left[\textbf{Hints: } \text{From Eqn. (5.24), we have } \frac{\partial^2 p}{dt^2} = -k\frac{\partial^3 \xi}{\partial t^2 dx} \text{ and } \frac{\partial p}{\partial x} = -K\frac{\partial^2 \xi}{\partial x^2} = -\rho_0\frac{\partial^2 \xi}{\partial t^2}\right].$$

18. Show that the speed of sound in a gas is independent of pressure (at constant temperature), and increases with the square root of the absolute temperature and also the speed is large in gases of low molar mass.

19. (*a*) Find the velocity of sound in air at 25°C.

(*b*) The human ear can perceive sounds over a frequency range of 30 Hz to 15 kHz. Find the wavelengths at these extremities. [Use the value of the speed of sound in air at 25°C.]

20. At 25°C the speed of sound in sea-water is 1531 ms^{-1}, and the density of sea-water is 1.025×10^3 kg m^{-3}. Calculate the bulk modulus of sea-water at 25°C.

21. Find the speed of sound in a diatomic ideal gas ($\gamma = 1.4$) that has a density of 3.5 kg/m^3 and a pressure of 215 kPa. [1 Pa = 1 N/m^2]

22. Determine the speed of sound in carbon dioxide (M = 44 g/mol, γ = 1.30) at a pressure of 0.5 atm and a temperature of 403°C.

23. An increase in pressure of 100 kPa causes a certain volume of water to decrease by 5×10^{-3} per cent of its original volume.

(*a*) What is the bulk modulus of water?

(*b*) What is the speed of sound in water?

[$\rho = 10^3$ kg/m^3]

24. A loud sound has an intensity of 0.54 W/m^2. Find the amplitude of such a sound wave if its frequency is 802 Hz.

[The density of air = 1.30 kg/m^3 and the speed of sound = 340 m/s]

25. Calculate the intensity of a sound wave in air at 0°C and 1 atm if its amplitude is 0.001 mm and its wavelength is 66 cm. The density of air at S.T.P. is 1.293 kg/m^3. [1 atm = 1.013×10^5 Pa, γ = 1.4]

26. Two monotomic ideal gases 1 and 2 molecular masses m_1 and m_2 are enclosed in separate containers kept at the same temperature. The ratio of the speed of sound in gas 1 to that in gas 2 is given by

(*a*) $\sqrt{\frac{m_1}{m_2}}$ (*b*) $\sqrt{\frac{m_2}{m_1}}$ (*c*) $\frac{m_1}{m_2}$ (*d*) $\frac{m_2}{m_1}$. (*I.I.T. 2000*)

27. The ratio of the speed of sound in nitrogen gas to that in helium gas at 300 K is

(*a*) $\sqrt{2/7}$ (*b*) $\sqrt{1/7}$ (*c*) $\sqrt{3}/5$ (*d*) $\sqrt{6}/5$ (*I.I.T. 1999*)

28. As a wave propagates

(*a*) the wave intensity remains constant for a plane wave.

(*b*) the wave intensity decreases as the inverse of the distance from the source for a spherical wave.

(*c*) the wave intensity decreases as the inverse square of the distance from the source for a spherical wave.

(*d*) total intensity of the spherical wave over the spherical surface centered at the source remains constant at all times. (*I.I.T. 1999*)

29. A given quantity of an ideal gas is at pressure P and absolute temperature T. The isothermal bulk modulus of the gas is

(*a*) $\frac{2}{3}P$ (*b*) P (*c*) $\frac{3}{2}P$ (*d*) 2 P (*I.I.T. 1998*)

30. The average translational energy and rms speed of molecules in a sample of oxygen gas at 300 K are 6.21×10^{-21} J and 484 m/s respectively. The corresponding values at 600 K are nearly (assuming ideal gas behaviour)

(*a*) 12.42×10^{-21} J, 968 m/s

(*b*) 8.78×10^{-21} J, 684 m/s

(*c*) 6.21×10^{-21} J, 968 m/s
(*d*) 12.42×10^{-21} J, 684 m/s

$$\left[\textbf{Hints: } v_{rms} = \sqrt{\frac{3kT}{m}}, E = \frac{1}{2}mv^2, \frac{v_1}{v_2} = \sqrt{\frac{T_1}{T_2}}, \frac{E_1}{E_2} = \frac{T_1}{T_2}.\right]$$ (*I.I.T. 1997*)

31. A plane progressive wave of frequency 25 Hz, amplitude 2.5×10^{-5} m and initial phase zero propagates along the negative *x*-direction with a velocity of 300 m/s. At any instant the phase difference between the oscillations at two points 6 m apart along the line of propagation is ...... and the corresponding amplitude is ..... *m*.
(*I.I.T. 1997*)

[**Hints :** Phase difference = $\frac{2\pi}{\lambda}$ × path difference
Amplitude does not change for a plane progressive wave.]

32. A source of sound of frequency 600 Hz is placed inside water. The speed of sound in water is 1500 m/s and in air it is 300 m/s. The frequency of sound recorded by an observer who is standing in air is
(*a*) 200 Hz (*b*) 3000 Hz (*c*) 120 Hz (*d*) 600 Hz. (*I.I.T. 2004*)
[**Hints:** The frequency, a characteristic of source, is independent of the medium.]

33. The temperature of a gas is 20°C and the pressure is changed from 1.01×10^5 Pa to 1.165×10^5 Pa. If the volume is decreased isothermally by 10%, Bulk modulus of the gas is (in Pa)
(*a*) 1.55×10^5 (*b*) 0.155×10^5 (*c*) 1.4×10^5 (*d*) 1.01×10^5. (*I.I.T. 2005*)

34. A monoatomic ideal gas, initially at temperature T_1 is enclosed in a cylinder fitted with a frictionless piston. The gas is allowed to expand adiabatically to a temperature T_2 by releasing the piston suddenly. If L_1 and L_2 are the lengths of the gas column before and after expansion respectively, then T_1/T_2 is given by

(*a*) $\left(\frac{L_1}{L_2}\right)^{2/3}$ (*b*) $\frac{L_1}{L_2}$ (*c*) $\frac{L_2}{L_1}$ (*d*) $\left(\frac{L_2}{L_1}\right)^{2/3}$. (*I.I.T. 2000*)

[**Hints:** $T_1 (A L_1)^{\gamma-1} = T_2 (A L_2)^{\gamma-1}$, $\gamma = 5/3$, A = Area of cross-section of the gas column.]

6 Superposition of Waves

6.1 SUPERPOSITION PRINCIPLE

If two or move waves of the same kind reach a point of the medium simultaneously, the resultant displacement $\vec{\xi}$ of the point is the vector sum of the displacements $\vec{\xi}_1, \vec{\xi}_2, \ldots,$ of that point due to the individual waves:

$$\vec{\xi} = \vec{\xi}_1 + \vec{\xi}_2 + \ldots \quad \ldots(6.1)$$

6.2 STATIONARY WAVES

When two waves of same amplitude and frequency travel in a medium in opposite directions with the same velocity, due to superposition of two waves there are some points of the medium which have no displacements (nodes) and there are some points which vibrate with maximum amplitudes (antinodes). The resultant wave is called a stationary wave or standing wave. The resultant wave remains confined in the region in which they are produced and is non-progressive in character.

6.3 WAVE REFLECTION

Standing waves on a string are produced by reflection of travelling waves from the ends of the string. If an end is fixed, it must be the position of a node; if it is free, it is the position of an antinode. These boundary conditions limit the frequencies of waves for which standing wave will occur on a given string. Each possible frequency is a resonant frequency. For a stretched string of length l with fixed ends the resonant frequencies are

$$\nu = \frac{v}{\lambda} = \frac{v}{2l}n, \; n = 1, 2, 3, \ldots$$

We can obtain standing sound waves in a fixed length of a pipe. The closed end of the pipe, like the fixed end of a string, is a displacement node. At the open end of a pipe, however, we find a displacement antinode. The allowed frequencies of an open pipe are

$$\nu = \frac{v}{\lambda} = \frac{v}{2l}\,n, \; n = 1, 2, 3, \ldots$$

where l = Length of the pipe.

For a pipe closed at one end the allowed frequencies are

$$\nu = \frac{v}{\lambda} = \frac{v}{4l}n, \; n = 1, 3, 5,...$$

The lowest frequency that may be excited, corresponding to $n = 1$ is called the fundamental or the first harmonic, the remaining frequencies being called the second ($n = 2$), third ($n = 3$) harmonics, and so on. Note that in a closed pipe only odd harmonics are excited.

The air particles at the end of a pipe have freedom of movement and hence the vibration of air particles at the open end of the pipe is extended a little more into the air outside the pipe. The antinode at the open end of the pipe is thus situated at a distance, say, x into the air outside. This distance is known as *end-correction.*

For $n = 1$, $\dfrac{\lambda}{4} = l + x$ (closed pipe) (Fig. 6.1)

For $n = 1$, $\dfrac{\lambda}{2} = l + x + x$ (open pipe), since two end corrections are required (Fig. 6.2).

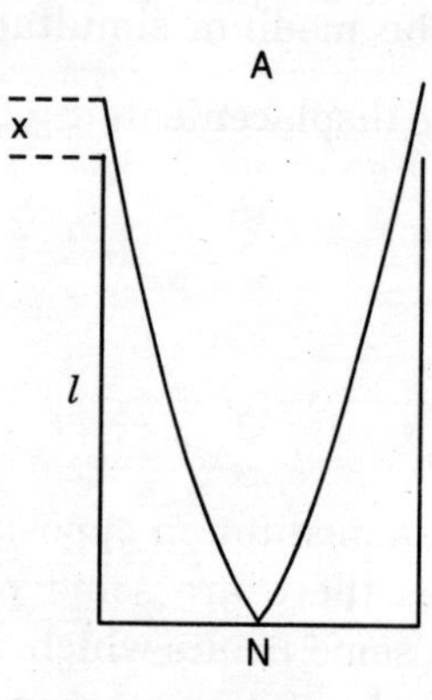

Fig. 6.1

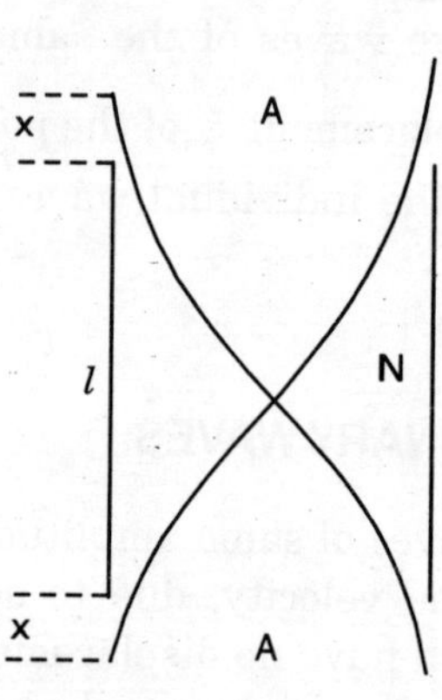

Fig. 6.2

According to Helmholtz and Rayleigh, $x = 0.6r$, where r is the radius of the pipe.

6.4 PHASE VELOCITY AND GROUP VELOCITY

The phase velocity v_p and group velocity v_g are defined as

$$v_p = \frac{\omega}{k} = \frac{\text{Angular frequency}}{\text{Angular wave number}} \qquad ...(6.2)$$

and

$$v_g = \frac{d\omega}{dk}. \qquad ...(6.3)$$

When two waves are superposed to form a wave group the energy is carried forward with the group velocity.

When ω/k or v_p depends on the wavelength (hence on the frequency), the waves are called dispersive. If v_p increases with wavelength in some wavelength range, the medium is said to possess normal dispersion in this range (Fig. 6.3). If v_p decreases with wavelength in some wavelength range, it is said to possess anomalous dispersion in this range (Fig. 6.4).

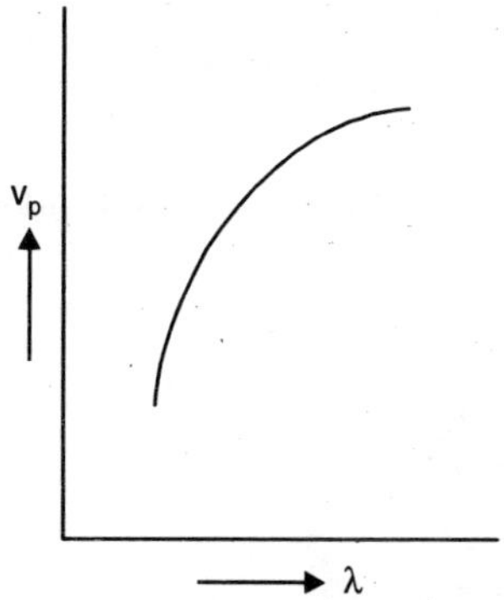

Fig. 6.3

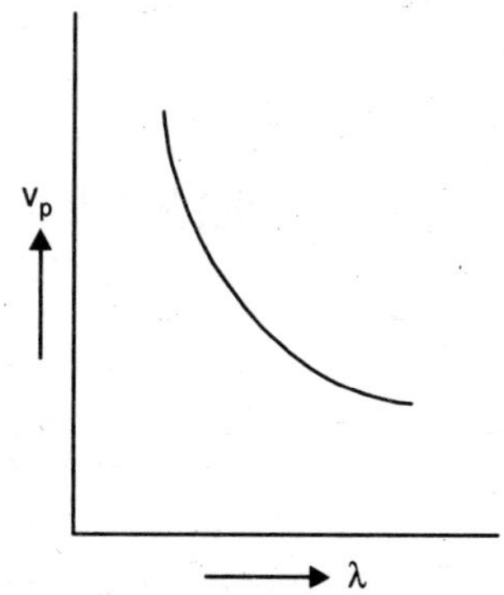

Fig. 6.4

SOLVED PROBLEMS

1. *Obtain the general expression for the standing wave solution of the wave equation*

$$\frac{\partial^2 y}{\partial t^2} = v^2 \frac{\partial^2 y}{\partial x^2}.$$

Solution

For the standing wave, the parts of the body oscillate in harmonic motion at the same angular frequency ω and with the same phase constant φ. Thus $y(x, t)$ should have the same time dependence, $\sin(\omega t + \phi)$, for all particles *i.e.,* for all x. In this case, the amplitude of vibration at x can be written as a continuous function of x denoted by $A(x)$. We can write the general expression for standing wave as

$$y(x, t) = A(x) \sin(\omega t + \phi).$$

Substituting this into the wave equation, we get

$$-\omega^2 A(x) = v^2 \frac{d^2 A(x)}{dx^2}.$$

The general solution of this equation is

$$A(x) = A \sin\left(\frac{\omega}{v}\right) x + B \cos\left(\frac{\omega}{v}\right) x$$

$$= A \sin kx + B \cos kx.$$

Thus, the general solution for the displacement in the standing wave is

$$y(x, t) = (A \sin kx + B \cos kx) \sin(\omega t + \phi).$$

For the standing waves the space and time variables are no longer associated together as $(x \pm vt)$.

2. *The vibration of a string fixed at both ends is represented by the equation*

$$y = 2 \sin \frac{\pi x}{3} \cos 50\pi t \text{ metre.}$$

If the above stationary wave is produced due to superposition of two waves of same frequency, velocity and amplitude travelling in opposite directions,

$$y_1 = A \sin \frac{2\pi}{\lambda}(x - vt) \text{ and } y_2 = A \sin \frac{2\pi}{\lambda}(x + vt);$$

(i) find the equations of the component waves, and

(ii) what is the distance between two consecutive nodes of the stationary wave?

Solution

(*i*) We have

$$y = y_1 + y_2 = 2A \sin \frac{2\pi x}{\lambda} \cos \frac{2\pi}{\lambda} vt$$

Thus, $A = 1$ m, $\lambda = 6$ m, $\frac{2\pi v}{\lambda} = 50\pi$ or $v = 150$ m/s.

Hence $$y_1 = \sin \frac{\pi}{3}(x - 150t) \text{ m}$$

and $$y_2 = \sin \frac{\pi}{3}(x + 150t) \text{ m}$$

(*ii*) Distance between two consecutive nodes = $\lambda/2 = 3$ m.

3. *A high frequency (HF) radio receiver receives simultaneously two signals from a transmitter 400 km away, one by a path along the surface of the earth, and the other by reflection from a portion of the ionospheric layer situated at a height of 200 km. We assume that the earth is flat and the ionospheric layer acts as a perfect horizontal reflector, which is moving slowly in the vertical direction. When the frequency of the transmitted wave is 10 MHz, it is observed that the combined signal strength varies from maximum to minimum and back to maximum 6 times per minute. With what slow vertical speed is the ionospheric layer moving?*

[Ignore atmospheric disturbance]

Solution

Let d = Length of the direct path along the earth's surface, h = Height of the reflecting ionospheric layer. The path difference p between the two routes (Fig. 6.5) is

$$p = 2(h^2 + d^2/4)^{1/2} - d \qquad ...(6.4)$$

Interference takes place between the signals arriving at the receiver by the two routes.

Fluctuation in intensity is due to motion of the ionospheric layer. We have from Eqn. (6.4)

$$\frac{dp}{dt} = 2h\left(h^2 + d^2/4\right)^{-1/2} \frac{dh}{dt}.$$

Each time p changes by λ (the wavelength of the signal) the received signal strength will vary through one cycle. The frequency f of the observed fluctuation will be given by

$$f = \frac{1}{\lambda}\frac{dp}{dt} = \frac{\nu}{c}\frac{dp}{dt}$$

where ν is the frequency of the radiation and c is the speed of light. Thus, we have

$$\frac{dh}{dt} = \frac{cf}{2\nu h}\left(h^2 + \frac{d^2}{4}\right)^{1/2} \qquad ...(6.5)$$

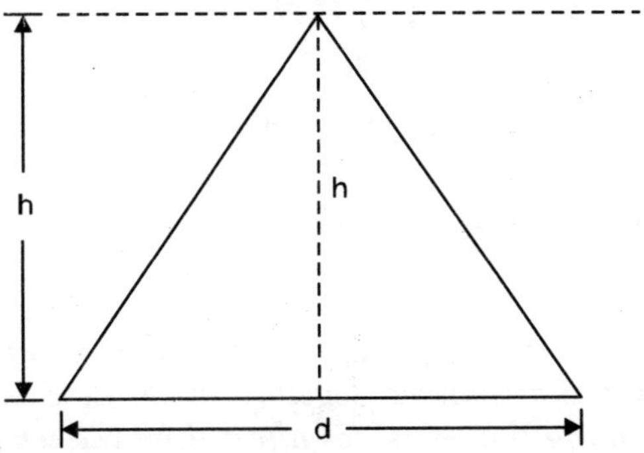

Fig. 6.5

where $v = \dfrac{dh}{dt}$ = Vertical velocity of the reflecting layer.

In our problem,
$$h = 200 \times 10^3 \text{ m},$$
$$d = 400 \times 10^3 \text{ m}, \quad \nu = 10 \times 10^6 \text{ Hz},$$
$$f = \frac{6}{60} = 0.1 \text{ Hz}, \; c = 3 \times 10^8 \text{ m/s}.$$

Substituting all these values in Eqn. (6.5), we get

$$v = \frac{dh}{dt} = 2.12 \text{ ms}^{-1}.$$

4. *An open organ pipe has a fundamental frequency of 300 Hz. The third harmonic of a closed organ pipe has the same frequency as the second harmonic of the open pipe. How long is each pipe? [speed of sound in air = 340 m/s]*

Solution

For an open organ pipe, $\nu = v/\lambda = v/2l_1$ for fundamental frequency (n = 1) without end-correction.

Thus,
$$l_1 = \frac{v}{2\nu} = \frac{340}{2 \times 300} \text{ m} = 0.57 \text{ m}$$

For third harmonic of the closed organ pipe,

$$\nu = \frac{v}{4l_2} \cdot 3 = \text{Second harmonic of the open pipe} = 600 \text{ Hz}.$$

Thus,
$$l_2 = \frac{3v}{4 \times 600} = \frac{3 \times 340}{4 \times 600} \text{ m} = 0.425 \text{ m}.$$

5. *An open pipe is suddenly closed at one end with the result that the frequency of third harmonic of the closed pipe is found to be higher by 100 Hz than the fundamental frequency of the open pipe. Find the fundamental frequency of the open pipe.* (I.I.T. 1996)

Solution

For the open pipe, the fundamental frequency is $\nu_1 = \dfrac{v}{2l}$, where v is the velocity of sound in air and l = Length of the pipe. For the third harmonic of the closed pipe, we have

$$\nu_3 = \frac{v}{4l} \cdot 3 = \frac{3\nu_1}{2}.$$

Now, $$\nu_3 = \nu_1 + 100$$

or $$\frac{3\nu_1}{2} = \nu_1 + 100$$

or $$\nu_1 = 200 \text{ Hz.}$$

6. *An organ pipe, open at both ends, sounds in unison with a tuning fork at 20°C. When the tuning fork and the pipe are sounded together at 30°C, 6 beats are heard. Find the frequency n of the fork assuming that it is not affected by the temperature change.*

Solution

The velocity of sound in air at t°C is given by

$$v_t = v_0 (1 + 0.00183t)$$

where v_0 = Velocity of sound in air at 0°C.

Thus,

$$n = \frac{v_0(1 + 0.00183 \times 20)}{2(l + 2x)}$$

where l = Length of the open pipe and

x = End-correction

At 30°C, we have

$$n + 6 = \frac{v_0(1 + 0.00183 \times 30)}{2(l + 2x)}$$

Hence, $$\frac{n+6}{n} = \frac{1 + 0.00183 \times 30}{1 + 0.00183 \times 20}$$

which gives n = 339.87 Hz.

7. *A metallic rod of length 1m is rigidly clamped at its mid-point. Longitudinal stationary waves are set-up in the rod in such a way that there are two nodes on either side of the mid-point. The amplitude of an antinode is 2 × 10^{-6} m. Write the equation of motion at a point 2 cm from the mid-point and those of the constituent waves in the rod.*

[Young's modulus = 2 × 10^{11} Nm^{-2}, density = 8000 kg m^{-3}] (I.I.T. 1994)

Solution

Velocity of propagation of longitudinal waves in the rod

$$= v = \sqrt{E/\rho} = [2 \times 10^{11}/8000]^{1/2} = 5 \times 10^3 \text{ m/s.}$$

The rod has node at the mid-point and antinodes at the two ends. There are two nodes on either side of the mid-point (Fig. 6.6).

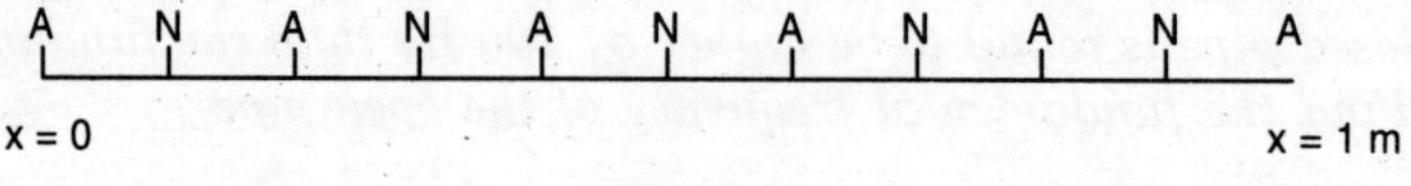

Fig. 6.6

Length of the rod = $5\lambda/2$ = 1 m.

Thus, λ = 0.4 m.

Frequency $\nu = \frac{v}{\lambda} = \frac{5000}{0.4} = 12500\text{Hz}$

The equation of the standing wave in the rod is

$$y = \left(A \sin\frac{2\pi}{\lambda}x + B\cos\frac{2\pi}{\lambda}x\right)\cos 2\pi\nu t$$

$$= (A \sin 5\pi x + B \cos 5\pi x) \cos 2\pi\nu t$$

Now, $y = 0$ at $x = \frac{1}{10}, \frac{3}{10}, \frac{5}{10}, \frac{7}{10}, \frac{9}{10}$ m.

Thus, $A = 0$.

The equation of the standing wave is

$$y = B \cos 5\pi x \cos 25000\pi t.$$

Since the amplitude at the antinode $= 2 \times 10^{-6}$ m,

$$B = 2 \times 10^{-6} \text{ m}.$$

Hence, $y = 2 \times 10^{-6} \cos 5\pi x \cos 25000\pi t$

which can be rewritten as

$$y = 1 \times 10^{-6} \cos (5\pi x + 25000\pi t) + 1 \times 10^{-6} \cos (5\pi x - 25000\pi t)$$

Hence the component waves are

$$y_1 = 1 \times 10^{-6} \cos(5\pi x + 25000\pi t)$$

$$y_2 = 1 \times 10^{-6} \cos(5\pi x - 25000\pi t)$$

At a point 2 cm from the mid-point $x = (0.5 \pm 0.02)$ m. The equation of motion at this point is given by

$$y = 2 \times 10^{-6} \cos 5\pi(0.5 \pm 0.02) \cos 25000\pi t$$

$$= \mp 2 \times 10^{-6} \sin \frac{\pi}{10} \cos 25000\pi t.$$

8. *Consider the superposition of two sinusoidal waves.*

$$y_1 = A \exp\left[i\left\{\left(k + \frac{1}{2}\Delta k\right)x - \left(\omega + \frac{1}{2}\Delta\omega\right)t\right\}\right]$$

$$y_2 = A \exp\left[i\left\{\left(k - \frac{1}{2}\Delta k\right)x - \left(\omega - \frac{1}{2}\Delta\omega\right)t\right\}\right]$$

which have identical amplitudes but have wave numbers differing by Δk and angular frequencies differing by $\Delta\omega$. Show that the beats occur with the frequency $\Delta\omega$ and the pattern of beats travels with the group velocity $v_g = \frac{\Delta\omega}{\Delta k}$ which becomes $\frac{d\omega}{dk}$ in the limit $\Delta k \to 0$.

Solution

The average phase velocity of two waves is

$$\frac{1}{2}\left[\frac{\omega + \frac{1}{2}\Delta\omega}{k + \frac{1}{2}\Delta k} + \frac{\omega - \frac{1}{2}\Delta\omega}{k - \frac{1}{2}\Delta k}\right] \approx \frac{\omega}{k}$$

where we neglect terms of second order of smalless *i.e.*, $(\Delta k)^2$, $\Delta k\ \Delta\omega$.

The sum of two waves according to the principle of superposition may be written as

$$y = y_1 + y_2 = A \exp [i(kx - \omega t)] \times \left[\exp\left\{ i\left(\frac{1}{2}\Delta kx - \frac{1}{2}\Delta\omega t\right)\right\} \right.$$

$$\left. + \exp\left\{ -i\left(\frac{1}{2}\Delta kx - \frac{1}{2}\Delta\omega t\right)\right\} \right]$$

$$= 2A \cos\left(\frac{1}{2}\Delta kx - \frac{1}{2}\Delta\omega t\right) \exp\left[i(kx - \omega t)\right].$$

The physical wave is represented by the real part:

$$y = 2A \cos\left(\frac{1}{2}\Delta kx - \frac{1}{2}\Delta\omega t\right)\cos(kx - \omega t) \qquad ...(6.6)$$

which is shown graphically in Fig. 2.18. The fast oscillating part of Eqn. (6.6) is cos $(kx - \omega t)$ having wave number k and angular frequency ω. The slowly varying part or the modulating wave of Eqn. (6.6) is $\cos\left(\frac{1}{2}\Delta kx - \frac{1}{2}\Delta\omega t\right)$ having wave number $\frac{1}{2}\Delta k$ and angular frequency $\frac{1}{2}\Delta\omega$. The amplitude of the fast oscillating wave is modulated by a wave envelope that has a wave number $\frac{1}{2}\Delta k$, which is half the difference of the wave numbers of the two component waves. Although the wavelength of the modulating wave is $\frac{2\pi}{\frac{1}{2}\Delta k} = \frac{4\pi}{\Delta k}$, the separation of successive beats is half this distance namely $\frac{2\pi}{\Delta k}$. The beats occur with the angular frequency $\Delta\omega$ and are separated in time by $\frac{2\pi}{\Delta\omega}$. The modulating wave has the velocity = $(\Delta\omega/2)/(\Delta k/2)$ = $\Delta\omega/\Delta k$. The pattern of beats travels with the group velocity in the limit $\Delta k \to 0$ *i.e.*

$$v_g = \frac{\Delta\omega}{\Delta k} \to \frac{d\omega}{dk} \text{ in the limit } \Delta k \to 0.$$

Due to superposition of sinusoidal waves of equal amplitude but slightly different frequencies a wave packet or wave group is formed. Individual wave travels with the phase velocity whereas the wave packet may travel with a different velocity known as the group velocity.

9. *Establish the relation*

$$v_g = v_p - \lambda\frac{dv_p}{d\lambda}$$

between the phase velocity v_p *and group velocity* v_g.

Solution

We have $v_p = \omega/k$ or, $\omega = kv_p$.

Now,
$$v_g = \frac{d\omega}{dk} = \frac{d}{dk}(kv_p) = v_p + k\frac{dv_p}{dk}$$

$$= v_p + k\frac{dv_p}{d\left(\frac{2\pi}{\lambda}\right)} = v_p + \frac{k}{2\pi}\frac{dv_p}{\left(-d\lambda/\lambda^2\right)}$$

$$= v_p - \lambda\frac{dv_p}{d\lambda}.$$

10. *The phase velocity of a surface wave on a liquid of density* ρ *and surfacce tension* T *is given by*

$$v_p = \left(\frac{g\lambda}{2\pi} + \frac{2\pi T}{\lambda\rho}\right)^{1/2},$$

where λ *is the wavelength of the wave and* g *is the acceleration due to gravity. Find the group velocity of the surface wave. Find the wavelength* λ *for which* v_p *is minimum. Evaluate the minimum value of* v_p *and the corresponding value of* v_g.

Solution

Since $v_p = \frac{\omega}{k}$ and $\lambda = \frac{2\pi}{k}$, we have

$$\frac{\omega}{k} = \left(\frac{g}{k} + \frac{kT}{\rho}\right)^{1/2}$$

or
$$\omega = \left(gk + \frac{k^3T}{\rho}\right)^{1/2}$$

which is the dispersion relation of the surface wave on a liquid.

Thus,
$$v_g = \frac{d\omega}{dk} = \frac{g + 3k^2T/\rho}{2\left(gk + k^3T/\rho\right)^{1/2}}$$

$$= \frac{g + 12\pi^2T/(\rho\lambda^2)}{2\left[2\pi g/\lambda + 8\pi^3T/\left(\rho\lambda^3\right)\right]^{1/2}}.$$

When v_p is minimum, so is v_p^2, and the conditions for this is

$$\frac{d}{d\lambda}\left(v_p^2\right) = 0$$

or
$$\frac{d}{d\lambda}\left(\frac{g}{k} + \frac{kT}{\rho}\right) = 0$$

or
$$\frac{d}{d\lambda}\left(\frac{\lambda g}{2\pi} + \frac{2\pi T}{\lambda\rho}\right) = 0$$

or
$$\frac{8}{2\pi} - \frac{2\pi T}{\lambda^2\rho} = 0$$

giving
$$\lambda = 2\pi\left(\frac{T}{\rho g}\right)^{1/2}$$

This gives the minimum value of v_p since $\frac{d^2}{d\lambda^2}(v_p^2) > 0$. Thus the minimum value of v_p is

$$(v_p)_{\min} = \sqrt{2}\left(\frac{Tg}{\rho}\right)^{1/4}$$

and the corresponding value of v_g is $\sqrt{2}\left(\frac{Tg}{\rho}\right)^{1/4}$.

11. *Prove that the group velocity v_g of electromagnetic waves in a dispersive medium is given by*

$$v_g = \frac{c}{n + \omega\frac{dn}{d\omega}}$$

where c is the velocity of light in vacuum and n is the refractive index of the medium for the angular frequency ω of the waves.

Solution

The refractive index n is defined by

$$n = \frac{c}{v_p} = \frac{ck}{\omega}.$$

Since,

$$v_g = \frac{d\omega}{dk},$$

$$\frac{1}{v_g} = \frac{dk}{d\omega} = \frac{1}{c}\frac{d}{d\omega}(n\omega) = \frac{1}{c}\left[n + \omega\frac{dn}{d\omega}\right]$$

Thus,

$$v_g = \frac{c}{n + \omega\frac{dn}{d\omega}}.$$

12. *The refractive index n of the interstellar medium is given by*

$$n^2 = 1 - \frac{Ne^2}{\varepsilon_0 m\omega^2}$$

where e and m are the charge and mass of an electron and N is the electron density. A pulsar emits very sharp pulses over a broad range of radio frequencies. The arrival time of a particular pulse at Earth measured at 400 MHz is 0.72 s later than the arrival time of the same pulse measured at 1420 MHz. Calculate the distance of the pulsar.

[$m = 9.1 \times 10^{-31}$ kg, $e = 1.6 \times 10^{-19}$ Coulomb, $N = 3 \times 10^4\ m^{-3}$, $\varepsilon_0 = 8.85 \times 10^{-12}\ Fm^{-1}$].

Solution

Since a pulse travels with group velocity, the time taken for a pulse to travel a distance D is

$$t = \frac{D}{v_g} = \frac{D}{c}\left(n + \omega\frac{dn}{d\omega}\right)$$

[see problem 11].

The quantity $\frac{Ne^2}{\varepsilon_0 m\omega^2}$ is very small compared to unity for the radio frequencies.

For ν = 400 MHz,

$$\frac{N e^2}{\varepsilon_0 m\omega^2} = \frac{3\times10^4\times(1.6)^2\times10^{-38}}{8.85\times10^{-12}\times9.1\times10^{-31}\times(2\pi\times4)^2\times10^{16}}$$

$$= 1.5 \times 10^{-11}$$

Thus, we can write

$$n = \left[1-\frac{Ne^2}{\varepsilon_0 m\omega^2}\right]^{1/2} \approx 1 - \frac{Ne^2}{2\,\varepsilon_0 m\omega^2}$$

and

$$\frac{dn}{d\omega} = \frac{Ne^2}{\varepsilon_0 m\,\omega^3}$$

Thus, we get

$$t = \frac{D}{c}\left(1+\frac{Ne^2}{2\varepsilon_0 m\omega^2}\right)$$

The difference in arrival times of the two pulses is

$$\Delta t = \frac{Ne^2 D}{2\varepsilon_0 mc}\left(\frac{1}{\omega_1^2}-\frac{1}{\omega_2^2}\right) = \frac{Ne^2 D\left(\omega_2^2-\omega_1^2\right)}{2\varepsilon_0 mc\omega_1^2\omega_2^2}$$

which gives

$$D = \frac{2\varepsilon_0 mc\omega_1^2\omega_2^2\Delta t}{Ne^2\left(\omega_2^2-\omega_1^2\right)} = 3.1 \times 10^{19}\text{ m.}$$

13. *(a) The phase velocity v_P of gravity waves in a liquid of depth h is given by*

$$v_p = \left[\frac{g}{k}\tanh kh\right]^{1/2}$$

where g is the acceleration due to gravity and k is the wave number. Find the dispersion relation for such waves and show that the group velocity is given by

$$v_g = v_p\left[\frac{1}{2}+\frac{kh}{\sin h(2kh)}\right].$$

(b) Find the phase and group velocities for gravity waves of frequency 1 Hz in a liquid of depth 0.1 m.

Solution

Since $v_p = \frac{\omega}{k}$, we have the dispersion relation

$$\omega = [gk \tanh kh\,]^{1/2}.$$

$$v_g = \frac{d\omega}{dk} = \frac{1}{2}\frac{g\tanh kh + \dfrac{gkh}{\cosh^2 kh}}{[gk\tanh kh]^{1/2}}$$

Thus $$\frac{v_g}{v_p} = \frac{1}{2}\frac{gk \tanh kh + \dfrac{gk^2h}{\cosh^2 kh}}{[gk \tanh kh]}$$

$$= \frac{1}{2}+\frac{kh}{\sinh(2kh)}$$

When $kh \to 0$, $\dfrac{kh}{\sinh(2kh)} \to \dfrac{1}{2}$ and $v_g = v_p$.

When $kh \to \infty$, $\dfrac{kh}{\sinh(2kh)} \to 0$ and $v_g = \dfrac{1}{2}v_p$.

(*b*) When $\nu = 1$ Hz, $\omega = 2\pi$ rad. s^{-1}. From the dispersion relation, we have

$$\frac{\omega^2 h}{g} = kh \tanh kh = x \tanh x$$

where $x = kh$.

Thus, $x \tanh x = \dfrac{4\pi^2 \times 0.1}{9.81} = 0.402$

which gives $x = 0.680$ so that $k = \dfrac{x}{h} = 6.80$ m^{-1},

and tanh 0.68 = 0.591.

The phase velocity $v_p = \left[\dfrac{9.81 \tanh 0.68}{6.8}\right]^{1/2} = 0.92$ m s^{-1}.

The group velocity $v_g = v_p\left[\dfrac{1}{2}+\dfrac{0.68}{\sinh 1.36}\right] = 0.80$ m s^{-1}.

14. *A string of negligible mass is fixed at x = 0 and x = 1. The N number of massive beads are fixed on the string at x = a, 2a,..., Na at equilibrium (Fig. 6.7). Each bead has mass m. The tension of the beaded string is T. Show that the dispersion relation for the normal modes of small transverse oscillations of the beads along the y-direction is*

$$\omega = \sqrt{\frac{4T}{ma}}\, sin\left(\frac{ka}{2}\right).$$

From this equation, show with proper limiting procedure that the dispersion relation for the massive continuous string is

$$\omega = \sqrt{\frac{T}{\mu}}\, k$$

where μ is the mass per unit length of the massive continuous string.

Solution

In order to find the equation of motion of the *n*th bead (Fig. 6.7),

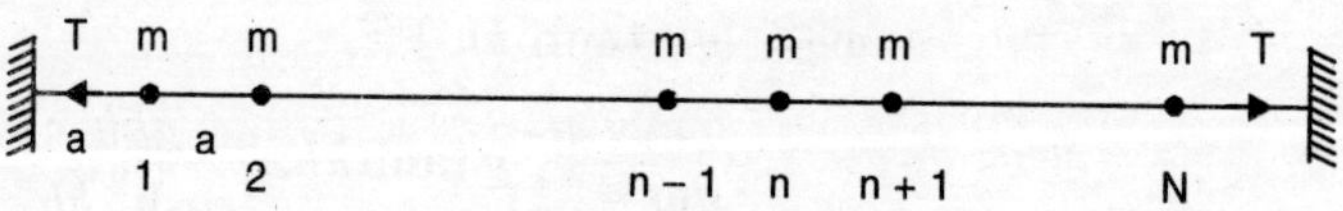

Fig. 6.7

we consider the nth bead and its neighbours $(n - 1)$ (to the left) and $(n + 1)$ (to the right). Let y_{n-1}, y_n and y_{n+1} be the transverse displacement of the $(n - 1)$, n and $(n + 1)$th bead respectively (Fig. 6.8).

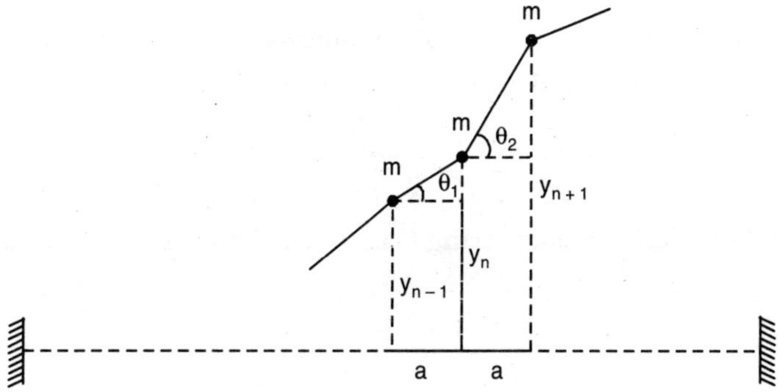

Fig. 6.8

Then the force on the nth bead along the y-direction is

$$F_y = T \sin \theta_2 - T \sin \theta_1$$

$$\approx T\frac{y_{n+1} - y_n}{a} - T\frac{y_n - y_{n-1}}{a}$$

where for small oscillation θ_1 and θ_2 are very small so that we may put $\sin \theta_2 \approx \tan \theta_2$ and $\sin \theta_1 \approx \tan \theta_1$.

Thus,
$$m\frac{d^2 y_n}{dt^2} = \frac{T}{a}[y_{n+1} - 2y_n + y_{n-1}] \qquad ...(6.7)$$

For normal modes of oscillations, each bead oscillates harmonically with the same frequency ω and with the same phase constant ϕ :

$$y_i = A_i \cos(\omega t + \phi),\ i = 1, 2 ....N.$$

Now, we have from Eqn. (6.7), the following difference equation

$$A_{n+1} + A_{n-1} = \left(2 - \frac{m\omega^2 a}{T}\right)A_n. \qquad ...(6.8)$$

Let us try a solution of Eqn. (6.8) in the form

$$A_n = B \sin(kna) \qquad ...(6.9)$$

where B is a constant and $k = 2\pi/\lambda$. Substituting Eqn. (6.9) into Eqn. (6.8), we get

$$\sin[k(n+1)a] + \sin[k(n-1)a] = \left(2 - \frac{m\omega^2 a}{T}\right)\sin(kna)$$

or
$$2\sin(kna)\cos ka = \left(2 - \frac{m\omega^2 a}{T}\right)\sin(kna)$$

Hence Eqn. (6.9) is a solution provided

$$2\cos ka = 2 - \frac{m\omega^2 a}{T}$$

or
$$\omega^2 = \frac{4T}{ma}\sin^2\left(\frac{ka}{2}\right).$$

Thus, the required dispersion relation is

$$\omega = \sqrt{\frac{4T}{ma}}\sin\left(\frac{ka}{2}\right). \qquad ...(6.10)$$

In a length 'a' total mass = m so that the mass per unit length = $\mu = m/a$. For the continuous string we take the limit $a \to 0$ in Eqn. (6.10) which gives

$$\omega = \sqrt{\frac{4T}{ma}}\cdot\frac{ka}{2} = \sqrt{\frac{Ta^2}{ma}}\cdot k = \sqrt{\frac{T}{\mu}}k \qquad ...(6.11)$$

The ω-k graph of the continuous string (1) and the beaded string (2) is shown in Fig. 6.9.

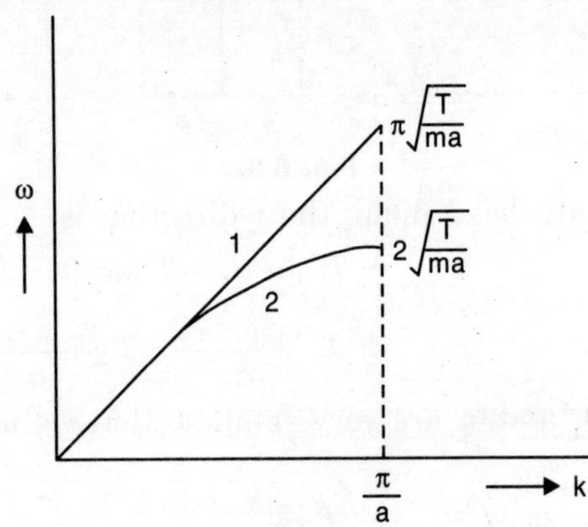

Fig. 6.9

15. *In the experiment for the determination of the speed of sound in air using the resonance column method, the length of the air column that resonates in the fundamental mode, with a tuning fork is 0.1 m. When the length is changed to 0.35 m, the same tuning fork resonates with the first overtone. Calculate the end correction.* (I.I.T. 2003)

Solution

For the fundamental mode

$$\nu = \frac{v}{\lambda} \text{ where } \frac{\lambda}{4} = l_1 + x$$

and x = End correction,

For the first overtone

$$\nu = \frac{v}{\lambda} \text{ where } \frac{\lambda}{4} + \frac{\lambda}{2} = l_2 + x$$

Thus,
$$\frac{v}{4(l_1+x)} = \frac{3v}{4(l_2+x)}$$

Here, $l_1 = 0.1$ m, $l_2 = 0.35$ m

$x = 0.025$ m.

16. *Two narrow cylindrical pipes A and B have the same length. Pipe A is open at both ends and is filled with a monoatomic gas of molar mass M_A. Pipe B is open at one end and closed at the other end, and is filled with a diatomic gas of molar mass M_B. Both gases are at the same temperature.*

(a) If the frequency of the second harmonic of the fundamental mode in pipe A is equal to the frequency of the third harmonic of the fundamental mode in pipe B, determine the value of M_A/M_B.

(b) Now the open end of pipe B is closed (so that the pipe is closed at both ends). Find the ratio of the fundamental frequency of pipe A to that in pipe B. (I.I.T. 2002)

Solution

(*a*) For an open pipe, $\nu_A = \dfrac{v_A}{2l}\, n,\ n = 1, 2, 3,...$

For a pipe closed at one end, $\nu_B = \dfrac{v_B}{4l}\, n,\ n = 1, 3, 5,...$

We have, $\dfrac{2v_A}{2l} = \dfrac{3v_B}{4l}$

Again, $v_A = \sqrt{\gamma_A RT/M_A}$ and $v_B = \sqrt{\gamma_B RT/M_B}$

$\gamma_A = 5/3,\ \gamma_B = 7/5.$

From these equations, we get $M_A/M_B = 400/189$

(*b*) Fundamental frequency in pipe $A = \dfrac{v_A}{2l}$

Fundamental frequency in pipe $B = \dfrac{v_B}{2l}$

The ratio of the fundamental frequencies $= \dfrac{v_A}{v_B} = \dfrac{3}{4}$.

17. *A 3.6 m long vertical pipe resonates with a source of frequency 212.5 Hz when water level is at certain heights in the pipe. Find the heights of water level (from the bottom of the pipe) at which resonances occur. Neglect end correction. Now, the pipe is filled to a height H ($\approx$ 3.6 m). A small hole is drilled very close to its bottom and water is allowed to leak. Obtain an expression for the rate of fall of water level in the pipe as a function of H. If the radii of the pipe and the hole are 2×10^{-2} m and 1×10^{-3} m respectively, calculate the time interval between the occurrence of first two resonances. Speed of sound in air is 340 m/s and g = 10 m/s^2.* (I.I.T. 2000)

Solution

The frequency of the source = 212.5 Hz

and wavelength in air $= \dfrac{340}{212.5} = 1.6$ m.

For a pipe closed at one end the length of the air column for resonance should be $(2n + 1)\ \lambda/4$, $n = 0, 1, 2,...$ *i.e.*, $\lambda/4, 3\lambda/4, 5\lambda/4,...$

Again the length of air column must be less than or equal to 3.6 m. Thus the possible lengths of air column are 0.4 m, 1.2 m, 2 m, 2.8 m, 3.6 m.

Possible heights of water columns are 0 m, 0.8 m, 1.6 m, 2.4 m, 3.2 m from the bottom of the pipe.

Suppose, dH = Change of water level in time dt,

A = Area of cross-section of the pipe,

v = Velocity with which the water is going out through the hole,

a = Area of cross-section of the hole,

then
$$AdH = -av\,dt.$$

The –ve sign is due to the fact that H decrease with t.

Since,
$$\frac{1}{2}mv^2 = mgh \text{ or } v = \sqrt{2gH},$$

$$dH = -\frac{a}{A}\sqrt{2gH}\,dt$$

or
$$\frac{dH}{\sqrt{H}} = -\frac{a}{A}\sqrt{2g}\,dt$$

or
$$\int_{H_1}^{H_2}\frac{dH}{\sqrt{H}} = -\frac{a}{A}\sqrt{2g}\int_{t_1}^{t_2}dt$$

or
$$2\left[\sqrt{H_2}-\sqrt{H_1}\right] = -\frac{a}{A}\sqrt{2g}\,(t_2-t_1)$$

or
$$\sqrt{H_1}-\sqrt{H_2} = \frac{a}{2A}\sqrt{2g}\,(t_2-t_1)$$

Here, $H_1 = 3.2$ m and $H_2 = 2.4$ m.

Time interval $= t_2 - t_1 = \dfrac{2A}{a}\dfrac{1}{\sqrt{2g}}\left[\sqrt{3.2}-\sqrt{2.4}\right]$

Here,
$$\frac{A}{a} = \frac{\pi\left(4\times10^{-4}\right)}{\pi\left(1\times10^{-6}\right)} = 4\times10^2$$

Thus,
$$t_2 - t_1 = 160\left[2-\sqrt{3}\right]\text{s.}$$

18. *The air column in a pipe closed at one end is made to vibrate in its* ***second overtone*** *by a tuning fork of frequency 440 Hz. The speed of sound in air is 330 ms^{-1}. End corrections may be neglected. Let P_0 denote the mean pressure at any point in the pipe, and ΔP_0 the maximum amplitude of pressure variation.*

(a) Find the length L of the air column.

(b) What is the amplitude of pressure variation at the middle of the column?

(c) What are the maximum and minimum pressures at the open end of the pipe?

(d) What are the maximum and minimum pressures at the closed end of the pipe?

(I.I.T. 1998)

Solution

(*a*) For the pipe closed at one end we have for frequency

$$\nu = \frac{v}{4L}\,n,\; n = 1, 3, 5, \ldots$$

where $n = 1$ for fundamental, $n = 3$ for first overtone and $n = 5$ for second overtone.

Thus for second overtone

$$\nu = \frac{5v}{4L} = \frac{v}{\lambda}$$

or $$\lambda = \frac{4L}{5}$$

$$440 = \frac{5\times 330}{4L}$$

or $$L = \frac{15}{16} \text{ m} = 0.9375 \text{ m}$$

(*b*) $$\text{Excess pressure} = p = -K\frac{\partial \xi}{\partial x}$$

where K = Bulk Modulus

ξ = Displacement of air particles in the pipe.

For standing wave

$$\xi = A \cos kx \cos \omega t$$

$$k = \frac{2\pi}{\lambda} = \frac{2\pi\times 5}{4L} = \frac{5\pi}{2L}.$$

At $x = 0$, there is antinode (maximum displacement of air particles) and at $x = L$, $kx = \frac{5\pi}{2}$, there is node.

$$p = AK\,k \sin kx \cos \omega t = \Delta p_0 \sin kx \cos \omega t$$

At the middle of the column, $x = \frac{L}{2}$

$$p = \Delta p_0 \sin\frac{5\pi}{4}\cos \omega t.$$

The amplitude of pressure variation at $x = \frac{L}{2}$ is $\frac{\Delta P_0}{\sqrt{2}}$.

(*c*) At $x = 0$ (at the open end) there is no variation of the pressure. The maximum and minimum pressure is equal to the atmospheric pressure.

(*d*) At x = L (at the closed end of the pipe) there is variation of pressure due to pressure amplitude variation $\pm \Delta p_0$. Hence the maximum pressure at the closed end = $p_0 + \Delta p_0$ and the minimum pressure = $p_0 - \Delta p_0$.

19. *The first overtone of an open organ pipe beats with the first overtone of a closed organ pipe with a beat frequency of 2.2 Hz. The fundamental frequency of the closed organ pipe is 110 Hz. Find the lengths of the pipes.* (I.I.T. 1997)

Solution

Allowed frequencies of an open organ pipe are

$$\nu_0 = \frac{v}{2l_1} n, \; n = 1, 2, 3,....$$

Allowed frequencies of a closed organ pipe are

$$\nu_c = \frac{v}{4l_2} n, \; n = 1, 3, 5,....$$

Thus, $$\frac{v}{2l_1}.2 = \frac{v}{4l_2}\cdot 3 \pm 2.2$$

Again, $$\frac{v}{4l_2}\cdot 1 = 110 \text{ Hz}$$

or $$l_2 = \frac{v}{440} = \frac{330}{440} = \frac{3}{4}\text{ m}$$

The length of the closed organ pipe $= l_2 = 0.75$ m

Thus, $$\frac{v}{l_1} = \frac{3v}{4l_2} \pm 2.2 = \frac{3\times 330}{4\times 3/4} \pm 2.2$$

$$= 330 \pm 2.2$$

$$l_1 = \frac{330}{330 \pm 2.2} = 0.9934 \text{ m or } 1.0067 \text{ m}$$

which is the length of the open organ pipe.

20. *The displacement of the medium in a sound wave is given by the equation*

$$y_1 = A \cos (ax + bt)$$

where A, a and b are positive constants. The wave is reflected by an obstacle situated at x = 0. The intensity of the reflected wave is 0.64 time that of the incident wave.

(a) What are the wavelength and frequency of incident wave?

(b) Write the equation for the reflected wave.

(c) In the resultant wave formed after reflection, find the maximum and minimum values of the particle speeds in the medium.

(d) Express the resultant wave as a superposition of a standing wave and travelling wave. What are the positions of the antinodes of the standing wave? What is the direction of propagation of travelling wave? (I.I.T. 1991)

Solution

(*a*) The general form of a travelling wave is given by

$$y = A \cos (kx + \omega t)$$

Comparing with the given equation, we get

$$k = a \text{ and } \omega = b$$

$$\frac{2\pi}{\lambda} = a \text{ or } \lambda = \frac{2\pi}{a} \text{ m}$$

$$2\pi\nu = b \text{ or } \nu = \frac{b}{2\pi} \text{ Hz.}$$

(*b*) The amplitude of the reflected wave $= 0.8A$ so that its intensity is $0.64A^2$.

It is moving in the opposite direction of the incident wave. The reflected and incident waves are 180° out of phase. Thus the equation of the reflected wave is

$$y_2 = -0.8A \cos (-ax + bt)$$
$$= -0.8A \cos (ax - bt)$$

(*c*) The resultant wave is given by

$$y = y_1 + y_2 = A \cos (ax + bt) - 0.8A \cos (ax - bt)$$

The particle velocity is

$$\frac{\partial y}{\partial t} = -Ab \sin (ax + bt) - 0.8Ab \sin (ax - bt)$$
$$= -1.8Ab \sin ax \cos bt - 0.2Ab \cos ax \sin bt$$

The maximum value of the particle velocity is obtained when sin ax = 1, cos bt = 1 and then cos ax = 0 and sin bt = 0.

$$\left|\frac{\partial y}{\partial t}\right|_{max} = |-1.8Ab| = 1.8Ab \text{ m/s.}$$

The minimum value of the particle velocity is obtained when sin ax = 0 and sin bt = 0,

$$\left|\frac{\partial y}{\partial t}\right|_{min} = 0$$

(*d*) The resultant wave is

$$y = y_1 + y_2 = A \cos (ax + bt) - 0.8A \cos (ax - bt)$$
$$= A \cos (ax + bt) - A \cos (ax - bt) + 0.2A \cos (ax - bt)$$
$$= -2A \sin ax \cos bt + 0.2A \cos (ax - bt).$$

The first term on the r.h.s. is the standing wave and the second term is the travelling wave. Thus the resultant wave is expressed as the superposition of standing wave and travelling wave.

The positions of the antinodes of the standing wave are obtained when sin ax = ± 1

or $$ax = n\pi + \frac{\pi}{2}$$

or $$x = \frac{\pi}{a}\left(n + \frac{1}{2}\right), n = 0, 1, 2,...$$

The travelling wave has the form $0.2A \cos (kx - \omega t)$. The direction of propagation of the wave is +ve in the x-direction.

21. *A closed organ pipe of length L and an open pipe contain gases of densities ρ_1 and ρ_2 respectively. The compressibility of gases is equal in both the pipes. Both the pipes are vibrating in their first overtone with the same frequency. The length of the open organ pipe is*

(*a*) $\frac{L}{3}$ (*b*) $\frac{4L}{3}$ (*c*) $\frac{4L}{3}\sqrt{\frac{\rho_1}{\rho_2}}$ (*d*) $\frac{4L}{3}\sqrt{\frac{\rho_2}{\rho_1}}$. (I.I.T. 2004)

Solution

Allowed frequencies of closed (ν_c) and open organ (ν_0) pipes are

$$\nu_c = \frac{v_1}{4L} n, n = 1, 3, 5, ...$$

$$\nu_0 = \frac{v_2}{2l} n, n = 1, 2, 3, ...$$

Thus, $$\frac{v_1}{4L}\cdot 3 = \frac{v_2}{2l}\cdot 2$$

or $$\frac{v_1}{v_2} = \frac{4L}{3l}$$

Again, $V_1 = \sqrt{\frac{K}{\rho_1}}$ and $v_2 = \sqrt{\frac{K}{\rho_2}}$

where K = Bulk modulus and $\frac{1}{K}$ = Compressibility

Thus, $\frac{v_1}{v_2} = \sqrt{\frac{\rho_2}{\rho_1}} = \frac{4L}{3l}$

$$l = \frac{4L}{3}\sqrt{\frac{\rho_1}{\rho_2}}$$

Correct Choice: c.

22. *In a resonance tube with tuning fork of frequency 512 Hz, first resonance occurs at water level equal to 30.3 cm and second resonance occurs at 63.7 cm. The maximum possible error in the speed of sound is: (a) 51.2 cm/s (b) 102.4 cm/s (c) 204.8 cm/s (d) 153.6 cm/s.* (I.I.T. 2005)

Solution

For a pipe closed at one end, we have

$$l_1 + x = \frac{v}{4f} \text{ and } l_2 + x = \frac{3v}{4f}$$

where
f = Frequency of the tuning fork = 512 Hz
v = Speed of sound
l_1 = Length of air column at first resonance
l_2 = Length of air column at second resonance
x = End correction.

Thus, $l_2 - l_1 = \frac{v}{2f}$

$$v = 2f\,(l_2 - l_1)$$

The error in v is

$$\Delta v = 2f\,\Delta\,(l_2 - l_1)$$
$$= 2f\,(\Delta l_2 + \Delta l_1) \text{ for maximum error}$$
$$= 2 \times 512 \times 0.2 = 204.8 \text{ cm/s}$$

Correct Choice : c.

23. *An open organ pipe resonated with frequency f_1 and 2nd harmonic. Now one end is closed and the frequency is slowly increased. It resonates with frequency f_2 and nth harmonic. Then*

(a) $n = 3,\ f_2 = \frac{3}{4} f_1$ *(b)* $n = 5,\ f_2 = \frac{3}{4} f_1$

(c) $n = 3,\ f_2 = \frac{5}{4} f_1$ *(d)* $n = 5,\ f_2 = \frac{5}{4} f_1$ (I.I.T. 2005)

Solution

For open organ pipe

$$f_1 = \frac{v}{2l}\cdot n,\ n = 2 \text{ for 2nd harmonic}$$

For closed organ pipe

$$f_2 = \frac{v}{4l}\cdot n,\ n = 1, 3, 5,....$$

Since $f_2 > f_1$ and f_2 is just greater than $f_1 = \frac{v}{l}$,

$$f_2 = \frac{v}{4l}\cdot 5,\ n=5$$

$$= \frac{5}{4}f_1$$

Correct Choice : *d.*

24. *In the experiment to determine the speed of sound using a resonance column*

(*a*) *prongs of the tuning fork are kept in a vertical plane.*

(*b*) *prongs of the tuning fork are kept in a horizontal plane.*

(*c*) *in one of the two resonances observed, the length of the resonating air column is close to the wavelength of sound in air.*

(*d*) *in one of the two resonances observed, the length of the resonating air column is close to half of the wavelength of sound in air.* (I.I.T. 2007)

Solution

$$l = \frac{\lambda}{4} \text{ for first resonance}$$

and

$$l = \frac{3\lambda}{4} \text{ for second resonance.}$$

To generate longitudinal wave propagating into the air column prongs of the tuning fork are kept in a vertical plane.

Correct Choice : *a.*

SUPPLEMENTARY PROBLEMS

1. Two waves represented by

$$\psi_1 = \cos(2x + 4t)$$

$$\psi_2 = 2\cos(2x + 4t + \pi/6)$$

interfere with each other. Show that the resultant of the two waves is given by

$$\psi = A\cos(2x + 4t + \phi)$$

where $A = 2.91$ and $\varphi = 20.1°$.

2. Show that standing waves are produced due to superposition of two waves of identical amplitude and frequency:

$$y_1 = \frac{A}{2}\cos(kx - \omega t)$$

$$y_2 = -\frac{A}{2}\cos(kx + \omega t)$$

which are travelling in opposite directions on a stretched string.

3. Two identical travelling waves moving in the same direction, are out of phase by 90°. What is the amplitude of the combined wave in terms of the common amplitude A of the two combining waves?

4. A source S and a detector D of radio waves are a distance d apart on the ground. The direct wave from S is found to be in phase at D with the wave from S that is reflected from a horizontal layer at an altitude H (Fig. 6.10). The incident and reflected rays make the same angle with the reflecting layer.

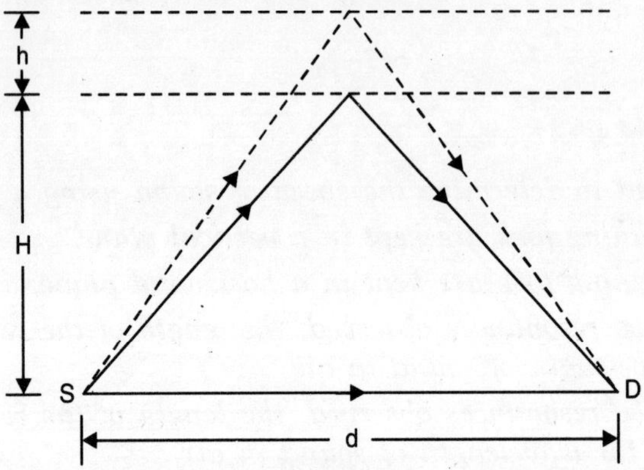

Fig. 6.10

When the layer rises a distance h, no signal is detected at D. Neglect absorption in the atmosphere and find the relation between d, h, H and the wavelength λ of the waves.

5. A sound wave in a fluid medium is reflected at a barrier so that a standing wave is formed. The distance between nodes is 3.8 cm and the speed of propagation is 1500 m/s. Find the frequency.

6. (*a*) If v_1 and v_2 are the speeds of sound wave in a gas at absoulte temperatures T_1 and T_2, then show that

$$\frac{v_1}{v_2} = \sqrt{\frac{T_1}{T_2}}.$$

(*b*) An organ pipe resonates to a frequency of 250 Hz when the temperature is 10°C. What will be its resonance frequency when the temperature is 20°C?

7. Consider two waves of same frequency, velocity and amplitude travelling in opposite directions which are represented by

$$y_1 = A \sin \frac{2\pi}{\lambda}(x - vt) \text{ and } y_2 = A \sin \frac{2\pi}{\lambda}(x + vt).$$

The vibration of a string fixed at both ends is represented by the equation

$$y = 4 \sin \frac{\pi x}{10} \cos 100\pi t.$$

Write down the equation of the component waves y_1 and y_2 whose superposition gives the above standing wave.

8. The water level in a vertical glass tube 1.0 m long can be adjusted to any position in the tube. A tuning fork vibrating at 680 Hz is held just over the open top end of the tube. At what positions of the water level will there be resonance?
[Speed of sound in air = 340 m/s]

9. A pipe 96 cm long is open at both ends. How long must a second pipe, closed at one end, be if it is to have the same fundamental resonance frequency as the open pipe?

10. Two tuning forks when sounded together give x beats per second. These forks when sounded individually with a closed pipe produce resonance with l_1 and l_2 lengths of air column. Find the frequencies of the forks and the velocity of sound in air.

11. A tube of certain diameter and length 50 cm is open at both ends. Its fundamental frequency of resonance is found to be 320 Hz. Estimate the diameter of the tube. If one end of the tube is now closed, find the fundamental frequency of resonance of this closed tube.
[Velocity of sound in air = 330 m/s]

12. A string fixed at both ends resonates in three segments to a frequency of 165 Hz. What frequency must be used if it is to resonate in four segments?

13. What must be the length of an iron rod that has the fundamental frequency 340 Hz when clamped at this centre? Assume longitudinal vibration at speed 5 km/s.

14. (*a*) Show that for normal dispersion $v_g < v_p$ and for anomalous dispersion $v_g > v_p$.
(*b*) In a graph of angular frequency ω against wave number k, show that the phase and group velocities are given, respectively by the slope of the straight line from the origin to a particular point on the curve and by the slope of the tangent to the curve at that point.

15. In solved problem 10 if $\rho = 1000 \text{ kg m}^{-3}$ and $T = 0.07 \text{ Nm}^{-1}$ show that the minimum value of v_p occurs at a wavelength of 0.017 m and has a value of 0.229 ms^{-1}. Find the value of v_g at this wavelength.

16. The dispersion relation for electromagnetic waves in the ionosphere is given by

$$\omega^2 = c^2k^2 + \omega_p^2$$

where c = The velocity of light in vacuum
and ω_p = Plasma oscillation frequency.

Show that the phase velocity exceeds c whereas the group velocity is always less than c.

17. Show that the dispersion relation for de-Broglie wave of a particle of momentum $\hbar k$ is given by

$$\omega^2 = c^2k^2 + \frac{m_0^2c^4}{\hbar^2}$$

where m_0 is the rest mass of the particle. Show that (*i*) the phase velocity exceeds c, (*ii*) the group velocity is equal to the velocity of the particle (*iii*) $v_pv_g = c^2$.

18. Consider a medium in which

$$v_p = A\omega^n$$

where A and n are constants. Show that

$$v_g = \frac{v_p}{1-n}, \; n \neq 1.$$

For what value of n is this dispersion normal?

19. In a resonance tube experiment to determine the speed of sound in air, a pipe of diameter 5 cm is used. The air column in pipe resonates with a tuning fork of frequency 480 Hz, when the minimum length of the air column is 16 cm. Find the speed of sound in air at room temperature. *(I.I.T. 2003)*

[Hints: $\nu = \dfrac{v}{4(l + 0.6\, r)}$]

20. A cylindrical resonance tube open at both ends has a fundamental frequency F in air. Half of the length of the tube is dipped vertically in water. The fundamental frequency of the air column now is .... *(I.I.T. 1992)*

21. Standing waves can be produced

(*a*) on a string clamped at both ends.

(*b*) on a string clamped at one end and free at the other end.

(*c*) when incident wave gets reflected from a wall.

(*d*) when two identical waves with a phase difference of π are moving in the same direction. *(I.I.T. 1999)*

[Hints: In case of (*b*) the wave is reflected at the clamped end and there is superposition of waves. In case of (*d*) there is no superposition of waves.]

7

Fourier Analysis

7.1 FOURIER'S THEOREM

If a function of x, $f(x)$, is integrable in $(-\pi, \pi)$ and it is periodic with period 2π outside of this interval *i.e.,*

$$f(x \pm 2k\pi) = f(x),\ k = 1, 2, 3, \ldots,$$

then the periodic function $f(x)$ can be analysed into a series

$$\frac{a_0}{2} + \sum_{n=1}^{\infty}(a_n \cos nx + b_n \sin nx). \qquad \ldots(7.1)$$

The series is called a Fourier series corresponding to $f(x)$, where the Fourier coefficients a_n and b_n are given by

$$a_n = \frac{1}{\pi}\int_{-\pi}^{\pi} f(x)\cos nx\, dx,\ n = 0,\ 1,\ 2, \ldots, \qquad \ldots(7.2)$$

$$b_n = \frac{1}{\pi}\int_{-\pi}^{\pi} f(x)\sin nx\, dx,\ n = 1,\ 2,\ 3, \ldots, \qquad \ldots(7.3)$$

7.2 DIRICHLET'S CONDITION OF CONVERGENCE OF FOURIER SERIES

If $f(x)$ is bounded and defined in the range $(-\pi, \pi)$ and $f(x)$ has only a finite number of maxima and minima and a finite number of finite discontinuities in this range and further if $f(x)$ is periodic with period 2π, *i.e.,* $f(x + 2\pi) = f(x)$, then the series (7.1) converges to $\frac{1}{2}[f(x + 0) + f(x - 0)]$. These are known as Dirichlet's conditions. If $f(t)$ is continuous at $t = x$, the series (7.1) converges to $f(x)$, *i.e.*

$$f(x) = \frac{a_0}{2} + \sum_{n=1}^{\infty}(a_n \cos nx + b_n \sin nx) \qquad \ldots(7.4)$$

7.3 FOURIER COSINE SERIES

If $f(x)$ is an even function of x, *i.e.*, $f(-x) = f(x)$, then

$$a_n = \frac{1}{\pi}\int_{-\pi}^{0} f(x)\cos nx\, dx + \frac{1}{\pi}\int_{0}^{\pi} f(x)\cos nx\, dx,$$

and
$$b_n = \frac{1}{\pi}\int_{-\pi}^{0} f(x)\sin nx\, dx + \frac{1}{\pi}\int_{0}^{\pi} f(x)\sin nx\, dx.$$

For the first integral of a_n or b_n, we put $y = -x$, we have

$$a_n = \frac{1}{\pi}\int_{\pi}^{0} f(-y)\cos ny.(-dy) + \frac{1}{\pi}\int_{0}^{\pi} f(x)\cos nx\, dx$$

$$= \frac{2}{\pi}\int_{0}^{\pi} f(x)\cos nx\, dx \qquad ...(7.5)$$

and
$$b_n = \frac{1}{\pi}\int_{\pi}^{0} f(-y)\sin ny\, dy + \frac{1}{\pi}\int_{0}^{\pi} f(x)\sin nx\, dx$$

$$= 0 \qquad ...(7.6)$$

We obtain only the Fourier cosine series. If the function is defined in the range $(0, \pi)$, it may be extended to the range $(-\pi, 0)$ by the equation $f(-x) = f(x)$.

7.4 FOURIER SINE SERIES

If $f(x)$ is an odd function of x, *i.e.*, $f(-x) = -f(x)$, then $a_n = 0$ and

$$b_n = \frac{2}{\pi}\int_{0}^{\pi} f(x)\sin nx\, dx. \qquad ...(7.7)$$

and the Fourier sine series is obtained. If the function is defined in the range $(0, \pi)$, it may be extended to the range $(-\pi, 0)$ by the equation $f(-x) = -f(x)$.

7.5 REPRESENTATION OF A FUNCTION BY FOURIER SERIES IN THE RANGE $a \le x \le b$

We write
$$x = \frac{1}{2}(a + b) - \frac{a-b}{2\pi}y,$$

or
$$y = \frac{\pi}{b-a}(2x - a - b),$$

so that when $x = a$, $y = -\pi$ and when $x = b$, $y = \pi$ and $f(x) = F(y)$.

The Fourier series of $F(y)$ is given by

$$\frac{1}{2}[F(y+0) + F(y-0)] = \frac{a_0}{2} + \sum_{n=1}^{\infty}(a_n \cos ny + b_n \sin ny)$$

and so
$$\frac{1}{2}[f(x+0)+f(x-0)] = \frac{a_0}{2}+\sum_{n=1}^{\infty}[a_n \cos\frac{n\pi}{b-a}(2x-a-b)$$
$$+ b_n \sin \frac{n\pi}{b-a}(2x - a - b)] \qquad ...(7.8)$$

where $a_n = \frac{1}{\pi}\int_{-\pi}^{\pi} F(y)\cos ny\, dy$

or
$$a_n = \frac{2}{b-a}\int_a^b f(x)\cos\frac{n\pi}{b-a}(2x - a - b)\, dx,\ n = 0, 1, 2,... \qquad ...(7.9)$$

and,
$$b_n = \frac{2}{b-a}\int_a^b f(x)\sin\frac{n\pi}{b-a}(2x - a - b)\, dx,\ n = 1, 2, 3,... \qquad ...(7.10)$$

(*i*) When $a = -\pi$, $b = \pi$, we reproduce Eqns. 7.1–7.3.

(*ii*) When $a = 0$, $b = 2\pi$, we can write the Fourier series as

$$\frac{1}{2}[f(x+0)+f(x-0)] = \frac{c_0}{2}+\sum_{n=1}^{\infty}(c_n \cos nx + d_n \sin nx), \qquad ...(7.11)$$

where
$$c_n = \frac{1}{\pi}\int_0^{2\pi} f(x)\cos nx\, dx, n = 0, 1, 2, ... \qquad ...(7.12)$$

$$d_n = \frac{1}{\pi}\int_0^{2\pi} f(x)\sin nx\, dx, n = 1, 2, 3, ... \qquad ...(7.13)$$

(*iii*) When $a = -L$ and $b = L$, we have

$$\frac{1}{2}[f(x+0)+f(x-0)] = \frac{a_0}{2}+\sum_{n=1}^{\infty}\left(a_n \cos\frac{n\pi x}{L} + b_n \sin\frac{n\pi x}{L}\right) \qquad ...(7.14)$$

where
$$a_n = \frac{1}{L}\int_{-L}^{L} f(x)\cos\frac{n\pi x}{L}\, dx \qquad ...(7.15)$$

$$b_n = \frac{1}{L}\int_{-L}^{L} f(x)\sin\frac{n\pi x}{L}\, dx. \qquad ...(7.16)$$

7.6 FOURIER INTEGRAL THEOREM

We consider the problem of representing a non-periodic function $f(x)$ over the infinite range $(-\infty, \infty)$. If $f(x)$ is piece-wise continuous in every finite interval and has a right- and left-hand derivative at every point and the integral

$$\int_{-\infty}^{+\infty} f(t)dt$$

exists, then $f(x)$ can be represented by the integral

$$f(x) = \frac{1}{2\pi}\int_{-\infty}^{+\infty} e^{-i\omega x}\,d\omega \int_{-\infty}^{+\infty} f(t)\,e^{i\omega t}\,dt. \qquad ...(7.17)$$

At a point where $f(x)$ is discontinuous, the value of the right hand side of Eqn. (7.17) equals the average of the left and right-hand limit of $f(x)$ at that point.

7.7 FOURIER TRANSFORM

We define the Fourier transform $g(\omega)$ of the function $f(t)$ by

$$g(\omega) = \frac{1}{\sqrt{2\pi}}\int_{-\infty}^{+\infty} f(t)\,e^{i\omega t}\,dt. \qquad ...(7.18)$$

The inverse relation from Eqn. (7.17) is

$$f(x) = \frac{1}{\sqrt{2\pi}}\int_{-\infty}^{+\infty} g(\omega)\,e^{-i\omega t}\,d\omega. \qquad ...(7.19)$$

7.8 FOURIER COSINE TRANSFORM

If $f(x)$ is an even function of x, we can define the Fourier cosine transform as

$$g_c(\omega) = \sqrt{\frac{2}{\pi}}\int_0^{\infty} f(t)\cos\omega t\,dt, \qquad ...(7.20)$$

and the inverse relation is

$$f(x) = \sqrt{\frac{2}{\pi}}\int_0^{\infty} g_c(\omega)\cos\omega x\,d\omega. \qquad ...(7.21)$$

7.9 FOURIER SINE TRANSFORM

If $f(x)$ is an odd function of x, we can define the Fourier sine transform as

$$g_s(\omega) = \sqrt{\frac{2}{\pi}}\int_0^{\infty} f(t)\sin\omega t\,dt \qquad ...(7.22)$$

and the inverse relation is

$$f(x) = \sqrt{\frac{2}{\pi}}\int_0^{\infty} g_s(\omega)\sin\omega x\,d\omega. \qquad ...(7.23)$$

SOLVED PROBLEMS

1. *Show that if m and n are positive integers*

(a) $\int_{-\pi}^{\pi} \cos mx \cos nx\,dx = \pi\delta_{mn}$, (b) $\int_{-\pi}^{\pi} \sin mx \sin nx\,dx = \pi\delta_{mn}$, (c) $\int_{-\pi}^{\pi} \sin mx \cos nx\,dx = 0.$

Solution

(*a*) If $m \neq n$,

$$\text{L.H.S.} = \frac{1}{2}\int_{-\pi}^{\pi}[\cos(m-n)\,x + \cos\,(m+n)x]\,dx = 0\,.$$

If $m = n$, L.H.S. $= \int_{-\pi}^{\pi}\cos^2 mx\,dx = \pi.$

(*b*) If $m \neq n$,

$$\text{L.H.S.} = \frac{1}{2}\int_{-\pi}^{\pi}[\cos(m-n)\,x - \cos\,(m+n)x]\,dx = 0.$$

If $m = n$, L.H.S. $= \int_{-\pi}^{\pi}\sin^2 mx\,dx = \pi.$

(*c*) If $m \neq n$,

$$\text{L.H.S.} = \frac{1}{2}\int_{-\pi}^{\pi}[\sin(m+n)\,x + \sin\,(m-n)x]\,dx = 0.$$

If $m = n$, L.H.S. $= \frac{1}{2}\int_{-\pi}^{\pi}\sin 2mx = 0.$

2. *The Fourier series corresponding to f (x) is given by*

$$f(x) = \frac{a_0}{2} + \sum_{n=1}^{\infty}(a_n \cos nx + b_n \sin nx) \qquad \text{...(7.24)}$$

where f (t) is continuous at t = x. Show that the Fourier coefficients a_n and b_n are given by

$$a_n = \frac{1}{\pi}\int_{-\pi}^{\pi} f(x)\cos nx\,dx,\ n = 0, 1, 2, \ldots$$

$$b_n = \frac{1}{\pi}\int_{-\pi}^{\pi} f(x)\sin nx\,dx,\ n = 1, 2, 3, \ldots$$

Solution

Multiplying (7.24) by cos mx with m = 1, 2, 3,..... and integrating from $-\pi$ to π, we get

$$\int_{-\pi}^{\pi} f(x)\cos mx\,dx = \frac{a_0}{2}\int_{-\pi}^{\pi}\cos mx\,dx + \sum_{n=1}^{\infty}\left(a_n\int_{-\pi}^{\pi}\cos nx\cos mx\,dx + b_n\int_{-\pi}^{\pi}\sin nx\cos mx\,dx\right)$$

$$= \sum_{n=1}^{\infty} a_n \pi \delta_{mn}, \quad \text{by using eqns. of problem 1.}$$

$$= \pi a_m.$$

Thus, $$a_m = \frac{1}{\pi}\int_{-\pi}^{\pi} f(x)\cos mx\, dx, m = 1, 2, 3, \ldots$$

Integrating Eqn. (7.24) from $-\pi$ to π, we get

$$\int_{-\pi}^{\pi} f(x)dx = \frac{a_0}{2}\int_{-\pi}^{\pi} dx + 0 = a_0\pi.$$

Thus, $$a_0 = \frac{1}{\pi}\int_{-\pi}^{\pi} f(x)dx.$$

Multiplying Eqn. (7.24) by sin mx with $m = 1, 2, 3, \ldots$ and integrating from $-\pi$ to π, we have

$$\int_{-\pi}^{\pi} f(x)\sin mx\, dx = \sum_{n=1}^{\infty} b_n \pi\delta_{mn} = \pi b_m.$$

Hence, $$b_m = \frac{1}{\pi}\int_{-\pi}^{\pi} f(x)\sin mx\, dx, m = 1, 2, 3, \ldots$$

3. *Let* $f(x) = \begin{cases} x, & \text{for } 0 < x < \pi \\ -x, & \text{for } -\pi < x < 0 \end{cases}$

where f (x) has period 2π. *Draw the graph of f (x) and obtain the Fourier series for f(x). Considering the point x = 0 show that*

$$\frac{1}{1^2} + \frac{1}{3^3} + \frac{1}{5^2} + \ldots = \frac{\pi^2}{8}.$$

Solution

The graph of $f(x)$ against x is shown in Fig. 7.1.

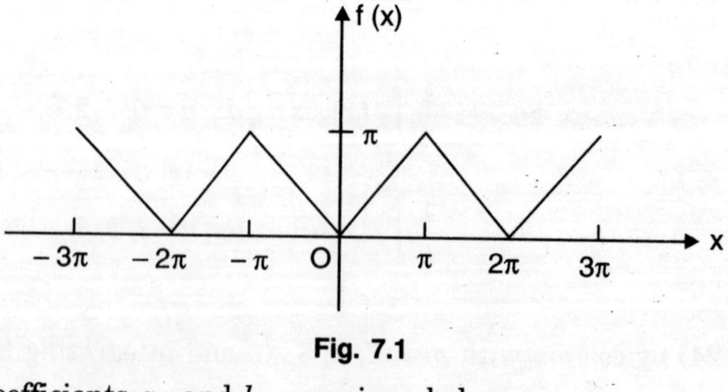

Fig. 7.1

The Fourier coefficients a_n and b_n are given below:

$$a_0 = \frac{1}{\pi}\int_{-\pi}^{\pi} f(x)dx = \frac{1}{\pi}\int_{-\pi}^{0}(-x)dx + \frac{1}{\pi}\int_{0}^{\pi} x\, dx = \pi$$

$$a_n = \frac{1}{\pi}\int_{-\pi}^{0}(-x)\cos nx\, dx + \frac{1}{\pi}\int_{0}^{\pi} x\cos nx\, dx$$

$$= \begin{cases} -\dfrac{4}{\pi n^2} & \text{for} \quad n = 1,\ 3,\ 5, \ldots \\ 0 & \text{for} \quad n = 2,\ 4,\ 6, \ldots \end{cases}$$

$$b_n = \frac{1}{\pi}\int_{-\pi}^{0}(-x)\sin nx\,dx + \frac{1}{\pi}\int_{0}^{\pi} x \sin nx\,dx = 0.$$

Since $f(x)$ is continuous for all values of x, the Fourier series for $f(x)$ is

$$f(x) = \frac{\pi}{2} - \frac{4}{\pi}\left[\cos x + \frac{\cos 3x}{3^2} + \frac{\cos 5x}{5^2} + \ldots\right]$$

At $x = 0$, $f(0) = 0$ (Fig. 7.1), and we have

$$0 = \frac{\pi}{2} - \frac{4}{\pi}\left[1 + \frac{1}{3^2} + \frac{1}{5^2} + \ldots\right]$$

or

$$\frac{1}{1^2} + \frac{1}{3^2} + \frac{1}{5^2} + \ldots = \frac{\pi^2}{8}.$$

4. (*a*) *Show that the Fourier series for*

$$f(x) = \cos px,\ -\pi \le x \le \pi$$

where the period is 2π *and* $p \ne 0, \pm 1, \pm 2, \ldots$ *is given by*

$$\frac{\sin p\pi}{\pi}\left(\frac{1}{p} - \frac{2p}{p^2 - 1}\cos x + \frac{2p}{p^2 - 2^2}\cos 2x - \frac{2p}{p^2 - 3^3}\cos 3x + \ldots\right)$$

(*b*) *Find also the sum of the series*

$$\frac{1}{p^2 - 1^2} - \frac{1}{p^2 - 2^2} + \frac{1}{p^2 - 3^2} - \frac{1}{p^2 - 4^2} + \ldots$$

where $p \ne 0, \pm 1, \pm 2, \ldots$

Solution

(*a*) The Fourier coefficients are given below:

$$a_0 = \frac{1}{\pi}\int_{-\pi}^{\pi}\cos px\,dx = \frac{2}{\pi}\frac{\sin p\pi}{p}$$

$$a_n = \frac{1}{\pi}\int_{-\pi}^{\pi}\cos px \cos nx\,dx = \frac{(-1)^n}{p^2 - n^2}\frac{2p\sin p\pi}{\pi}$$

$$b_n = \frac{1}{\pi}\int_{-\pi}^{\pi}\cos px \sin nx\,dx = 0$$

The Fourier series of $f(x)$ is thus

$$\frac{1}{\pi}\frac{\sin p\pi}{p} + \sum_{n=1}^{\infty}\frac{(-1)^n}{p^2 - n^2}\frac{2p\sin p\pi}{\pi}\cos nx.$$

(*b*) At $x = 0$, we have

$$1 = \frac{\sin p\pi}{p}\left[\frac{1}{p} - \frac{2p}{p^2-1^2} + \frac{2p}{p^2-2^2} - \frac{2p}{p^2-3^2} + \ldots\right]$$

or
$$\frac{1}{p^2-1^2} - \frac{1}{p^2-2^2} + \frac{1}{p^2-3^2} - \ldots = \frac{1}{2p}\left(\frac{1}{p} - \frac{\pi}{\sin p\pi}\right).$$

5. (*a*) *Obtain the Fourier series of the function*

$$f(x) = e^x \text{ for } -\pi < x < \pi$$

and
$$f(x + 2\pi) = f(x).$$

(*b*) *Find the sum of the series*

(*i*) $\sum_{m=1}^{\infty} \frac{1}{1+m^2}$ (*ii*) $\sum_{m=1}^{\infty} \frac{(-1)^m}{1+m^2}$

Solution

(*a*) The graph of $f(x)$ against x is shown in Fig. 7.2.

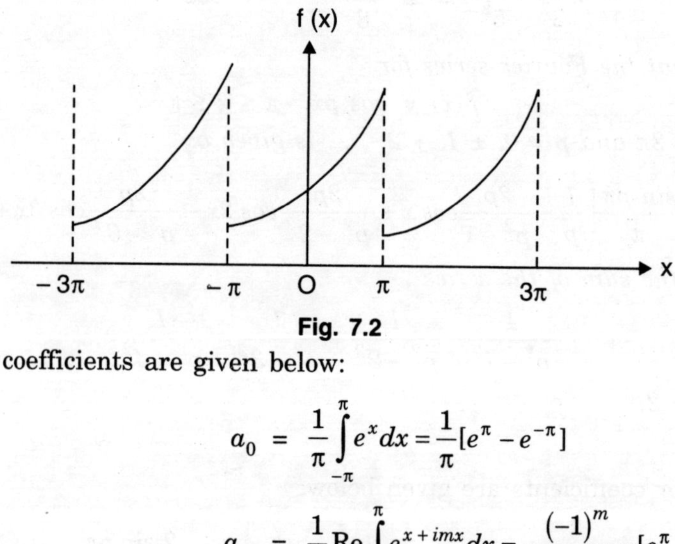

Fig. 7.2

The Fourier coefficients are given below:

$$a_0 = \frac{1}{\pi}\int_{-\pi}^{\pi} e^x dx = \frac{1}{\pi}[e^{\pi} - e^{-\pi}]$$

$$a_m = \frac{1}{\pi}\text{Re}\int_{-\pi}^{\pi} e^{x+imx} dx = \frac{(-1)^m}{\pi(1+m^2)}[e^{\pi} - e^{-\pi}]$$

$$b_m = \frac{1}{\pi}\text{Im}\int_{-\pi}^{\pi} e^{x+imx} dx = \frac{(-1)^{m+1} m}{\pi(1+m^2)}[e^{\pi} - e^{-\pi}]$$

The Fourier series is thus

$$\frac{1}{2}[f(x+0) + f(x-0)] = \frac{1}{\pi}(e^{\pi} - e^{-\pi})\left[\frac{1}{2} + \sum_{m=1}^{\infty}\left\{\frac{(-1)^m \cos mx}{1+m^2} + \frac{(-1)^{m+1} m}{1+m^2}\sin mx\right\}\right].$$

(*b*) (*i*) At $x = \pi$, we have

$$\frac{1}{2}\left[e^{\pi} + e^{-\pi}\right] = \frac{1}{\pi}(e^{\pi} - e^{-\pi})\left[\frac{1}{2} + \sum_{m=1}^{\infty}\frac{1}{1+m^2}\right]$$

or
$$\sum_{m=1}^{\infty} \frac{1}{1+m^2} = \frac{1}{2}[\pi \coth(\pi) - 1].$$

(*ii*) At $x = 0$, we have

$$1 = \frac{1}{\pi}\left(e^{\pi} - e^{-\pi}\right)\left[\frac{1}{2} + \sum_{m=1}^{\infty} \frac{(-1)^m}{1+m^2}\right]$$

or
$$\sum_{m=1}^{\infty} \frac{(-1)^m}{1+m^2} = \frac{1}{2}\left[\frac{\pi}{\sinh(\pi)} - 1\right].$$

6. (*a*) *Find the Fourier series corresponding to the function* $f(x) = x^2$, $0 < x < 2\pi$, *where* $f(x)$ *has period* 2π *outside of the interval* $(0, 2\pi)$. (*b*) *Show that*

$$\frac{1}{1^2} + \frac{1}{2^2} + \frac{1}{3^2} + \frac{1}{4^2} + \ldots = \frac{\pi^2}{6}.$$

Solution

(*a*) The Fourier series for the function $f(x)$ [Fig. 7.3] is given by

$$\frac{1}{2}[f(x+0) + f(x-0)] = \frac{a_0}{2} + \sum_{n=1}^{\infty}(a_n \cos nx + b_n \sin nx),$$

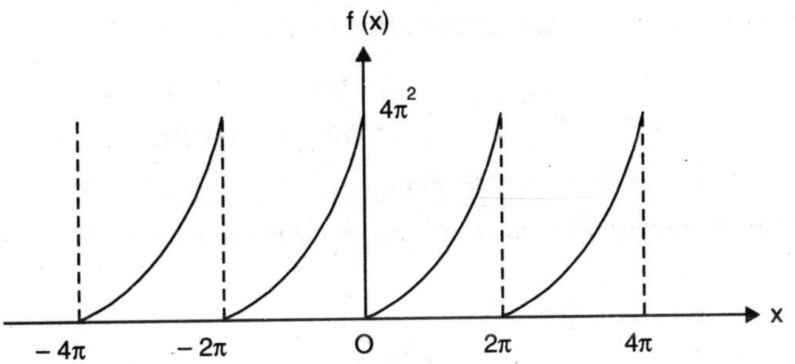

Fig. 7.3

where
$$a_0 = \frac{1}{\pi}\int_0^{2\pi} x^2\, dx = \frac{8\pi^2}{3}$$

$$a_n = \frac{1}{\pi}\int_0^{2\pi} x^2 \cos nx\, dx = \frac{4}{n^2},\quad n = 1, 2, 3, \ldots$$

$$b_n = \frac{1}{\pi}\int_0^{2\pi} x^2 \sin nx\, dx = -\frac{4\pi}{n},\quad n = 1, 2, 3, \ldots$$

(*b*) At $x = 0$, we have

$$2\pi^2 = \frac{4\pi^2}{3} + \sum_{n=1}^{\infty} \frac{4}{n^2}$$

or
$$\sum_{n=1}^{\infty} \frac{1}{n^2} = \frac{\pi^2}{6}.$$

7. *Find the Fourier series corresponding to the function*

$$f(x) = \begin{cases} 1 & \text{for } 0 < x < l \\ 0 & \text{for } -l < x < 0 \end{cases}$$

where the period is 2l.

Solution

$$a_0 = \frac{1}{l}\int_0^l dx = 1$$

$$a_n = \frac{1}{l}\int_0^l \cos\frac{n\pi x}{l}\,dx = 0$$

$$b_n = \frac{1}{l}\int_0^l \sin\frac{n\pi x}{l}\,dx = \frac{1}{n\pi}\left[1-(-1)^n\right]$$

The Fourier series is

$$\frac{1}{2}+\frac{2}{\pi}\left[\sin\frac{\pi x}{l}+\frac{1}{3}\sin\frac{3\pi x}{l}+\frac{1}{5}\sin\frac{5\pi x}{l}+\ldots\right].$$

8. *(a) Find the Fourier series corresponding to the function*

$$f(x) = \begin{cases} +1 & \text{for } 0 < x < \pi \\ -1 & \text{for } \pi < x < 2\pi \end{cases}$$

and
$$f(x + 2\pi) = f(x).$$

(b) Show that the value of the sum of the Fourier series just to the right of x = 0 is about 1.18.

Solution

(a)
$$a_0 = \frac{1}{\pi}\int_0^{2\pi} f(x)dx = 0$$

$$a_n = \frac{1}{\pi}\int_0^{2\pi} f(x)\cos nx\,dx = 0$$

$$b_n = \frac{1}{\pi}\int_0^{2\pi} f(x)\sin nx\,dx = \frac{2}{n\pi}\left[1-(-1)^n\right].$$

Thus the Fourier series is

$$\frac{4}{\pi}\left[\sin x + \frac{1}{3}\sin 3x + \frac{1}{5}\sin 5x+\ldots\right].$$

(b) The sum of the first *n* terms of the Fourier series is

$$S_n(x) = \frac{4}{\pi}\left[\sin x + \frac{\sin 3x}{3}+\ldots+\frac{\sin(2n-1)x}{2n-1}\right].$$

$S_n(x)$ has n maxima and $(n - 1)$ minima in the interval $0 < x < \pi$. Differentiating $S_n(x)$ with respect to x, we have

$$S_n'(x) = \frac{4}{\pi}\left[\cos x + \cos 3x + \ldots + \cos(2n-1)x\right]$$

$$= \frac{4}{\pi}\,\text{Re}\left[e^{ix} + e^{3ix} + \ldots + e^{(2n-1)ix}\right]$$

$$= \frac{4}{\pi}\,\text{Re}\left[\frac{e^{ix}\left(e^{2nix}-1\right)}{e^{2ix}-1}\right]$$

$$= \frac{2}{\pi}\frac{\sin 2nx}{\sin x}$$

Thus, $S'_n(x) = 0$ when $2nx = \pi, 2\pi, 3\pi, \ldots, (2n - 1)\pi$,

or $$x = \frac{\pi}{2n}, \frac{2\pi}{2n}, \frac{3\pi}{2n}, \ldots, \frac{(2n-1)\pi}{2n}$$

in the interval $0 < x < \pi$.

Now, $x = \dfrac{\pi}{2n}, \dfrac{3\pi}{2n}, \ldots, \dfrac{(2n-1)\pi}{2n}$ will correspond to maxima and $x = \dfrac{2\pi}{2n}, \dfrac{4\pi}{2n}, \ldots, \dfrac{(2n-2)\pi}{2n}$ will correspond to minima of $S_n(x)$.

At the first maximum point $\left(x = \dfrac{\pi}{2n}\right)$, the partial sum $S_n(x)$ has the value

$$S_n\left(x = \frac{\pi}{2n}\right) = \frac{4}{\pi}\left[\sin\frac{\pi}{2n} + \frac{1}{3}\sin\frac{3\pi}{2n} + \ldots\right]$$

or $$\frac{\pi}{2}S_n\left(x = \frac{\pi}{2n}\right) = \left[\frac{\sin(\pi/2n)}{(\pi/2n)}.\frac{\pi}{n} + \frac{\sin(3\pi/2n)}{3\pi/2n}.\frac{\pi}{n} + \ldots + \frac{\sin(2n-1)\pi/2n}{(2n-1)\pi/2n}\frac{\pi}{n}\right]. \qquad \ldots(7.25)$$

Now $$\int_0^{\pi}\frac{\sin x}{x}dx = \int_0^{\pi/n}\frac{\sin x}{x}dx + \int_{\pi/n}^{2\pi/n}\frac{\sin x}{x}dx + \ldots + \int_{(n-1)\pi/n}^{\pi}\frac{\sin x}{x}dx. \qquad \ldots(7.26)$$

For large n, the upper and lower limits of each integral of the r.h.s of Eqn. (7.26) are very close and their difference is π/n. For large n we can evaluate each integrand at the mid-point of the interval so that the r.h.s. of Eqn. (7.26) becomes equal to the r.h.s of Eqn. (7.25).

Thus, $$\lim_{n\to\infty}\frac{\pi}{2}S_n\left(x = \frac{\pi}{2n}\right) = \int_0^{\pi}\frac{\sin x}{x}dx$$

or $$\lim_{n\to\infty}S_n\left(x = \frac{\pi}{2n}\right) = \frac{2}{\pi}\int_0^{\pi}\left[1 - \frac{x^2}{3!} + \frac{x^4}{5!} - \frac{x^6}{7!} + \ldots\right]dx$$

$$\approx 1.18$$

Hence, just to the right of $x = 0$, the value of the sum of the Fourier series overshoots by about 18%. This is known as Gibbs' phenomenon or Gibbs' overshoot.

9. *(a) A displacement curve is given by*

$$f(t) = A\frac{t}{T}, \text{ for } 0 < t < T$$

and $$f(t + T) = f(t),$$

where A is a constant. Draw the graph of f (t) and obtain the Fourier series expansion for f (t).

(b) Considering the point t = T/4, show that

$$1-\frac{1}{3}+\frac{1}{5}-\frac{1}{7}+... = \frac{\pi}{4}.$$

Solution

(*a*) The graph of $f(t)$ against t is shown in Fig. 7.4 (saw-tooth wave).

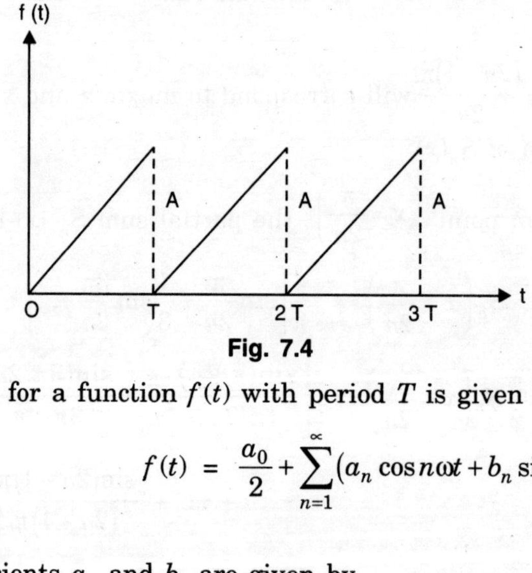

Fig. 7.4

The Fourier series for a function $f(t)$ with period T is given by

$$f(t) = \frac{a_0}{2}+\sum_{n=1}^{\infty}(a_n \cos n\omega t + b_n \sin n\omega t)$$

where $T = 2\pi/\omega$.

The Fourier coefficients a_n and b_n are given by

$$a_n = \frac{2}{T}\int_0^T f(t) \cos n\,\omega t\, dt, \quad n = 0, 1, 2, ...$$

$$b_n = \frac{2}{T}\int_0^T f(t) \sin n\,\omega t\, dt, \quad n = 1, 2, 3, ...$$

In the present problem

$$a_0 = \frac{2}{T}\int_0^T \frac{At}{T} dt = A$$

$$a_n = \frac{2}{T}\int_0^T \frac{At}{T} \cos n\,\omega t\, dt = 0$$

$$b_n = \frac{2}{T}\int_0^T \frac{At}{T}\sin n\,\omega t\,dt = -\frac{A}{n\pi}$$

Thus, the Fourier series for the function $f(t)$ is

$$\frac{A}{2}-\frac{A}{\pi}\left[\sin\omega t+\frac{1}{2}\sin 2\omega t+\frac{1}{3}\sin 3\omega t+\ldots\right]$$

(*b*) When $t = \frac{T}{4}$, $f(T/4) = \frac{A}{T}\cdot\frac{T}{4}=\frac{A}{4}$,

and $$\frac{A}{4} = \frac{A}{2}-\frac{A}{\pi}\left[1-\frac{1}{3}+\frac{1}{5}-\frac{1}{7}+\ldots\right]$$

or, $$1-\frac{1}{3}+\frac{1}{5}-\frac{1}{7}+\ldots = \frac{\pi}{4}.$$

This series is known as Gregory's series.

10. (*a*) *Obtain the Fourier series of* $|\sin x|$

(*b*) *Show that*

(*i*) $$\frac{1}{2^2-1}+\frac{1}{4^2-1}+\frac{1}{6^2-1}+\ldots=\frac{1}{2}$$

(*ii*) $$\frac{1}{2^2-1}-\frac{1}{4^2-1}+\frac{1}{6^2-1}-\frac{1}{8^2-1}+\ldots=\frac{\pi}{4}-\frac{1}{2}$$

(*iii*) $$\frac{1}{2^2-1}+\frac{1}{6^2-1}+\frac{1}{10^2-1}+\ldots=\frac{\pi}{8}.$$

Solution

(*a*) Since $|\sin x|$ is an even function of *x*, we get Fourier cosine series. We see that

$$a_0 = \frac{2}{\pi}\int_0^\pi \sin x\,dx = \frac{4}{\pi}$$

$$a_n = \frac{2}{\pi}\int_0^\pi \sin x \cos x\,dx$$

$$a_n = \begin{cases} -\dfrac{4}{\pi(n^2-1)}, & \text{when } n \text{ is even} \\ 0 & \text{when } n \text{ is odd}\end{cases}$$

$$b_n = 0$$

Thus,

$$|\sin x| = \frac{2}{\pi}-\frac{4}{\pi}\left[\frac{\cos 2x}{2^2-1}+\frac{\cos 4x}{4^2-1}+\frac{\cos 6x}{6^2-1}+\ldots\right].$$

(*b*) (*i*) Putting $x = 0$, we get

$$\frac{1}{2^2-1}+\frac{1}{4^2-1}+\frac{1}{6^2-1}+\ldots = \frac{1}{2}$$

(*ii*) Putting $x = \pi/2$, we get

$$\frac{1}{2^2-1}-\frac{1}{4^2-1}+\frac{1}{6^2-1}-\frac{1}{8^2-1}+\ldots = \frac{\pi}{4}-\frac{1}{2}$$

(*iii*) By adding (*i*) and (*ii*) we obtain this relation.

11. *Expand* $\frac{1}{8}\pi x(\pi - x)$ *in Fourier sine series, for* $0 \le x \le \pi$. *Hence show that*

$$1-\frac{1}{3^3}+\frac{1}{5^3}-\frac{1}{7^3}+\ldots=\frac{\pi^3}{32}.$$

Solution

For Fourier sine series, $a_n = 0$ and

$$b_n = \frac{2}{\pi}\int_0^{\pi}\frac{1}{8}\pi x(\pi - x)\sin nx\,dx$$

$$= \begin{cases} 0 & \text{if } n \text{ is even} \\ \dfrac{1}{n^3} & \text{if } n \text{ is odd} \end{cases}$$

Thus, the required series is

$$\frac{1}{8}\pi x(\pi - x) = \sin x + \frac{\sin 3x}{3^3}+\frac{\sin 5x}{5^3}+\ldots$$

Putting $x = \frac{\pi}{2}$, we obtain

$$\frac{1}{8}\pi\cdot\frac{\pi}{2}\cdot\frac{\pi}{2} = 1 - \frac{1}{3^3}+\frac{1}{5^3}-\frac{1}{7^3}+\ldots$$

or,

$$\frac{\pi^3}{32} = 1 - \frac{1}{3^3}+\frac{1}{5^3}-\frac{1}{7^3}+\ldots$$

12. *Expand* $\frac{1}{2}(\pi - x)$ *sin x in a cosine series in the range* $0 \le x \le \pi$.

Solution

Here

$$a_0 = \frac{2}{\pi}\int_0^{\pi}\frac{1}{2}(\pi - x)\sin x\,dx = 1$$

$$a_n = \frac{2}{\pi}\int_0^{\pi}\frac{1}{2}(\pi - x)\sin x\cos nx\,dx = \frac{1}{1-n^2} \text{ when } n \ne 1$$

$$a_1 = \frac{2}{\pi}\int_0^{\pi}\frac{1}{2}(\pi - x)\sin x\cos x\,dx = \frac{1}{4}$$

The required series is

$$\frac{1}{2}+\frac{1}{4}\cos x - \frac{\cos 2x}{2^2-1}-\frac{\cos 3x}{3^2-1}-\frac{\cos 4x}{4^2-1}-\ldots$$

13. *The Fourier series for the function $f(x)$ in the interval $[-L, L]$ is given by*

$$f(x) = \frac{a_0}{2} + \sum_{n=1}^{\infty}\left(a_n \cos\frac{n\pi x}{L} + b_n \sin\frac{n\pi x}{L}\right)$$

where $f(t)$ is continuous at $t = x$. The Fourier coefficients are given by

$$a_n = \frac{1}{L}\int_{-L}^{L} f(t)\cos\frac{n\pi t}{L}\,dt, \quad n = 0, 1, 2, \ldots$$

$$b_n = \frac{1}{L}\int_{-L}^{L} f(t)\sin\frac{n\pi t}{L}\,dt, \quad n = 1, 2, 3, \ldots$$

By taking the limit $L \to \infty$ with $\frac{n\pi}{L} = \omega$ and $\frac{\pi}{L} = \Delta\omega$, obtain the Fourier integral equation

$$f(x) = \frac{1}{2\pi}\int_{-\infty}^{\infty} e^{-i\omega x}\,d\omega \int_{-\infty}^{\infty} f(t)e^{i\omega t}\,dt.$$

Solution

For the Fourier series in the interval $[-L, L]$, we have

$$f(x) = \frac{1}{2L}\int_{-L}^{L} f(t)dt + \frac{1}{L}\sum_{n=1}^{\infty}\cos\frac{n\pi x}{L}\int_{-L}^{L} f(t)\cos\frac{n\pi t}{L}\,dt$$

$$+ \frac{1}{L}\sum_{n=1}^{\infty}\sin\frac{n\pi x}{L}\int_{-L}^{L} f(t)\sin\frac{n\pi t}{L}\,dt$$

$$= \frac{1}{2L}\int_{-L}^{L} f(t)dt + \frac{1}{L}\sum_{n=1}^{\infty}\int_{-L}^{L} f(t)\cos\left[\frac{n\pi}{L}(t-x)\right]dt.$$

We now let the parameter L approach infinity. Also, we set $\frac{n\pi}{L} = \omega, \frac{\pi}{L} = \Delta\omega$.

Thus we have

$$f(x) \to \frac{1}{\pi}\sum_{n=1}^{\infty}\Delta\omega\int_{-\infty}^{\infty} f(t)\cos[\omega(t-x)]\,dt.$$

Note that the term corresponding to a_0 has vanished, assuming that $\int_{-\infty}^{\infty} f(t)\,dt$ exists.

Replacing the infinite sum by integration over ω, from 0 to ∞, we have

$$f(x) = \frac{1}{\pi}\int_{0}^{\infty} d\omega \int_{-\infty}^{\infty} f(t)\cos[\omega(t-x)]dt$$

$$= \frac{1}{2\pi}\int_{-\infty}^{\infty} d\omega \int_{-\infty}^{\infty} f(t)\cos[\omega(t-x)]dt$$

Since $\cos[\omega\,(t - x)]$ is an even function of ω.

Now, $\frac{1}{2\pi}\int_{-\infty}^{\infty} d\omega \int_{-\infty}^{\infty} f(t)\sin[\omega(t-x)]dt = 0.$

Adding the first equation to i times the second equation, we get

$$f(x) = \frac{1}{2\pi}\int_{-\infty}^{\infty} d\omega \int_{-\infty}^{\infty} f(t)e^{i\omega(t-x)}\,dt$$

or

$$f(x) = \frac{1}{2\pi}\int_{-\infty}^{\infty} d\omega e^{-i\omega x}\int_{-\infty}^{\infty} f(t)e^{i\omega t}\,dt. \qquad ...(7.27)$$

In many physical problems, ω corresponds to the angular frequency. We thus interpret Eqn. (7.27) as representation of $f(x)$ in terms of a distribution of infinitely long sinusoidal wave trains of angular frequency ω.

Note: Dirac delta function is given by

$$\delta(t - x) = \frac{1}{2\pi}\int_{-\infty}^{\infty} e^{i\omega(t-x)}\,d\omega.$$

This equation may be used to evaluate the right hand side of Eqn. (7.27):

$$\text{R.H.S.} = \int_{-\infty}^{\infty} f(t)\delta(t-x)dt = f(x)$$

14. *Find the Fourier transform of the function*

$$f(x) = 1 \text{ for } |x| < 1,$$
$$f(x) = 0 \text{ for } |x| > 1,$$

Solution

This is an even function of x [Fig. 7.5]

Thus, we may find the Fourier cosine transform of $f(x)$

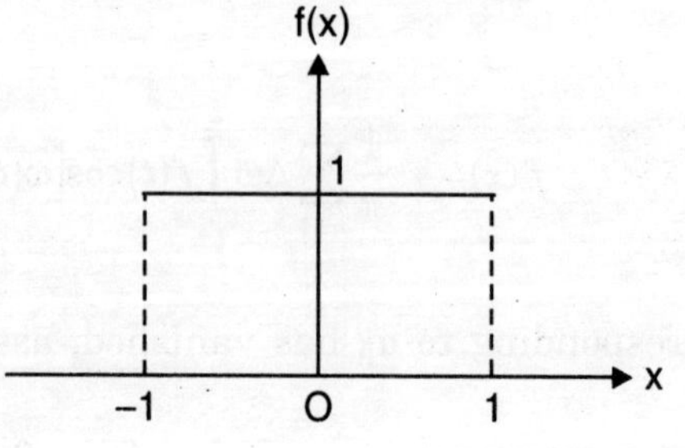

Fig. 7.5

$$g_c(\omega) = \sqrt{\frac{2}{\pi}}\int_0^1 f(t)\cos\omega t\,dt = \sqrt{\frac{2}{\pi}}\frac{\sin\omega}{\omega}$$

15. *By taking inverse Fourier transform of $g_c(\omega)$ of problem 14 show that*

$$\int_0^\infty \frac{\sin y}{y}\,dy = \frac{\pi}{2}.$$

Solution

The inverse Fourier transform of $g_c\,(\omega)$ is

$$f_c(x) = \sqrt{\frac{2}{\pi}}\int_0^\infty g_c(\omega)\cos\omega x\,d\omega = \frac{2}{\pi}\int_0^\infty \frac{\sin\omega\cos\omega x}{\omega}\,d\omega.$$

At $x = \pm 1$, $f_c\,(x) = \dfrac{1}{2} = \dfrac{2}{\pi}\displaystyle\int_0^\infty \frac{\sin\omega\cos\omega}{\omega}\,d\omega$

or,
$$\int_0^\infty \frac{\sin 2\omega}{\omega}\,d\omega = \frac{\pi}{2}.$$

With $2\omega = y$, we get $\displaystyle\int_0^\infty \frac{\sin y}{y}\,dy = \frac{\pi}{2}.$

16. *(a) Find the Fourier transform of the triangular pulse*

$$f(x) = \begin{cases} a(1 - a|x|) & \text{for } |x| < \dfrac{1}{a} \\ 0 & \text{for } |x| > \dfrac{1}{a} \end{cases}$$

where $a > 0$.

(b) By taking the Fourier inverse transform show that, in the limit $a \to \infty$,

$$\delta(x) = \frac{1}{2\pi}\int_{-\infty}^{\infty} e^{-ikx}\,dk.$$

So defined, $\delta\,(x)$ is called the Dirac delta function.

Solution

(*a*) $f(x)$ is an even function of x [Fig.7.6].

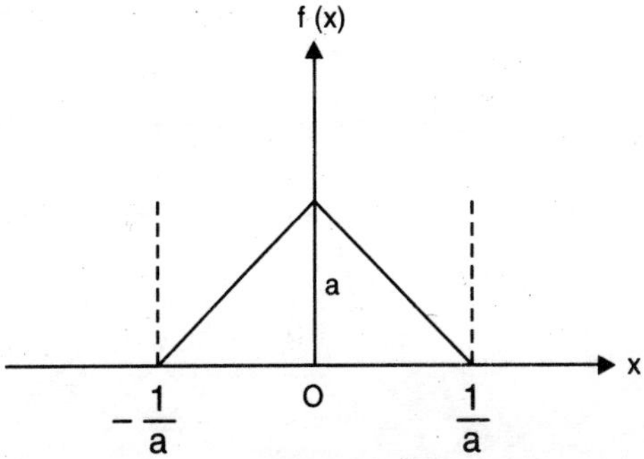

Fig. 7.6

$$g_c(\omega) = \sqrt{\frac{2}{\pi}} \int_0^{1/a} a(1 - ax) \cos \omega x \, dx$$

$$= \sqrt{\frac{2}{\pi}} \frac{2a^2}{\omega^2} \sin^2\left(\frac{\omega}{2a}\right).$$

(*b*) Area of the triangle in Fig.7.6 is $\frac{1}{a} \times a = 1$.

As $a \to \infty$, $f(x) \to \delta(x)$, and

$$g_c(\omega) = \sqrt{\frac{2}{\pi}} \cdot \frac{1}{2} \left(\frac{\sin\left(\frac{\omega}{2a}\right)}{\frac{\omega}{2a}} \right)^2 \to \frac{1}{\sqrt{2\pi}}$$

By taking Fourier inverse transform, we get

$$f(x) = \frac{1}{\sqrt{2\pi}} \int_{-\infty}^{\infty} g(\omega) e^{-i\omega x} d\omega.$$

In the time $a \to \infty$, this equation becomes

$$\delta(x) = \frac{1}{\sqrt{2\pi}} \int_{-\infty}^{\infty} \frac{1}{\sqrt{2\pi}} e^{-i\omega x} d\omega = \frac{1}{2\pi} \int_{-\infty}^{\infty} e^{-ikx} dk.$$

This is an integral representation of the δ-function.

17. *Find the Fourier transform of signal*

$$f(t) = \begin{cases} A \cos \omega_0 t & \text{for } |t| \le T \\ 0 & \text{for } |t| > T \end{cases}$$

Solution

Since $f(t)$ is even, we use Fourier cosine transform

$$g_c(\omega) = \sqrt{\frac{2}{\pi}} \int_0^T A \cos \omega_0 t \cos \omega t \, dt$$

$$= \frac{A}{\sqrt{2\pi}} \left[\frac{\sin(\omega_0 - \omega)T}{\omega_0 - \omega} + \frac{\sin(\omega_0 + \omega)T}{\omega_0 + \omega} \right].$$

18. *Find the Fourier transform of finite wave train*

$$f(t) = \begin{cases} \sin \omega_0 t & \text{for } |t| < \frac{N\pi}{\omega_0} \\ 0 & \text{for } |t| > \frac{N\pi}{\omega_0} \end{cases}$$

Solution

The function $f(t)$ represents a sine wave for the time $t = -\frac{N\pi}{\omega_0}$ to $t = +\frac{N\pi}{\omega_0}$ or, total time $= N.\frac{2\pi}{\omega_0}$ (waves of N cycles). Since $f(t)$ is odd, we use Fourier sine transform,

$$g_s(\omega) = \sqrt{\frac{2}{\pi}} \int_0^{N\pi/\omega_0} \sin \omega_0 t \sin \omega t \, dt$$

$$= \sqrt{\frac{1}{2\pi}} \left[\frac{\sin[N\pi(\omega_0 - \omega)/\omega_0]}{\omega_0 - \omega} - \frac{\sin[N\pi(\omega_0 + \omega)/\omega_0]}{\omega_0 + \omega} \right]. \quad ...(7.28)$$

19. *(a) Discuss the nature of the frequency pulse g(ω) of problem 18 when ω_0 is large. (b) Show that the spread in frequency of wave pulse g(ω) may be given by $\Delta\omega = \omega_0/N$. (c) Obtain the relation $\Delta E \cdot \Delta t \approx h$ for our waves, where ΔE = uncertainty in the energy of the pulse and Δt = uncertainty in time.*

Solution

(*a*) For large ω_0 only the first term of Eqn. (7.28) will be of any importance when $\omega \approx \omega_0$. This is shown in Fig. 7.7.

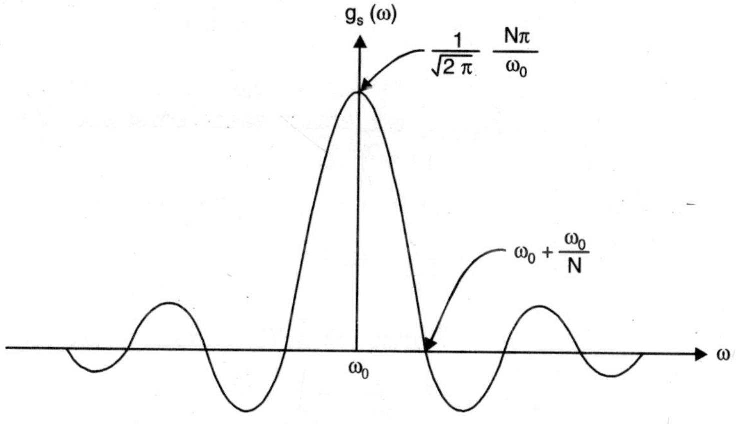

Fig. 7.7

Note that this is the amplitude curve for the single slit diffraction pattern. There are zeroes at

$$\frac{\omega_0 - \omega}{\omega_0} = \frac{\Delta\omega}{\omega_0} = \pm\frac{1}{N}, \pm\frac{2}{N}, \pm\frac{3}{N}, \text{etc.,}$$

where $$\Delta\omega = \omega_0 - \omega.$$

(*b*) From Fig. 7.7 we find that the value of $g_s(\omega)$ outside the central maximum is small. We may take

$$\Delta\omega = \frac{\omega_0}{N}$$

as a good measure of the spread in frequency of the wave pulse. If N is large [a long pulse], the frequency spread = $\Delta\omega$ is very small. If N is small, the frequency distribution is wide.

(*c*) For an electromagnetic wave, $E = \hbar\omega$ or, $\Delta E = \hbar\,\Delta\omega$, where ΔE represents the uncertainty in the energy of the pulse. There is also uncertainty in the time. A cycle of the wave requires a time $2\pi/\omega_0$ to pass through a point. Thus, the wave of N cycles requires a time $N2\pi/\omega_0$ to pass through a point. We may take

$$\Delta t = N\frac{2\pi}{\omega_0}.$$

Then, the product $\Delta E\Delta t$ is of the order

$$\hbar\Delta\omega\,\Delta t = \hbar\frac{\omega_0}{N}.N\frac{2\pi}{\omega_0} = 2\pi\hbar = h.$$

20. *In a resonant cavity, an electromagnetic oscillation with angular frequency ω_0 is given by*

$$f(t) = Ae^{-\frac{\omega_0 t}{2\theta}}e^{-i\omega_0 t}, t > 0$$

and

$$f(t) = 0, \text{ for } t < 0.$$

Find the frequency distribution $|g(\omega)|^2$ of the electromagnetic oscillation. Show schematically on a diagram how $|g(\omega)|^2$ vs ω behaves as θ is increased from small values to very high values.

Solution

We have

$$g(\omega) = \frac{1}{\sqrt{2\pi}}\int_0^\infty Ae^{-\frac{\omega_0 t}{2\theta}}e^{-i\omega_0 t}e^{i\omega t}dt$$

$$= \frac{A}{\sqrt{2\pi}}\frac{\frac{\omega_0}{2\theta} - i(\omega_0-\omega)}{\left(\frac{\omega_0}{2\theta}\right)^2 + (\omega_0-\omega)^2}$$

and

$$|g(\omega)|^2 = \frac{A^2}{2\pi}\frac{1}{\left(\frac{\omega_0}{2\theta}\right)^2 + (\omega_0-\omega)^2}$$

$|g(\omega)|^2$ has a peak at $\omega = \omega_0$ and the peak value is $\dfrac{2A^2\theta^2}{\pi\omega_0^2} \propto \theta^2$.

The value of ω at which $|g(\omega)|^2$ becomes half of its peak value is given by

$$\frac{A^2}{2\pi}.\frac{1}{\left(\frac{\omega_0}{2\theta}\right)^2 + (\omega_0-\omega)^2} = \frac{A^2\theta^2}{\pi\omega_0^2}$$

or

$$\omega = \omega_0 \pm \frac{\omega_0}{2\theta}.$$

Thus $\Delta\omega$ = Full width of the frequency distribution curve at half maximum = ω_0/θ [Fig. 7.8].

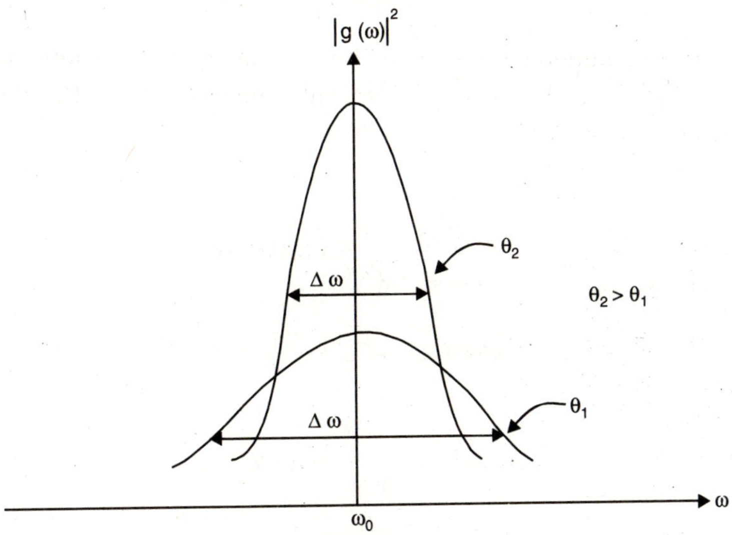

Fig. 7.8

For small θ (damping of the electromagnetic wave is large), the value of the peak of the frequency distribution is small and $\Delta\omega$ is large. As θ increases (damping of the electromagnetic wave decreases), the peak value increases and $\Delta\omega$ decreases *i.e.*, the frequency distribution curve becomes more peaked and sharper.

21. *If $f(x)$ vanishes as $x \to \pm \infty$ show that the Fourier transform of the derivative of $f(x)$ (i.e. $\frac{d}{dx} f(x)$) is given by $-i\omega g(\omega)$.*

Solution

The Fourier transform of the derivative $\frac{d}{dx} f(x)$ is

$$g_1(\omega) = \frac{1}{\sqrt{2\pi}} \int_{-\infty}^{+\infty} \frac{d}{dx} f(x) e^{i\omega x} dx$$

$$= \frac{1}{\sqrt{2\pi}} e^{i\omega x} f(x) \Big|_{-\infty}^{\infty} - \frac{i\omega}{\sqrt{2\pi}} \int_{-\infty}^{\infty} f(x) e^{i\omega x} dx$$

$$= -i\omega g(\omega),$$

since $f(x)$ vanishes as $x \to \pm \infty$.

Note: The transform of the derivative is $(-i\omega)$ times the transform of the original function. This may be generalized to the nth derivative to yield the result

$$g_n(\omega) = (-i\omega)^n g(\omega),$$

provided all the integrated parts vanish as $x \to \pm \infty$.

22. *Derive Parseval's relation*

$$\int_{-\infty}^{+\infty} f(t)g^*(t)dt = \int_{-\infty}^{+\infty} F(\omega)G^*(\omega)d\omega$$

where F (ω) and G (ω) are Fourier transforms of f (t) and g (t) respectively.

Solution

We have

$$F(\omega) = \frac{1}{\sqrt{2\pi}}\int_{-\infty}^{+\infty} f(t)e^{i\omega t}dt,$$

and
$$G(\omega) = \frac{1}{\sqrt{2\pi}}\int_{-\infty}^{+\infty} g(t)e^{i\omega t}dt.$$

Hence,
$$g(t) = \frac{1}{\sqrt{2\pi}}\int_{-\infty}^{+\infty} G(\omega)e^{-i\omega t}d\omega,$$

and
$$g^*(t) = \frac{1}{\sqrt{2\pi}}\int_{-\infty}^{+\infty} G^*(\omega)e^{i\omega t}d\omega.$$

Thus,

$$\int_{-\infty}^{+\infty} f(t)g^*(t)dt = \frac{1}{2\pi}\int_{-\infty}^{+\infty}\int_{-\infty}^{+\infty}\int_{-\infty}^{+\infty} F(\omega)e^{-i\omega t}d\omega G^*(x)e^{ixt}dx\,dt$$

$$= \frac{1}{2\pi}\int_{-\infty}^{+\infty}\int_{-\infty}^{+\infty} F(\omega)G^*(x)\,d\omega\,dx\int_{-\infty}^{+\infty} e^{it(x-\omega)}dt$$

$$= \int_{-\infty}^{+\infty}\int_{-\infty}^{+\infty} F(\omega)G^*(x)\,\delta(x-\omega)\,d\omega\,dx$$

$$= \int_{-\infty}^{+\infty} F(\omega)G^*(\omega)\,d\omega$$

In the special case when $g(t) \equiv f(t)$, we obtain

$$\int_{-\infty}^{+\infty} f^*(t)\,f(t)\,dt = \int_{-\infty}^{+\infty} F^*(\omega)F(\omega)\,d\omega.$$

23. *For one-dimensional wave function ψ (x), the momentum function g (p) is defined by*

$$g(p) = \frac{1}{\sqrt{2\pi\hbar}}\int_{-\infty}^{+\infty}\psi(x)e^{-i\pi x/\hbar}\,dx$$

[Technically we have employed the inverse Fourier transform].

Derive the Parseval's relation

$$\int_{-\infty}^{+\infty} g^*(p)\, g(p)\, dp = \int_{-\infty}^{+\infty} \psi^*(x)\, \psi(x)\, dx.$$

Solution

We have

$$\int_{-\infty}^{+\infty} g^*(p)\, g(p)\, dp = \frac{1}{2\pi\hbar}\int_{-\infty}^{+\infty}\int_{-\infty}^{+\infty}\int_{-\infty}^{+\infty} \psi^*(x)\, e^{ipx/\hbar} dx\, \psi(x') e^{-ipx'/\hbar} dx'\, dp$$

$$= \int_{-\infty}^{+\infty} \psi^*(x)\, \psi(x) dx.$$

Thus, if function $\psi(x)$ is normalised, then so is $g(p)$ and vice versa.

SUPPLEMENTARY PROBLEMS

1. Let $f(x) = x$ for $-\pi < x < \pi$ and $f(x + 2\pi) = f(x)$.
   Draw the graph of $f(x)$ and obtain the Fourier series for $f(x)$. Considering the point $x = \frac{\pi}{2}$ show that

$$1 - \frac{1}{3} + \frac{1}{5} - \frac{1}{7} + \ldots = \frac{\pi}{4}$$

2. (*a*) Obtain the Fourier series of the function (square wave)

$$f(x) = \begin{cases} 0 & \text{for } -\pi < x < 0 \\ 1 & \text{for } 0 < x < \pi \end{cases}$$

and $$f(x + 2\pi) = f(x)$$

(*b*) Considering the point $x = \frac{\pi}{2}$, find the sum of the series

$$1 - \frac{1}{3} + \frac{1}{5} - \frac{1}{7} + \ldots\ .$$

3. Obtain the Fourier series expansion for the function $f(x)$ defined as follows:

$$f(x) = \begin{cases} -\frac{2}{\pi}(x + \pi) & \text{for } -\pi \le x \le -\frac{\pi}{2} \\ \frac{2}{\pi}x & \text{for } -\frac{\pi}{2} \le x \le \frac{\pi}{2} \\ -\frac{2}{\pi}(x - \pi) & \text{for } \frac{\pi}{2} \le x \le \pi \end{cases}$$

and $$f(x + 2\pi) = f(x).$$

Considering the point $x = \frac{\pi}{2}$, show that

$$\frac{1}{1^2}+\frac{1}{3^2}+\frac{1}{5^2}+... = \frac{\pi^2}{8}$$

4. Show that the Fourier series corresponding to

$$f(x) = 2x \text{ for } 0 \le x < 3$$
$$f(x) = 0 \text{ for } -3 < x < 0$$

where the period is 6, is given by

$$\frac{3}{2}+\sum_{n=1}^{\infty}\left[\frac{6(\cos n\pi - 1)}{n^2\pi^2}\cos\frac{n\pi x}{3} - \frac{6\cos n\pi}{n\pi}\sin\frac{n\pi x}{3}\right].$$

5. Find the Fourier series corresponding to the function

$$f(x) = \begin{cases} 0 & \text{for } -5 < x < 0 \\ 3 & \text{for } 0 < x < 5, \end{cases} \quad \text{Period} = 10.$$

6. Find the Fourier series corresponding to the function

$$f(x) = 4x,\ 0 < x < 10,\ \text{Period} = 10.$$

7. The Fourier series corresponding to the periodic function $f(t)$ with period T [*i.e.*, $f(t + T) = f(t)$] is given by

$$f(t) = \frac{a_0}{2}+\sum_{n=1}^{\infty}(a_n \cos n\,\omega t + b_n \sin n\,\omega t)$$

where $T = \frac{2\pi}{\omega}$ and $f(x)$ is continuous at $x = t$. Show that the Fourier coefficients a_n and b_n are given by

$$a_n = \frac{2}{T}\int_0^T f(t)\cos n\,\omega t\,dt, \quad n = 0, 1, 2, ...$$

$$b_n = \frac{2}{T}\int_0^T f(t)\sin n\,\omega t\,dt, \quad n = 1, 2, 3, ...$$

8. A displacement curve is given by

$$y(t) = \left(1-\frac{t}{T}\right)A \text{ for } 0 < t < T$$

and $y(t + T) = y(t)$.

Obtain the Fourier series expansion of $y(t)$.

9. A displacement curve is given by

$$f(t) = \sin \omega t \text{ for } 0 < t < T/2,$$
$$f(t) = -\sin \omega t \text{ for } T/2 < t < T$$

where $T = \frac{2\pi}{\omega}$ and $f(t + T) = f(t)$.

Give a rough sketch of the displacement curve and obtain the Fourier series expansion for $f(t)$.

10. A square wave is defined as

$$f(t) = B \text{ for } -\frac{\alpha T}{2} < t < \frac{\alpha T}{2}$$

$$= 0 \text{ elsewhere in the range } -\frac{T}{2} \text{ to } \frac{T}{2},$$

where $0 < \alpha < 1$ and $f(t)$ is periodic with period $T = \frac{2\pi}{\omega}$. Obtain the Fourier series of the function.

11. For a periodic function $f(t)$ of fundamental frequency f_0 (*i.e.*, $f_0 = \frac{1}{T}$, where T is the period) we have the Fourier series expansion

$$f(t) = \frac{a_0}{2} + \sum_{n=1}^{\infty}(a_n \cos 2\pi n f_0 t + b_n \sin 2\pi n f_0 t),$$

where $f(t)$ is defined for the time $-\frac{1}{2f_0} < t < \frac{1}{2f_0} \left(\text{or}, -\frac{T}{2} < t < \frac{T}{2}\right)$ and $f(t + T) = f(t)$.

Show that the Fourier coefficients are given by

$$a_n = 2f_0 \int_{-\frac{1}{2f_0}}^{\frac{1}{2f_0}} f(t)\cos(2\pi n f_0 t)\, dt, \quad n = 0, 1, 2, \ldots$$

and

$$b_n = 2f_0 \int_{-\frac{1}{2f_0}}^{\frac{1}{2f_0}} f(t)\sin(2\pi n f_0 t)\, dt, \quad n = 1, 2, 3, \ldots .$$

12. Show that for values of x between 0 and π, the function $\frac{1}{8}\pi(\pi - 2x)$ can be expanded in the cosine series

$$\cos x + \frac{\cos 3x}{3^2} + \frac{\cos 5x}{5^2} + \ldots .$$

13. Find a Fourier sine series corresponding to the function

$$f(x) = \cos x, \; 0 < x < \pi.$$

14. Expand $f(x) = x$, $0 < x < 2$ in a half range in (*a*) sine series (*b*) cosine series.

15. (*a*) Prove that for $0 \le x \le \pi$,

$$x(\pi - x) = \frac{\pi^2}{6} - \left[\frac{\cos 2x}{1^2} + \frac{\cos 4x}{2^2} + \frac{\cos 6x}{3^2} + \ldots\right]$$

(*b*) Show that

(*i*) $\displaystyle\sum_{n=1}^{\infty} \frac{1}{n^2} = \frac{\pi^2}{6}$ (*ii*) $\displaystyle\sum_{n=1}^{\infty} \frac{(-1)^{n-1}}{n^2} = \frac{\pi^2}{12}.$

16. Show that when $0 \le x \le \pi$,

$$\frac{1}{96}\pi(\pi - 2x)\left(\pi^2 + 2\pi x - 2x^2\right) = \cos x + \frac{\cos 3x}{3^4} + \frac{\cos 5x}{5^4} + \ldots$$

and

$$1 + \frac{1}{3^4} + \frac{1}{5^4} + \ldots = \frac{\pi^4}{96}.$$

17. Find the Fourier cosine and sine transforms of exp $(-at)$, $t > 0$. By taking inverse Fourier transform show that

(*i*) $$\int_0^\infty \frac{\cos \omega x}{\omega^2 + a^2} d\omega = \frac{\pi}{2a} e^{-ax}, \quad x > 0$$

(*ii*) $$\int_0^\infty \frac{\omega \sin \omega x}{\omega^2 + a^2} d\omega = \frac{\pi}{2} e^{-ax}, \quad x > 0$$

18. (*a*) Find the Fourier transform of the rectangular pulse

$$f(x) = \begin{cases} 0 & \text{for } |x| > X \\ \dfrac{1}{2X} & \text{for } |x| < X \end{cases}$$

(*b*) By taking Fourier inverse transform show in the limit $X \to 0$ that

$$\delta(x) = \frac{1}{2\pi} \int_{-\infty}^{+\infty} e^{-ikx} dk.$$

19. By taking the Fourier transform of the differential equation

$$\frac{d^2 y}{dx^2} + xy = 0$$

obtain the differential equation satisfied by $g(\omega)$, and find its solution. Inverting the transform show that

$$y(x) = \frac{A}{\sqrt{2\pi}} \int_{-\infty}^{\infty} d\omega \exp\left[-i\left(\omega x - \frac{\omega^3}{3}\right)\right]$$

[The integral is known as Airy integral].

20. Using the definition of momentum function $g(p)$ [see problem 23] find the momentum function representation for the one-dimensional Schrödinger equation for harmonic oscillator:

$$-\frac{\hbar^2}{2m}\frac{d^2\psi(x)}{dx^2} + \frac{1}{2}kx^2\psi(x) = E\psi(x).$$

21. A linear quantum oscillator is its ground state has the normalised wave function

$$\psi(x) = a^{-1/2}\pi^{-1/4}\exp\left(-x^2/2a^2\right).$$

Show that the corresponding momentum function is

$$g(p) = a^{-1/2}\,\pi^{-1/4}\,\hbar^{-1/2}\,\exp\left(-a^2p^2/2\hbar^2\right).$$

$$\left[\text{Use the relation } \int_{-\infty}^{+\infty} \exp\left(-\alpha x^2 - \beta x\right)dx = \sqrt{\frac{\pi}{\alpha}}\exp\left(\beta^2/4\alpha\right)\right]$$

22. Using the relation

$$< p^2 > = \int_{-\infty}^{+\infty} g^*(p)p^2 g(p)dp,$$

Find $< p^2 >$ in the ground state of quantum harmonic oscillator of problem 21.

23. Squaring both sides of the expression of problem 16 of supplementary problems and integrating x from 0 to π show that

$$1+\frac{1}{3^8}+\frac{1}{5^8}+\ldots = \frac{17\pi^8}{161280}.$$

[**Hints:** $\int_0^{\pi} \cos(2n+1)x\cos(2m+1)x\,dx = \frac{\pi}{2}\delta_{mn}$]

8 Vibrations of Strings and Membranes

8.1 TRANSVERSE VIBRATION OF A STRING FIXED AT TWO ENDS

The normal mode frequencies of transverse waves on a string of length l which is under tension T and is fixed at two ends are given by

$$\nu_n = \frac{n}{2l}\sqrt{\frac{T}{\mu}}, \; n = 1, 2, 3, \ldots, \qquad \ldots(8.1)$$

where μ is the mass per unit length of the string. For the fundamental mode, $n = 1$ and higher values of n correspond to higher harmonics (overtones).

8.2 PLUCKED STRING

A string which is fixed at both ends and is under some tension, is plucked at some point to a small height and then released from rest. The string then executes small transverse vibration.

8.3 STRUCK STRING

A perfectly flexible string which is fixed at both ends and is under some tension, is struck by a hammer at a point, the time of contact between the string and the hammer being very very small. The force given by the hammer is of the nature of an impulse and it imparts initially ($t = 0$) a velocity to the point struck but all other points have zero velocity. We investigate the motion of the string at later times.

8.4 BOWED STRING

We study the motion of the violin string when bowed at some point. The string is fixed at both ends and is under some tension.

Characteristics of a bowed string: For maintained vibration of the bowed string, Helmholtz observed the following characteristics:

(*i*) The bowed point moves with the same velocity as that of the bow.

(*ii*) All points of the string vibrate in a plane at any instant. The motion of any point on the string consists of an ascent with uniform forward velocity followed by a descent with another uniform backward velocity. The two velocities are equal in magnitude at the middle point of the string.

The displacement-time graph of a point on the string can be represented as two step zig-zag straight lines (Fig. 8.1). Here, the point under observation moves forward with constant velocity for the time $AB = T_1$ and moves backward with another constant velocity for the time $BC = T_2$. The time period of vibration $\tau = T_1 + T_2 = AC$.

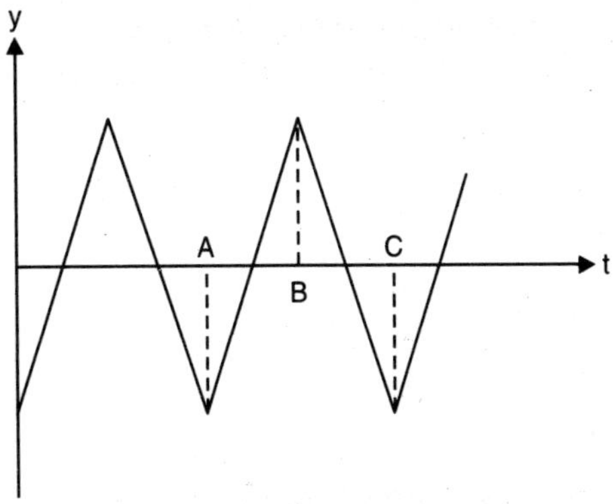

Fig. 8.1

8.5 TRANSVERSE VIBRATION OF MEMBRANES

A perfectly flexible thin membrane is stretched and then fixed along its entire boundary in the *xy*-plane. The tension *T* per unit length caused by the stretching of the membrane is same in all directions. The deflection $u\ (x, y, t)$ for small transverse vibration of the membrane satisfies the two-dimensional wave equation

$$\frac{\partial^2 u}{\partial t^2} = v^2\ \nabla^2 u \qquad ...(8.2)$$

where $v = \sqrt{T/\sigma}, \sigma$ being mass per unit area of the membrane and $\nabla^2 = \frac{\partial^2}{\partial x^2} + \frac{\partial^2}{\partial y^2}$.

SOLVED PROBLEMS

1. *Derive the formula given in Eqn. (8.1).*

Solution

Since the string is fixed at its ends, each end must be stationary and therefore nodes are produced at the two ends. Again, the string must be an integral number of half-wavelengths in length (Fig. 8.2):

$$l = n\frac{\lambda_n}{2},\ n = 1, 2, 3, ...$$

where λ_n is the wavelength of the *n*th normal mode.

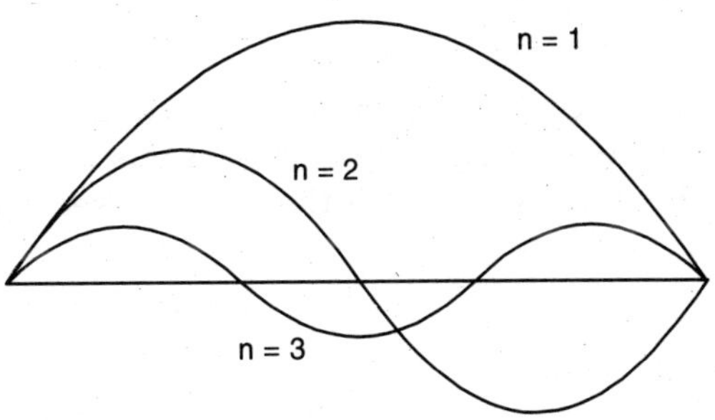

Fig. 8.2

The frequency of the *n*th mode is

$$\nu_n = \frac{v}{\lambda_n} = \frac{n}{2l}\sqrt{\frac{T}{\mu}}$$

where $v = \sqrt{T/\mu}$ is the velocity of propagation of transverse wave along the string.

2. *A string of length $l = 0.5$ m and mass per unit length 0.01 kgm^{-1} has a fundamental frequency of 250 Hz. What is the tension in the string?*

Solution

We have

$$\nu = \frac{1}{2l}\sqrt{\frac{T}{\mu}}$$

or

$$T = 4l^2 \nu^2 \mu = 4 \times (0.5)^2 (250)^2 \times 0.01$$
$$= 625 \text{ N}.$$

3. *Two wires of radii r and 2r respectively are welded together end to end. This combination is used as a sonometer wire kept under tension T. The welded point is midway between the two bridges. What would be the ratio of the number of loops formed in the wires such that the joint is a node when stationary vibrations are set up in the wires.* (I.I.T. 1976)

Solution

Let n_1 and n_2 be the number of loops formed in the wires of radii r and $2r$ respectively. Then

$$\nu_1 = \frac{n_1}{2l}\sqrt{\frac{T}{\mu_1}} \text{ and } \nu_2 = \frac{n_2}{2l}\sqrt{\frac{T}{\mu_2}}.$$

Since the welded wire is continuous, $\nu_1 = \nu_2$ and

$$\frac{n_1}{n_2} = \sqrt{\frac{\mu_1}{\mu_2}} = \left[\frac{\pi r^2 \times 1 \times \rho}{\pi (2r)^2 \times 1 \times \rho}\right]^{\frac{1}{2}} = \frac{1}{2}$$

where ρ = density of the material of the wires.

4. *A metal wire of diameter 1 mm is held on two knife edges separated by a distance of 50 cm. The tension in the wire is 100 N. The wire vibrating with its fundamental frequency and a vibrating tuning fork together produce 5 beats/sec. The tension in the wire is then reduced to 81 N. When the two are excited, beats are heard at the same rate. Calculate (i) the frequency of the fork and (ii) the density of the material of wire.* (I.I.T. 1980)

Solution

Let the frequency of the tuning fork be n. When tension in the wire is 100 N, the fundamental frequency of the wire is $n + 5$, and when $T = 81$ N, the fundamental frequency of the wire is $n - 5$, so that 5 beats/sec are produced in both the cases. Thus,

$$n + 5 = \frac{1}{2 \times 0.5}\sqrt{\frac{100}{\mu}} = \frac{10}{\sqrt{\mu}}$$

$$n - 5 = \frac{1}{2 \times 0.5}\sqrt{\frac{81}{\mu}} = \frac{9}{\sqrt{\mu}}.$$

On solving these two equations, we get

$$n = 95 \text{ Hz}, \mu = 0.01 \text{ kg/m}.$$

Now, ρ = Density of the material of wire

$$= \frac{\mu}{\pi r^2 \times 1} = \frac{0.01}{\pi \times \left(0.5 \times 10^{-3}\right)^2}$$

$$= 12732.4 \text{ kg/m}^3.$$

5. *A string 25 cm long and having a mass of 2.5 g is under tension. A pipe closed at one end is 40 cm long. When the string is vibrating in its first overtone and the air in the pipe in its fundamental frequency, 8 beats per second are heard. It is observed that decreasing the tension in the string decreases the beat frequency. If the speed of sound in air is 320 m/s, find the tension in the string.* (I.I.T. 1982)

Solution

The fundamental frequency of the closed pipe is

$$\nu_1 = \frac{v}{4l} = \frac{320}{4 \times 0.4} = 200 \text{ Hz.}$$

The frequency of the vibrating string is 208 Hz so that with decrease in tension of the string the frequency of the string decreases and the beat frequency also decreases.

The first overtone of the string is

$$208 = \frac{2}{2l}\sqrt{\frac{T}{\mu}}$$

or $$T = 208 \times 208 \times l^2 \mu$$

$$= 208 \times 208 \times l^2 \times (m/l)$$

where $$l = 25 \text{ cm} = 0.25 \text{ m}$$

and $$m = 2.5 \text{ g} = 2.5 \times 10^{-3} \text{ kg.}$$

Thus, $$T = 27.04 \text{ N.}$$

6. *A sonometer wire fixed at one end has a solid mass M hanging from its other end to produce tension in it. It is found that a 70 cm length of the wire produces a certain fundamental frequency when plucked. When the same mass M is hanging in water, completely submerged in it, it is found that the length of the wire has to be changed by 5 cm in order that it will produce the same fundamental frequency. Calculate the density of the material of the mass M hanging from the wire.* (I.I.T 1972)

Solution

The fundamental frequency of the sonometer wire is

$$\nu = \frac{1}{2l}\sqrt{\frac{T}{\mu}} = \frac{1}{2 \times 70}\sqrt{\frac{Mg}{\mu}}.$$

When M is submerged in water completely, the tension in the wire decreases to $[M - M/\rho]\,g$, where ρ is the density of the material of the mass M. The frequency remains the same for a length of 65 cm. Hence,

$$\nu = \frac{1}{2 \times 65}\sqrt{\frac{[M - M/\rho]g}{\mu}}$$

From these two equations, we have

$$\frac{M}{M - M/\rho} = \left(\frac{70}{65}\right)^2$$

or $$\rho = 7.26 \text{ g/cm}^3 = 7.26 \times 10^3 \text{ kg/m}^3.$$

7. *A steel wire of length 1 m, mass 0.1 kg and uniform cross sectional area 10^{-6} m^2 is rigidly fixed at both ends. The temperature of the wire is lowered by 20°C. If transverse waves are set up by plucking the string in the middle, calculate the frequency of the fundamental mode of vibration.*

Young's modulus of steel = 2×10^{11} N/m^2,

Coefficient of linear expansion of steel = 1.21×10^{-5} /°C. (I.I.T. 1984)

Solution

Change in length dl of the wire when the temperature is lowered by 20°C is

$$dl = l \times \alpha \times 20 = 1 \times 1.21 \times 10^{-5} \times 20$$
$$= 2.42 \times 10^{-4} \text{ m}.$$

Now, $$Y = \frac{T/A}{dl/l}$$

or $$T = YA\, dl/l = 2 \times 10^{11} \times 10^{-6} \times 2.42 \times 10^{-4}$$
$$= 48.4 \text{ N}.$$

The fundamental frequency is

$$\nu = \frac{1}{2l}\sqrt{\frac{T}{\mu}} = \frac{1}{2}\left[\frac{48.4}{0.1}\right]^{\frac{1}{2}} = 11 \text{ Hz}.$$

8. *If y be the displacement at x of an elementary segment δx of a uniform string under tension T at any instant, then show that the kinetic energy (K.E.) and potential energy (P.E.) of the element at that instant are given by*

$$K.E. = \frac{1}{2}\mu\left(\frac{\partial y}{\partial t}\right)^2 \delta x$$

$$P.E. = \frac{1}{2}T\left(\frac{\partial y}{\partial x}\right)^2 \delta x$$

where μ is the mass per unit length of the string.

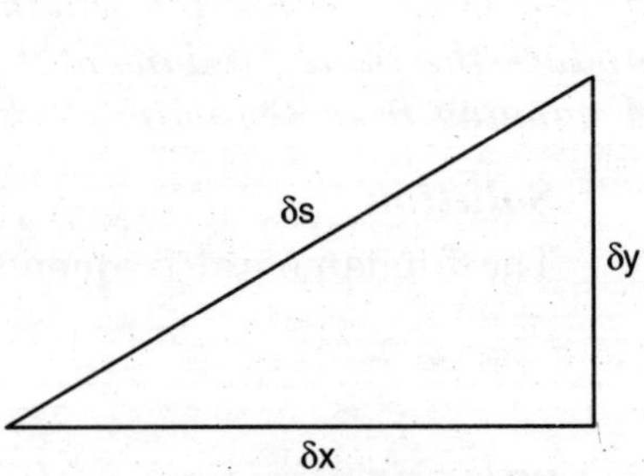

Fig. 8.3

Solution

The mass of the element δx is $\mu\delta x$ and the velocity is $\frac{\partial y}{\partial t}$. Thus, the kinetic energy is $\frac{1}{2}\mu\delta x\left(\frac{\partial y}{\partial t}\right)^2$.

Let δS be the element in the displaced position (Fig. 8.3). The work done against the tension T when the element is stretched is the potential energy of the element.

Since, $$(\delta S)^2 = (\delta x)^2 + (\delta y)^2,$$

$$\delta S = \delta x\left[1 + \left(\frac{\delta y}{\delta x}\right)^2\right]^{1/2}$$

$$\approx \delta x\left[1+\frac{1}{2}\left(\frac{\partial y}{\partial x}\right)^2\right] \text{in the limit } \delta x \to 0.$$

We assume that the y-displacement of the string is small so that $\left(\frac{\partial y}{\partial x}\right)$ is small.

Thus, $$\text{P.E.} = \frac{1}{2}T\left(\frac{\partial y}{\partial x}\right)^2 \delta x.$$

Here, we neglect the increase in tension of the string when it is stretched.

9. *If a wave of the form $y = f(u)$ with $u = x - vt$ moves along the string under tension T with velocity v, show that the instantaneous power passing any position x is given by*

$$P = vT\,[f'(u)]^2.$$

Solution

At any position and time, the kinetic energy density (K.E. per unit length of the string)

$$= \frac{1}{2}\mu\left(\frac{\partial y}{\partial t}\right)^2 = \frac{1}{2}\mu\left(\frac{\partial y}{\partial u}\frac{\partial u}{\partial t}\right)^2 = \frac{1}{2}\mu v^2[f'(u)]^2$$

$$= \frac{1}{2}T[f'(u)]^2.$$

The potential energy density $= \frac{1}{2}T\left(\frac{\partial y}{\partial x}\right)^2 = \frac{1}{2}T[f'(u)]^2$

Total energy density $= E_1 = T[f'(u)]^2$.

Since the wave moves with velocity v, the instantaneous power passing any position x is

$$P = vE_1 = vT\,[f'(u)]^2.$$

10. *Using the following general expression for the power that passes any position x along the string*

$$P = FV,$$

where $F = y$ *– component of the force* $= -T\dfrac{\partial y}{\partial x}$,

V = *the transverse velocity* $= \dfrac{\partial y}{\partial t}$,

and the general form of the travelling wave

$$y = f(x - vt) = f(u) \text{ with } u = x - vt,$$

show that

(*i*) $\dfrac{F}{V} = \sqrt{\mu T}$ (*ii*) $P = vT\,[f'(u)]^2$.

Solution

We have, $$\frac{\partial y}{\partial x} = f'(u) \text{ and } \frac{\partial y}{\partial t} = -vf'(u).$$

Thus $$\frac{F}{V} = \frac{T}{v} = \sqrt{\mu T}$$

and $$FV = P = vT\,[f'(u)]^2.$$

[*Note* : The ratio $F/V = \sqrt{\mu T}$ is called the *wave impedance* or, the *characteristic impedance* (z) of a transverse wave on the string. This expression is analogous with the electrical impedance = voltage/current, while electric power = voltage × current. Voltage and current are the electrical analogous of the mechanical force and displacement velocity.]

11. *Show that the mean power required to maintain a travelling wave of amplitude A and angular frequency ω on a long string is*

$$P = \frac{1}{2}\mu v A^2\omega^2 = \frac{1}{2}zA^2\omega^2,$$

where μ is the mass per unit length of the string, v is the speed of transverse waves on the string and z (= $\sqrt{\mu T}$ *) is the characteristic impedance of the string for transverse waves.*

Solution

For the displacement of the string let us take

$$y = A \sin (kx - \omega t) = A \sin ku = f(u).$$

where $u = x - vt$.

The instantaneous power passing any position is

$$P = vT[f'(u)^2] = vT A^2 k^2 \cos^2 ku.$$

Since the average value of $\cos^2 ku$ is $\frac{1}{2}$, the average power is

$$\bar{P} = \frac{1}{2}vTA^2k^2 = \frac{1}{2}\mu v A^2\omega^2 = \frac{1}{2}zA^2\omega^2.$$

12. *A long string of mass per unit length 0.1 kgm*$^{-1}$ *is stretched to a tension of 250 N. Find the speed of transverse waves on the string and the mean power required to maintain a travelling wave of amplitude 5mm and wavelength 0.5 m.*

Solution

$$v = \sqrt{T/\mu} = 50\ \text{ms}^{-1}$$

$$\text{Mean power} = \frac{1}{2}\mu v A^2\omega^2 = \frac{1}{2}\times 0.1\times 50\times\left(5\times 10^{-3}\right)^2\times\left(\frac{2\pi\times 50}{0.5}\right)^2 = 24.67\ \text{W}.$$

13. *Consider the motion of the transverse waves on a long string consisting of two parts. The left part has a linear mass density* μ_1 *and the right part a different linear mass density* μ_2 *with both parts under the same tension T Fig. 8.4*

For convenience, we place the x-origin at the discontinuity. Suppose that a source of sinusoidal waves on the negative x-axis is sending waves toward the discontinuity and that the waves continue past it are absorbed with no reflection by a distant sink. Find the power reflection and transmission coefficients at the point of discontinuity.

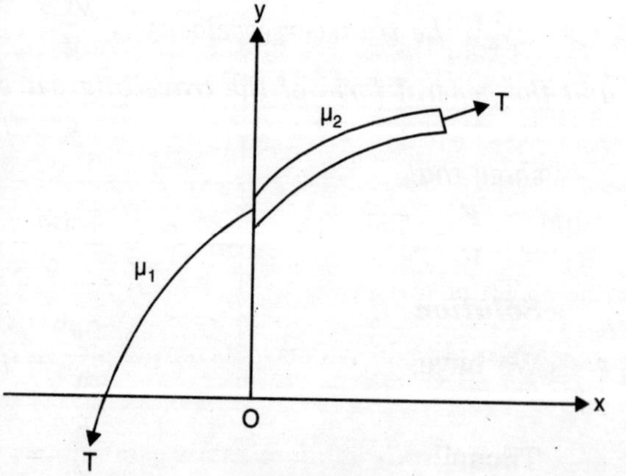

Fig. 8.4

Solution

There are two independent boundary conditions at the point of discontinuity where the two strings are joined;

(*i*) Continuity of the displacement of the string:

$$y_{\text{left}} = y_{\text{right}} \text{ at } x = 0 \text{ for all times.}$$

(otherwise they would not be joined together)

(*ii*) Continuity of the transverse force in the string: If the force is not continuous at the boundary, an infinitesimal mass therefore would be subject to a finite force, resulting in an infinite acceleration, which is not possible. Thus, we have

$$-T\frac{\partial y}{\partial x}\bigg|_{\text{left}} = -T\frac{\partial y}{\partial x}\bigg|_{\text{right}} \quad \text{at } x = 0 \text{ for all times,}$$

or,

$$\frac{\partial y}{\partial x}\bigg|_{\text{left}} = \frac{\partial y}{\partial x}\bigg|_{\text{right}} \quad \text{at } x = 0 \text{ for all times,}$$

Let us take the *incident wave* coming from the left to be the real part of

$$y_1 = A_1 \exp\,[i(k_1x - \omega t)],\ -\infty < x < 0.$$

which has the amplitude A_1 and the velocity $v_1 = \omega/k_1 = \sqrt{T/\mu_1}$.

The wave transmitted past the discontinuity is assumed to be the real part of

$$y_2 = A_2 \exp\,[i(k_2x - \omega t)],\ 0 < x < \infty,$$

which has the amplitude A_2 and the velocity $v_2 = \omega/k_2 = \sqrt{T/\mu_2}$.

Both the waves must necessarily have the same frequency.

There must exist a third wave that is reflected from the boundary. We assume that the reflected wave travelling to the left is the real part of

$$y_1' = B_1 \exp\,[i(-k_1x - \omega t)],\ -\infty < x < 0,$$

which is moving towards the negative direction. The reflected wave has the amplitude B_1 and the wave number k_1 which is appropriate to the string on the left side of the boundary.

The boundary conditions (*i*) and (*ii*) now give

$$A_1 + B_1 = A_2$$

$$-T(ik_1A_1) + T(ik_1B_1) = -T(ik_2A_2)$$

From these two equations, we get

$$A_2 = \frac{2k_1}{k_1 + k_2}A_1$$

and

$$B_1 = \frac{k_1 - k_2}{k_1 + k_2}A_1$$

Thus, we have

$$\text{Amplitude reflection coefficient} = R_a = \frac{B_1}{A_1} = \frac{k_1 - k_2}{k_1 + k_2},$$

$$\text{Amplitude transmission coefficient} = T_a = \frac{A_2}{A_1} = \frac{2k_1}{k_1 + k_2}.$$

The characteristic impedances of the two parts of the string are:

$$z_1 = \mu_1 v_1 = \sqrt{T\mu_1}$$

$$z_2 = \mu_2 v_2 = \sqrt{T\mu_2}$$

and the wave numbers k_1 and k_2 are given by

$$k_1 = \frac{\omega}{v_1} = \frac{\omega z_1}{T}$$

$$k_2 = \frac{\omega}{v_2} = \frac{\omega z_2}{T}$$

Thus, we get

$$R_a = \frac{z_1 - z_2}{z_1 + z_2} = \frac{\sqrt{\mu_1} - \sqrt{\mu_2}}{\sqrt{\mu_1} + \sqrt{\mu_2}}$$

$$T_a = \frac{2z_1}{z_1 + z_2} = \frac{2\sqrt{\mu_1}}{\sqrt{\mu_1} + \sqrt{\mu_2}}.$$

We find that both R_a and T_a are real. If $\mu_1 > \mu_2$, R_a is positive which implies that the reflected wave has the same phase as the incident wave. If $\mu_1 < \mu_2$, R_a is negative showing that the reflected and incident waves are 180° out of phase. T_a is always positive showing that the transmitted wave has the same phase as the incident wave.

Next, we define the power reflection coefficient R_p and power transmission coefficient T_p. The power carried by a travelling wave is $\frac{1}{2}\mu v\omega^2 A^2$. Thus, we may define

$$\text{Power reflection coefficient} = R_p = \frac{\frac{1}{2}\mu_1 v_1 \omega^2 B_1^2}{\frac{1}{2}\mu_1 v_1 \omega^2 A_1^2}$$

$$= \frac{B_1^2}{A_1^2} = \left(\frac{z_1 - z_2}{z_1 + z_2}\right)^2 \qquad \text{...(8.3)}$$

$$\text{Power transmission coefficient} = T_p = \frac{\frac{1}{2}\mu_2 v_2 \omega^2 A_2^2}{\frac{1}{2}\mu_1 v_1 \omega^2 A_1^2}$$

$$= \frac{\mu_2 v_2 A_2^2}{\mu_1 v_1 A_1^2} = \frac{4z_1 z_2}{(z_1 + z_2)^2}. \qquad \text{...(8.4)}$$

We see that $R_p + T_p = 1$, showing that the incident power equals the reflected power plus the transmitted power. Since R_p and T_p depend only on the properties of the string and not on the frequency and amplitude of the waves, the expressions (8.3) and (8.4) must hold for waves of arbitrary shape.

14. *A perfectly elastic string of length l which is under tension T and is fixed at both ends, has the linear mass density (i.e., mass per unit length)* μ. *The string is given initial deflection and initial velocity at its various points and is released at time t = 0. The string executes small transverse vibrations. The initial deflection and the initial velocity of the string at any point x are denoted by h (x) and V (x) respectively. Find the different normal modes of vibrations and the deflection of the string at any point x and at any time t > 0.*

Solution

We have to solve the wave equation

$$\frac{\partial^2 y}{\partial t^2} = v^2 \frac{\partial^2 y}{\partial x^2}, \quad \text{...(8.5)}$$

with
$$v = \sqrt{T/\mu}$$

under the following boundary conditions:

(*i*) $y(0, t) = 0$ ...(8.6)

(*ii*) $y(l, t) = 0$ ...(8.7)

(*iii*) $y(x, 0) = h(x)$ ...(8.8)

(*iv*) $\left.\frac{\partial y}{\partial x}\right|_{t=0} = V(x)$ ...(8.9)

(*v*) $|y(x, t)| < M$ ...(8.10)

where M is a fixed number *i.e.* the motion is bounded.

Solution of Eqn. (8.5) by the method of separation of variables: We assume that y (x, t) can be written as

$$y(x, t) = F(x)\, G(t)$$

where F (x) is a function of x only and G (t) is a function of t only. Substituting in Eqn. (8.5), we get

$$F\frac{d^2G}{dt^2} = v^2 G\frac{d^2F}{dx^2}$$

or
$$\frac{1}{v^2G}\frac{d^2G}{dt^2} = \frac{1}{F}\frac{d^2F}{dx^2} \quad \text{...(8.11)}$$

L.H.S. of Eqn. (8.11) is a function of t only and the R.H.S of Eqn. (8.11) is a function of x only. Since Eqn. (8.11) is true for all values of x and t, the two sides of Eqn. (8.11) must be equal to a constant, independent of x and t. This constant should also be negative on physical grounds, $|y(x, t)| < M$ (*i.e.*, the displacement is bounded). If it were a positive constant q, then G (t) would be G (t) ~ exp $(\pm \sqrt{v^2 q}\, t)$. The positive sign in the exponential is not allowed since it would mean growing displacement and the negative sign is not acceptable since there is no damping force in the system. Thus, we have

$$\frac{1}{v^2G}\frac{d^2G}{dt^2} = \frac{1}{F}\frac{d^2F}{dx^2} = -p^2$$

where $-p^2$ is the separation constant.

The solution for $F(x)$ is

$$F(x) = A \cos px + B \sin px.$$

The boundary conditions of Eqns. (8.6) and (8.7) give

$$A = 0 \text{ and } B \sin pl = 0,$$

or $$pl = n\pi,\ n = 1, 2, 3,...$$

or $$p = n\pi/l$$

For any particular n, we have

$$F_n(x) = B_n \sin(n\pi x/l).$$

For different values of n we obtain different solutions. In fact there are infinitely many solutions. For a particular n, the differential equation for $G(t)$ is

$$\frac{d^2G_n}{dt^2} + \omega_n^2 G_n = 0$$

where $\omega_n = N\pi v/l$.

The general solution of this equation is

$$G_n(t) = D'_n \cos \omega_n t + E'_n \sin \omega_n t.$$

Thus, the displacement for the nth mode is

$$y_n(x, t) = F_n(x)G_n(t) = B_n \sin \frac{n\pi x}{l}(D'_n \cos \omega_n t + \text{E}'_n \sin \omega_n t.)$$

or $$y_n(x, t) = \sin \frac{n\pi x}{l}(D_n \cos \omega_n t + E_n \sin \omega_n t) \qquad ...(8.12)$$

where $$D_n = B_n D'_n \text{ and } E_n = B_n E'_n.$$

We have obtained the solutions $y_n(x, t)$ of the partial differential Eqn. (8.5) satisfying the boundary conditions (8.6), (8.7) and (8.10). The functions $y_n(x, t)$ are called the eigenfunctions or characteristic functions and the values $\omega_n = n\pi v/l$ are called the eigen frequencies or characteristic frequencies of the vibrating string. Each y_n represents a harmonic motion having the angular frequency $\omega_n = 2\pi\nu_n$, where

$$\nu_n = \frac{nv}{2l} = \frac{n}{2l}\sqrt{\frac{T}{\mu}}.$$

The motion is called the nth normal mode of the string. The first normal mode ($n = 1$) is known as the fundamental mode, and higher modes ($n = 2, 3, 4, ...$) as overtones.

Form Eqn. (8.12), we have $y_n(x, t) = 0$ for all time when

$$\sin \frac{n\pi x}{l} = 0 \text{ or, } \frac{n\pi x}{l} = k\pi,$$

or $$x = k\frac{l}{n},\ k = 0, 1, 2,....., n$$

These are the points of the string which do not move (nodes). For $k = 0, n$ we have $x = 0, l$ which are the two fixed end points of the string.

When $n = 1$, the nodes are at $x = 0, l$ (Fig. 8.5). The fundamental frequency is

$$\nu_1 = \frac{1}{2l}\sqrt{T/\mu}$$

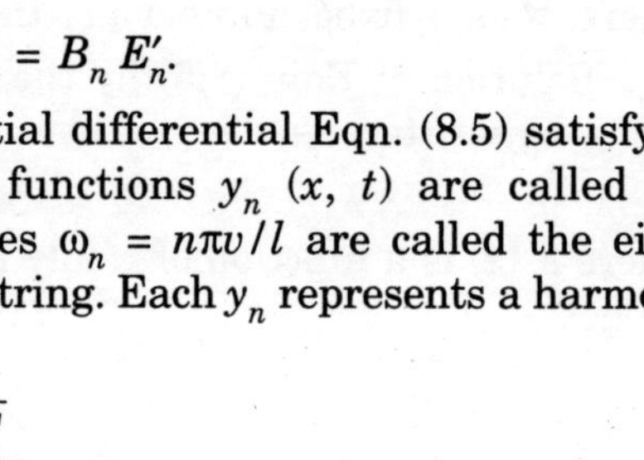

Fig. 8.5

and the fundamental wavelength is

$$\lambda_1 = \frac{v}{\nu_1} = 2l.$$

When $n = 2$, the nodes are at $x = 0, l/2, l$ (Fig. 8.6). The corresponding frequency and wavelength are

$$\nu_2 = 2\nu_1 \text{ and } \lambda_2 = l = \lambda_1/2.$$

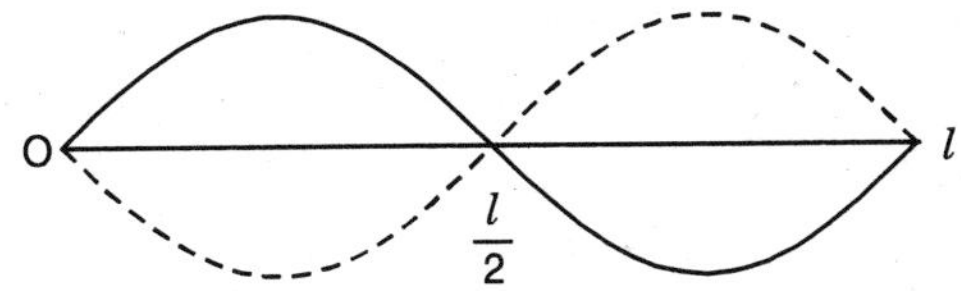

Fig. 8.6

When $n = 3$, the nodes are at $x = 0, \frac{l}{3}, \frac{2l}{3}, l$ (Fig. 8.7). The corresponding frequency and wavelength are

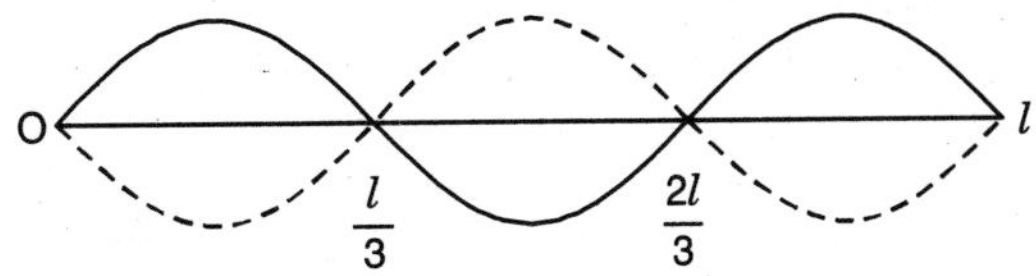

Fig. 8.7

$$\nu_3 = 3\nu_1 \text{ and } \lambda_3 = \frac{1}{3}\lambda_1.$$

Equation (8.12) gives the nth normal mode solution of Eqn. (8.5) satisfying the boundary conditions (8.6), (8.7) and (8.10). The sum of infinitely many solutions $y_n(x, t)$ is also a solution. Therefore, the general solution is

$$y(x, t) = \sum_{n=1}^{\infty} (D_n \cos \omega_n t + E_n \sin \omega_n t) \sin \frac{n\pi x}{l}. \quad \text{...(8.13)}$$

The boundary condition (8.8) gives

$$y(x, 0) = \sum_{n=1}^{\infty} D_n \sin \frac{n\pi x}{l} = h(x)$$

which is the Fourier sinc series of $h(x)$. Thus, we have

$$D_n = \frac{2}{l}\int_0^l h(x) \sin \frac{n\pi x}{l} dx, \quad n = 1, 2, 3, \ldots \quad \text{...(8.14)}$$

By applying the boundary condition (8.9), we get

$$\left.\frac{\partial y}{\partial t}\right|_{t=0} = \sum_{n=1}^{\infty} E_n \omega_n \sin \frac{n\pi x}{l} = V(x)$$

or

$$E_n \omega_n = \frac{2}{l}\int_0^l V(x) \sin \frac{n\pi x}{l} dx$$

or

$$E_n = \frac{2}{n\pi v}\int_0^l V(x) \sin \frac{n\pi x}{l} dx, \quad n = 1, 2, 3, \ldots \quad \text{...(8.15)}$$

The deflection of the string at any point x and at any time t is given by Eqn. (8.13) where the coefficients D_n and E_n are obtained from Eqns. (8.14) and (8.15).

15. *(a) A string of length $l = \pi$ which is under tension T and is fixed at both ends has mass per unit length μ. The initial deflection at any point x is given by*

$$h(x) = 0.01\, x(\pi - x).$$

The initial velocity is zero at any point x. Find the deflection y(x, t) of the string at point x and at any time t > 0.

(b) What is the ratio of the amplitudes of the fundamental mode and the next non-zero overtone (i.e., D_1/D_3)?

(c) Find the ratio

$$D_1^2/\left(D_1^2+D_3^2+D_5^2+....\right).$$

Solution

(*a*) The deflection $y(x, t)$ of the string is given by

$$y(x,\ t) = \sum_{n=1}^{\infty} D_n \cos\omega_n t \sin\frac{n\pi x}{l}$$

where
$$D_n = \frac{2}{l}\int_0^l h(x)\sin nx\, dx$$

$$D_n = \frac{2}{\pi}\int_0^\pi 0.01\, x\,(\pi - x)\sin nx\, dx$$

$$= -\frac{0.04}{\pi n^3}[(-1)^n - 1]$$

Thus,
$$D_2 = D_4 = D_6 = ... = 0.$$

(*b*) $D_1/D_3 = 27$

(*c*) The deflection at time $t = 0$ is

$$0.01x\ (\pi - x) = \sum_{k=0}^{\infty}\frac{0.08}{\pi(2k+1)^3}\sin(2k+1)x.$$

Squaring this expression and integrating from 0 to π, we get

$$\int_0^\pi [0.01x(\pi - x)]^2 dx = \sum_k\sum_l \frac{(0.08)^2}{\pi^2(2k+1)^3(2l+1)^3}$$

$$\times \int_0^\pi \sin(2k+1)x\ \sin(2l+1)x\, dx$$

or
$$1 \times 10^{-4}\frac{\pi^5}{30} = \sum_k \frac{64\times 10^{-4}}{2\pi(2k+1)^6}$$

Thus,
$$\sum_{k=0}^{\infty}\frac{1}{(2k+1)^6} = \frac{\pi^6}{960}$$

Now,
$$\frac{D_1^2}{D_1^2+D_3^2+D_5^2+...} = \left(\sum_{k=0}^{\infty}\frac{1}{(2k+1)^6}\right)^{-1}$$

$$= \frac{960}{\pi^6}.$$

16. *A perfectly elastic string of length l which is under tension T and is fixed at both ends, has mass per unit length μ. It is plucked at the point $x = a$ to a height h (Fig. 8.8) and then released from rest. The string executes small transverse vibration. Find the different normal modes of vibrations and the deflection of the string at any point x at any later time.*

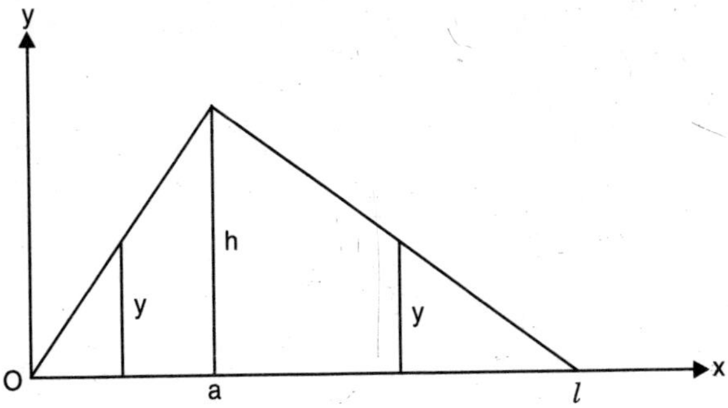

Fig. 8.8

Solution

Initially we have triangular deflection:

For $0 \le x \le a$, $\quad \dfrac{y}{h} = \dfrac{x}{a}$ or, $y = \dfrac{hx}{a}$,

For $a \le x \le l$, $\dfrac{y}{h} = \dfrac{l-x}{l-a}$, or $y = \dfrac{h(l-x)}{l-a}$.

Thus, at time $t = 0$, the deflection of the string is given by

$$h(x) = \begin{cases} \dfrac{xh}{a} & \text{for} \quad 0 \le x \le a \\ \dfrac{h(l-x)}{l-a} & \text{for} \quad a \le x \le l \end{cases}$$

We have also $\left.\dfrac{\partial y}{\partial t}\right|_{t=0} = 0$

Thus, from Eqn. (8.13), we obtain

$$y(x, t) = \sum_{n=1}^{\infty} D_n \cos\frac{n\pi vt}{l}\sin\frac{n\pi x}{l}$$

where

$$D_n = \frac{2}{l}\int_0^l h(x)\sin\frac{n\pi x}{l}dx$$

$$= \frac{2}{l}\left[\int_0^a \frac{xh}{a}\sin\frac{n\pi x}{l}dx + \int_a^l \frac{h(l-x)}{l-a}\sin\frac{n\pi x}{l}dx\right]$$

$$= \frac{2hl^2}{a(l-a)\pi^2 n^2}\sin\frac{n\pi a}{l}$$

For nth harmonic we have the vibration mode

$$y_n(x, t) = \frac{2hl^2}{\pi^2 a(l-a)}\left[\frac{1}{n^2}\sin\frac{n\pi a}{l}\sin\frac{n\pi x}{l}\cos\frac{n\pi vt}{l}\right]$$

At the antinode of this particular mode, $\sin\dfrac{n\pi x}{l} = 1$ and the maximum amplitude of vibration for the nth mode is

$$A_n = \frac{2hl^2}{\pi^2 a(l-a)}\frac{\sin\dfrac{n\pi a}{l}}{n^2}$$

The amplitude of higher harmonics decreases very fast due to appearance of n^2 in the denominator.

When $a = \frac{l}{2}$ (plucked at the mid-point of the string)

$$y_n(x, t) = \frac{8h}{\pi^2}\left[\frac{1}{n^2}\sin\frac{n\pi}{l}\sin\frac{n\pi x}{l}\cos\frac{n\pi vt}{l}\right]$$

and

$$y(x, t) = \frac{8h}{\pi^2}\left[\frac{1}{1^2}\sin\frac{\pi x}{l}\cos\frac{\pi vt}{l} - \frac{1}{3^2}\sin\frac{3\pi x}{l}\cos\frac{3\pi vt}{l} + \frac{1}{5^2}\sin\frac{5\pi x}{l}\cos\frac{5\pi vt}{l} - ...\right]$$

We see that the 2nd, 4th, 6th and all the even harmonics are absent.

If $a = \frac{l}{3}$, we see that 3rd, 6th, 9th, ..., harmonics will be absent. In general, if $a = \frac{l}{p}$, where p is an integer, pth, $(2p)$th, $(3p)$th, ..., harmonics will be absent in the vibrations.

17. *A string of length l which is fixed at both ends is under tension T. It is plucked at the point $x = a$ to a height h (Fig. 8.8) and then released from rest. The string executes small transverse vibrations.*

(i) Show that the initial potential energy of the string is $\frac{\mu v^2 h^2 l}{2a(l-a)}$.

(ii) Find the total energy for the nth harmonic of the string.

(iii) Show that the total energy is the sum of the energies of the harmonics.

(iv) Show that the total energy at any instant is equal to the initial potential energy of the string.

Solution

(*i*) At time $t = 0$, the deflection of the string is given by, (see problem 16),

$$y(x) = \begin{cases} \frac{xh}{a} & \text{for } 0 \le x \le a, \\ \frac{h(l-x)}{l-a} & \text{for } a \le x \le l \end{cases}$$

From problem 8, we have for the total potential energy of the string,

$$\text{Total P.E.} = \int_0^l \frac{1}{2}T\left(\frac{\partial y}{\partial x}\right)^2 dx$$

$$= \frac{1}{2}\mu v^2\left[\int_0^a \left(\frac{h}{a}\right)^2 dx + \int_a^l \left(\frac{h}{l-a}\right)^2 dx\right]$$

$$= \frac{\mu v^2 h^2 l}{2a(l-a)}.$$

(*ii*) For the *n*th harmonic the total energy is given by

$$E_n = \frac{\mu}{2}\int_0^l\left[\left(\frac{\partial y_n}{\partial t}\right)^2 + v^2\left(\frac{\partial y_n}{\partial x}\right)^2\right]dx \quad \text{(see problems 8 and 16)}$$

where
$$y_n = A_n \sin\frac{n\pi x}{l}\cos\frac{n\pi v t}{l}$$

with
$$A_n = \frac{2hl^2}{\pi^2 a(l-a)}\frac{1}{n^2}\sin\frac{n\pi a}{l}.$$

After performing the integrations, we get

$$E_n = \frac{\mu v^2 h^2 l^3}{\pi^2 a^2 (l-a)^2}\frac{1}{n^2}\sin^2\frac{n\pi a}{l}.$$

Thus, the energy of higher harmonics decreases very fast with increase in n.

(*iii*) The total energy of the string is given by

$$E = \frac{\mu}{2}\int_0^l\left(\frac{\partial y}{\partial t}\right)^2 dx + \frac{\mu}{2}v^2\int_0^l\left(\frac{\partial y}{\partial x}\right)^2 dx$$

where $y = \sum y_n$ is the deflection of the string.

Now,
$$\left(\frac{\partial y}{\partial t}\right)^2 = \sum_n\sum_m A_n A_m \frac{n\pi v}{l}\frac{m\pi v}{l}\sin\frac{n\pi x}{l}\sin\frac{m\pi x}{l} \times \sin\frac{n\pi v t}{l}\sin\frac{m\pi v t}{l}$$

Since
$$\int_0^l \sin\frac{n\pi x}{l}\sin\frac{m\pi x}{l}dx = \frac{l}{2}\delta_{mn},$$

$$\frac{\mu}{2}\int_0^l\left(\frac{\partial y}{\partial t}\right)^2 dx = \frac{\mu}{4}\frac{A_n^2 n^2\pi^2 v^2}{l}\sin^2\frac{n\pi v t}{l}$$

Similarly,
$$\frac{\mu}{2}v^2\int_0^l\left(\frac{\partial y}{\partial x}\right)^2 dx = \sum_n\frac{\mu}{4}\frac{A_n^2 n^2\pi^2 v^2}{l}\cos^2\frac{n\pi v t}{l}.$$

Thus,
$$E = \sum_n\frac{\mu}{4}\frac{A_n^2\pi^2 v^2 n^2}{l}$$

$$= \sum_n E_n$$

(*iv*)
$$E = \frac{\mu v^2 h^2 l^3}{\pi^2 a^2(l-a)^2}\sum_n\frac{1}{n^2}\sin^2\frac{n\pi a}{l}$$

$$= \frac{\mu v^2 h^2 l^3}{\pi^2 a^2(l-a)^2}\frac{\pi^2 a(l-a)}{2l^2}$$

[See Supplementary problem 11]

Thus, $$E = \frac{\mu v^2 h^2 l}{2a(l-a)}.$$

18. *A perfectly flexible string of length l and linear mass density μ, which is fixed at both ends and is under tension T, is struck by a pointed hammer at the point x = a, the time of contact between the string and the hammer being very very small. Write the wave equation and the proper boundary conditions of this problem of struck string. Find the deflection of the string at any point x at a later time.*

Solution

Since the string is perfectly flexible, the point, say $x = a$, where the hammer strikes is not at rest, though other points are at rest initially *i.e.*, $\left(\frac{\partial y}{\partial t}\right)_{(t=0)} = \dot{y}_0 \neq 0$ at $x = a$ and $\dot{y}_0 = 0$ when $x \neq a$. In this case, there is initial motion, but no initial displacement *i.e.*, $y\,(x,\,0) = 0$. Thus, we have to solve the wave equation.

$$\frac{\partial^2 y}{\partial x^2} = \frac{1}{v^2}\frac{\partial^2 y}{\partial t^2}$$

under the following boundary conditions:

(*i*) $y\,(0,\,t) = 0$

(*ii*) $y\,(l,\,t) = 0$

(*iii*) $|y\,(x,\,t)| < M$

(*iv*) $y(x,\,0) = 0$

(*v*) $\dot{y}_0 = \dot{y}_0(x)\,\delta\,(x-a)$

where M is a fixed number *i.e.*, the motion is bounded, $v = \sqrt{T/\mu}$ is the velocity of propagation of transverse wave on the string and $\delta\,(x-a)$ is the Dirac delta function having the definition

$$\delta(x-a) = 0 \text{ when } x \neq a$$
$$\neq 0 \text{ when } x = a.$$

The general equation for deflection of the vibrating string satisfying the boundary conditions (*i*), (*ii*) and (*iii*) is [see problem 14].

$$y(x,\,t) = \sum_{n=1}^{\infty}\left(A_n \cos\frac{n\pi vt}{l} + B_n \sin\frac{n\pi vt}{l}\right)\sin\frac{n\pi x}{l}$$

From the condition (*iv*) we have

$$y(x,\,0) = \sum_{n=1}^{\infty} A_n \sin\frac{n\pi x}{l} = 0$$

or, $A_n = 0$, since the above equation is true for all values of x. Thus,

$$y(x,\,t) = \sum_{n=1}^{\infty} B_n \sin\frac{n\pi vt}{l}\sin\frac{n\pi x}{l}$$

and $$\left(\frac{\partial y}{\partial t}\right)_{t=0} = \dot{y}_0 = \sum_{n=1}^{\infty} B_n \frac{n\pi v}{l}\sin\frac{n\pi x}{l}$$

which is nothing but Fourier sine series with coefficients

$$\frac{n\pi v}{l}B_n = \frac{2}{l}\int_0^l \dot{y}_0 \sin\frac{n\pi x}{l}dx$$

or

$$B_n = \frac{2}{n\pi v}\int_0^l \dot{y}_0(x)\delta(x-a)\sin\frac{n\pi x}{l}dx$$

$$= \frac{2}{n\pi v}\dot{y}_0(a)\sin\frac{n\pi a}{l}$$

Hence,

$$y(x, t) = \frac{2\dot{y}_0(a)}{\pi v}\sum_{n=1}^{\infty}\frac{1}{n}\sin\frac{n\pi a}{l}\sin\frac{n\pi x}{l}\sin\frac{n\pi vt}{l}.$$

The amplitude of the nth mode of vibration is given by

$$\frac{2\dot{y}_0(a)}{\pi v}\frac{1}{n}\sin\frac{n\pi a}{l}\sin\frac{n\pi x}{l}$$

which decreases as $\frac{1}{n}$. When $a = \frac{l}{2}$, 2nd, 4th, 6th, . . . harmonics are absent. When $a = \frac{l}{3}$, 3rd, 6th, 9th, . . . harmonics are absent.

Note: In practice, the results for the vibration of struck string are to be modified because of finite time of contact between the hammer and the string and also because the area under the hammer is finite.

19. *A violin string of length l and linear mass density μ is fixed at both ends and is under tension T. The string is bowed at some point. It is observed that a point x on the string has a constant forward velocity v_1 from $t = 0$ to $t = T_1$ and a constant backward velocity v_2 (in magnitude) from $t = T_1$ to $t = \tau$ where τ is the period of vibration. Show that the deflection of the string is given by*

$$y(x, t) = \frac{\tau(v_1+v_2)}{\pi^2}\sum_{n=1}^{\infty}\frac{1}{n^2}\sin\frac{n\pi vT_1}{2l}\sin\frac{n\pi v}{l}\left(t-\frac{T_1}{2}\right)$$

where $v = \sqrt{T/\mu}$ = velocity of transverse wave along the string.

Solution

We have to solve the wave equation

$$\frac{\partial^2 y}{\partial x^2} = \frac{1}{v^2}\frac{\partial^2 y}{\partial t^2}$$

under the following boundary conditions:

(*i*) $y(0, t) = 0$

(*ii*) $y(l, t) = 0$

(*iii*) $|y(x, t)| < M$

(*iv*) $$\frac{\partial y}{\partial t} = \begin{cases} v_1 & \text{for } 0<t<T_1 \\ -v_2 & \text{for } T_1<t<\tau \end{cases}$$

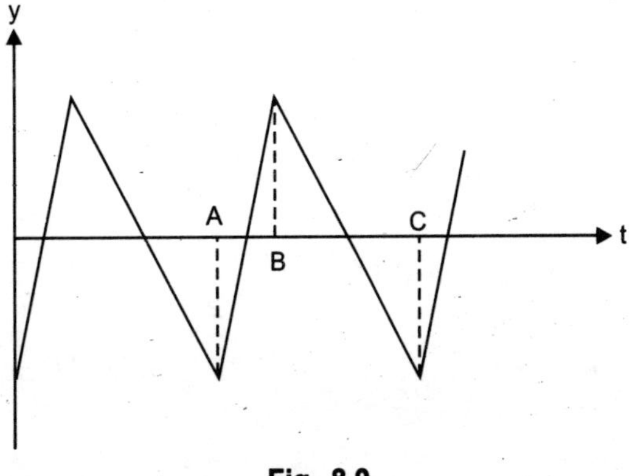

Fig. 8.9

We know that the general solution for finite displacement of a string fixed at $x = 0$ and $x = l$ is

$$y(x, t) = \sum_{n=1}^{\infty}\left(A_n \cos\frac{n\pi vt}{l} + B_n \sin\frac{n\pi vt}{l}\right)\sin\frac{n\pi x}{l},$$

where the boundary conditions (*i*), (*ii*) and (*iii*) are satisfied.

The above equation can be rewritten as

$$y(x, t) = \sum_{n=1}^{\infty}(A_n \cos n\omega t + B_n \sin n\omega t)\sin\frac{n\pi x}{l} \qquad ...(8.16)$$

where $\omega = \pi\dfrac{v}{l}$ and $\tau = \dfrac{2\pi}{\omega} = \dfrac{2l}{v}$.

Now,
$$\frac{\partial y}{\partial t} = \sum_{n=1}^{\infty}(-n\omega A_n \sin n\omega t + n\omega B_n \cos n\omega t)\sin\frac{n\pi x}{l}$$

According to the Fourier series, we have

$$-n\omega\, A_n \sin\frac{n\pi x}{l} = \frac{2}{\tau}\int_0^{\tau}\left(\frac{\partial y}{\partial t}\right)\sin n\omega t\, dt,$$

$$n\omega\, B_n \sin\frac{n\pi x}{l} = \frac{2}{\tau}\int_0^{\tau}\left(\frac{\partial y}{\partial t}\right)\cos n\omega t\, dt$$

which gives

$$-n\omega A_n \sin\frac{n\pi x}{l} = \frac{2}{\tau}\left[\int_0^{T_1} v_1 \sin n\omega t\, dt + \int_{T_1}^{\tau}(-v_2)\sin n\,\omega t\, dt\right]$$

$$= \frac{2}{n\omega\tau}[-v_1 \cos n\omega T_1 + v_1 + v_2 \cos n\omega\tau - v_2 \cos\, n\omega T_1]$$

$$= \frac{(v_1 + v_2)}{\pi n}[1 - \cos n\omega T_1],$$

$$n\omega B_n \sin\frac{n\pi x}{l} = \frac{2}{\tau}\left[\int_0^{T_1} v_1 \cos n\omega t\, dt + \int_{T_1}^{\tau}(-v_2)\cos n\omega t\, dt\right]$$

$$= \frac{2}{n\omega\tau}[v_1 \sin n\omega T_1 + v_2 \sin n\omega T_1]$$

$$= \frac{v_1 + v_2}{\pi n}\sin n\omega T_1.$$

Thus, we get

$$y(x, t) = \sum_{n=1}^{\infty}\frac{(v_1 + v_2)}{\pi\omega n^2}[(\cos n\omega T_1 - 1)\cos n\omega t$$
$$+ \sin n\omega T_1 \sin n\omega t]$$

$$= \sum_{n=1}^{\infty} \frac{\tau(v_1 + v_2)}{2\pi^2 n^2} \left[-2\sin^2 \frac{n\omega T_1}{2} \cos n\omega t \right.$$

$$\left. + 2\sin \frac{n\omega T_1}{2} \cos \frac{n\omega T_1}{2} \sin n\omega t \right]$$

$$= \sum_{n=1}^{\infty} \frac{\tau(v_1 + v_2)}{2\pi^2 n^2} . 2 \sin \frac{n\omega T_1}{2} \sin\left(n\omega t - \frac{n\omega T_1}{2} \right)$$

$$y(x, t) = \frac{\tau(v_1 + v_2)}{\pi^2} \sum_{n=1}^{\infty} \frac{1}{n^2} \sin \frac{n\pi v T_1}{2l} \sin n\pi \frac{v}{l}\left(t - \frac{T_1}{2} \right). \quad ...(8.17)$$

Hence, the amplitude of the nth harmonic decreases as $\frac{1}{n^2}$.

20. *Derive the partial differential equation for small transverse vibrations of a thin stretched membrane.*

Solution

We make the following assumptions regarding the vibrations of the membrane:

(*i*) The mass of the membrane σ per unit area is constant.

(*ii*) The membrane is perfectly flexible.

(*iii*) The membrane is stretched and then fixed along its entire boundary in the *xy*-plane. The tension T per unit length caused by stretching of the membrane is the same at all points and in all directions.

(*iv*) The deflection u (x, y, t) of the membrane during its motion is small compared with the size of the membrane, and all angles of inclination are small. The tension T does not change appreciably during the motion.

We consider small transverse vibrations of a thin membrane. We find the forces acting on a small portion (Fig. 8.10) whose sides are approximately equal to Δx and Δy.

The forces acting on the edges of the portion are $T\Delta x$ and $T\Delta y$ which are tangent to the membrane. The horizontal components of the forces are obtained by multiplying the forces by the cosines of the angles of inclination. The cosines of the angles are close to one. Hence the horizontal components of the forces at opposite edges are approximately equal and opposite. Thus, there will be no appreciable motion of the particles of the membrane in the horizontal direction.

The resultant of the vertical components of the forces along the edges parallel to the *yu*-plane is (Fig. 8.10).

$$T\Delta y \ (\sin \beta - \sin \alpha)$$

$$\approx T\Delta y \ (\tan \beta - \tan \alpha)$$

$$= T\Delta y \left[\frac{\partial u}{\partial x}(x + \Delta x, y_1) - \frac{\partial u}{\partial x}(x, y_2) \right]$$

where y_1 and y_2 are values between y and $y + \Delta y$. Similarly, the resultant of the vertical components of forces along the edges parallel to *xu*-plane is

$$T\Delta x = \left[\frac{\partial u}{\partial y}(x_1, y + \Delta y) - \frac{\partial u}{\partial y}(x_2, y) \right]$$

where x_1 and x_2 are values between x and $x + \Delta x$. Hence, the equation of motion of this portion of the membrane is

$$(\sigma\, \Delta x \Delta y)\frac{\partial^2 u}{\partial t^2} = T\Delta y\left[\frac{\partial u}{\partial x}(x+\Delta x, y_1) - \frac{\partial u}{\partial x}(x, y_2)\right]$$

$$+ T\Delta x\left[\frac{\partial u}{\partial y}(x_1, y+\Delta y) - \frac{\partial u}{\partial y}(x_2, y)\right]$$

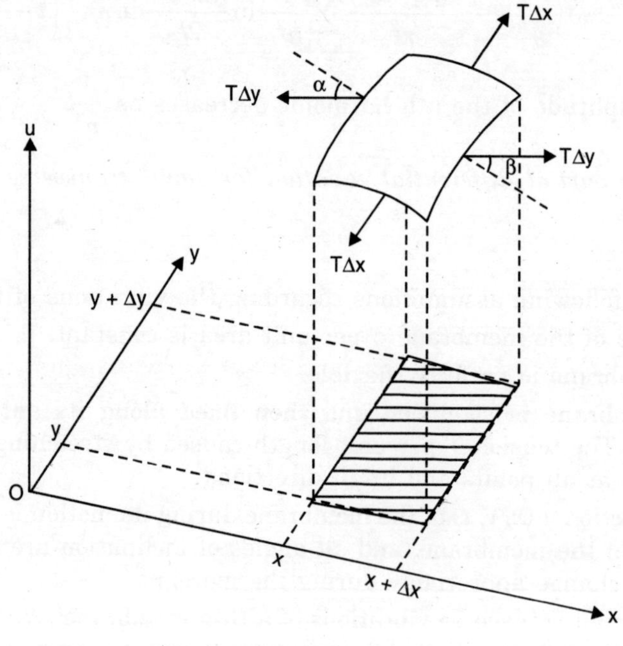

Fig. 8.10

Dividing by $\Delta x \Delta y$ and taking the limits $\Delta x \to 0$ and $\Delta y \to 0$, we obtain

$$\frac{\partial^2 u}{\partial t^2} = v^2\left[\frac{\partial^2 u}{\partial x^2} + \frac{\partial^2 u}{\partial y^2}\right] \quad ...(8.18)$$

where $v = \sqrt{T/\sigma}$.

Eqn. (8.18) is the two-dimensional wave equation. We may write Eqn. (8.18) in the form

$$\frac{\partial^2 u}{\partial t^2} = v^2\, \nabla^2 u.$$

21. (*a*) *Consider a stretched thin rectangular membrane of sides a and b which is fixed along its entire boundary. Write the differential equation for small transverse vibrations of the membrane with proper boundary conditions. Solve the problem and discuss the nature of vibrations.*

(*b*) *Consider the vibrations of a square membrane for which a = b = 1 and discuss the nature of the nodal lines.*

Solution

(*a*) We consider small transverse vibrations of the rectangular membrane (Fig. 8.11). The initial deflection and the initial velocity of the membrane at any point (x, y) are denoted by $f(x, y)$ and $g(x, y)$ respectively.

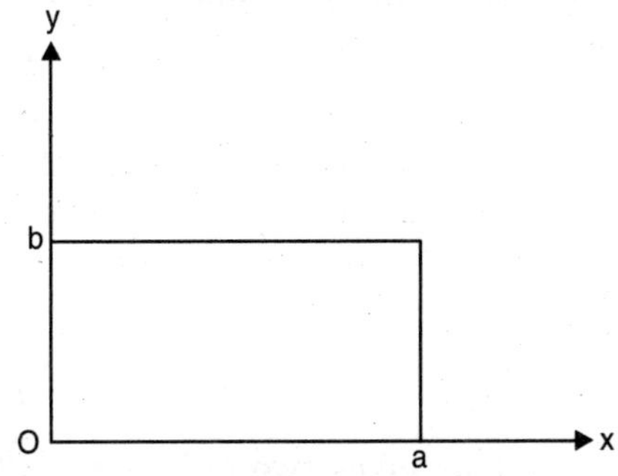

Fig. 8.11

We have to solve the two-dimensional wave equation

$$\frac{\partial^2 u}{\partial t^2} = v^2\left(\frac{\partial^2 u}{\partial x^2}+\frac{\partial^2 u}{\partial y^2}\right)$$

under the following boundary conditions:

(*i*) $u = 0$ on the boundary of the membrane for all $t \geq 0$

(*ii*) $u(x, y, 0) = f(x, y)$

(*iii*) $\left.\frac{\partial u}{\partial t}\right|_{t=0} = g(x, y)$

(*iv*) $|u(x, y, t)| < M$

where M is a fixed number *i.e.*, the motion is bounded.

We apply the method of separation of variables to the wave equation (8.18) and assume

$$u(x, y, t) = F(x, y)\, G(t). \qquad ...(8.19)$$

Substituting Eqn. (8.19) into Eqn. (8.18), we get

$$F\frac{d^2G}{dt^2} = v^2\left[G\frac{\partial^2 F}{\partial x^2}+G\frac{\partial^2 F}{\partial y^2}\right].$$

Dividing both sides by $v^2 FG$, we obtain

$$\frac{1}{v^2 G}\frac{d^2G}{dt^2} = \frac{1}{F}\left[\frac{\partial^2 F}{\partial x^2}+\frac{\partial^2 F}{\partial y^2}\right] \qquad ...(8.20)$$

Since L.H.S. of Eqn. (8.20) is a function of t only and the R.H.S. is a function of x and y, the two sides must be equal to a constant. This constant should also be negative which gives solution with proper boundary conditions. Thus, we may write

$$\frac{1}{v^2 G}\frac{d^2G}{dt^2} = \frac{1}{F}\left[\frac{\partial^2 F}{\partial x^2}+\frac{\partial^2 F}{\partial y^2}\right] = -p^2. \qquad ...(8.21)$$

The equation for F is

$$\frac{\partial^2 F}{\partial x^2}+\frac{\partial^2 F}{\partial y^2} + p^2 F = 0 \qquad ...(8.22)$$

We apply the method of separation of variables and write

$$F(x, y) = H(x)\, W(y) \qquad ...(8.23)$$

Substituting Eqn. (8.23) into Eqn. (8.22), we get

$$W\frac{d^2H}{dx^2} = -H\frac{d^2W}{dy^2} - p^2 HW.$$

Dividing both sides by HW, we obtain

$$\frac{1}{H}\frac{d^2H}{dx^2} = -\frac{1}{W}\left(\frac{d^2W}{dy^2}+p^2W\right) \quad ...(8.24)$$

Since L.H.S. is a function of x only and R.H.S. is a function of y only, the two sides must be equal to a constant. This constant should be negative on physical grounds.

$$\frac{1}{H}\frac{d^2H}{dx^2} = -\frac{1}{W}\left(\frac{d^2W}{dy^2}+p^2W\right) = -q^2.$$

This gives two ordinary differential equations:

$$\frac{d^2H}{dx^2}+q^2H = 0 \quad ...(8.25)$$

and

$$\frac{d^2W}{dy^2}+r^2W = 0 \quad ...(8.26)$$

where

$$r^2 = p^2 - q^2.$$

The general solutions of Eqns. (8.25) and (8.26) are

$$H(x) = A\cos qx + B\sin qx,$$

$$W(y) = C\cos ry + D\sin ry,$$

where A, B, C and D are constants. From the boundary condition (i) it follows that $F(x, y) = H(x)\,W(y)$ must be zero on the boundary. Thus, we have

$$H(0) = 0,\ H(a) = 0,\ W(0) = 0 \text{ and } W(b) = 0.$$

Hence,

$$H(0) = A = 0 \text{ and } H(a) = B\sin qa = 0.$$

We must take $B \neq 0$ since otherwise $H(x) \equiv 0$ and $F \equiv 0$ which corresponds to no deflection of membrane for all time. Hence $\sin qa = 0$ or, $q = \frac{m\pi}{a}$ where m is an integer. Similarly, $C = 0$ and $r = \frac{n\pi}{b}$ where n is an integer. Thus, the solutions are

$$H_m(x) = \sin\frac{m\pi x}{a} \text{ and } W_n(y) = \sin\frac{n\pi y}{b}$$

with $m = 1, 2, \ldots$, and $n = 1, 2, \ldots$. It is not necessary to consider $m, n = -1, -2, \ldots$, because the corresponding solutions are essentially the same as for positive m and n, except for a factor -1. Thus the functions

$$F_{mn}(x, y) = H_m(x)\,W_n(y) = \sin\frac{m\pi x}{a}\sin\frac{n\pi y}{b}$$

with $m, n = 1, 2, 3, \ldots$

are solutions of Eqn. (8.22) which satisfy the boundary conditions (i) and (iv). Since $p^2 = q^2 + r^2$, for a particular m and n, we have

$$p_{mn} = \left[\frac{m^2\pi^2}{a^2}+\frac{n^2\pi^2}{b^2}\right]^{1/2}$$

and the differential equation for $G(t)$ is

$$\frac{d^2G_{mn}}{dt^2} + v^2 p_{mn}^2 G_{mn} = 0$$

which has the general solution

$$G_{mn}(t) = K_{mn} \cos \omega_{mn} t + L_{mn} \sin \omega_{mn} t$$

where
$$\omega_{mn} = v p_{mn} = v\pi \left[\frac{m^2}{a^2} + \frac{n^2}{b^2}\right]^{1/2}$$

with $m = 1, 2, 3,.....$ and $n = 1, 2, 3,...$

The deflection for a particular value of m and n is

$$u_{mn}(x, y, t) = (K_{mn} \cos \omega_{mn} t + L_{mn} \sin \omega_{mn} t) \sin \frac{m\pi x}{a} \sin \frac{n\pi y}{b}.$$

The functions $u_{mn}(x, y, t)$ are called the eigenfunctions or characteristic functions, and the numbers ω_{mn} are called the eigenvalues or characteristic values of the vibrating membrane. The corresponding frequency is $\nu_{mn} = \omega_{mn}/2\pi$.

The general solution of the problem is

$$u(x, y, t) = \sum_{m=1}^{\infty} \sum_{n=1}^{\infty} u_{mn}(x, y, t)$$

$$= \sum_{m=1}^{\infty} \sum_{n=1}^{\infty} (K_{mn} \cos \omega_{mn} t + L_{mn} \sin \omega_{mn} t) \sin \frac{m\pi x}{a} \sin \frac{n\pi y}{b}. \quad ...(8.27)$$

From the boundary condition (*ii*), we have

$$u(x, y, 0) = \sum_{m} \sum_{n} K_{mn} \sin \frac{m\pi x}{a} \sin \frac{n\pi y}{b} = f(x, y).$$

This is a double Fourier series. The coefficients K_{mn} are obtained from the generalized Euler formula:

$$K_{mn} = \frac{4}{ab} \int_0^a \int_0^b f(x, y) \sin \frac{m\pi x}{a} \sin \frac{n\pi y}{b} dx\, dy \quad ...(8.28)$$

with $m = 1, 2, 3,....$ and $n = 1, 2, 3,.......$

From the boundary condition (*iii*), we obtain

$$\left.\frac{\partial u}{\partial t}\right|_{t=0} = \sum_{m} \sum_{n} \omega_{mn} L_{mn} \sin \frac{m\pi x}{a} \sin \frac{n\pi y}{b} = g(x, y).$$

The coefficients L_{mn} are obtained from the double Fourier series.

$$L_{mn} = \frac{4}{ab\omega_{mn}} \int_0^a \int_0^b g(x, y) \sin \frac{m\pi x}{a} \sin \frac{n\pi y}{b} dx\, dy \quad ...(8.29)$$

with $m = 1, 2, 3,.....$, and $n = 1, 2, 3,.....$

The deflection at any point (x, y) and at any time t is given by Eqn. (8.27) where the coefficients K_{mn} and L_{mn} are obtained from Eqns. (8.28) and (8.29).

(*b*) Here $a = b = 1$ and the eigenvalues are

$$\omega_{mn} = v\pi\,[m^2 + n^2]^{1/2} = \omega_{nm}.$$

For $m \neq n$, the corresponding functions.

$F_{mn} = \sin m\pi x \sin n\pi y$ and $F_{nm} = \sin n\pi x \sin m\pi y$ are different. For example, $\omega_{12} = \omega_{21} = v\pi\sqrt{5}$, but the corresponding two functions.

$F_{12} = \sin \pi x \sin 2\pi y$ and $F_{21} = \sin 2\pi y \sin \pi x$ are different. Now, $F_{12} = 0$ when $y = \frac{1}{2}$ and $F_{21} = 0$ when $x = \frac{1}{2}$. Hence, the corresponding eigenfunctions

$$u_{12} = (K_{12} \cos v\pi\,\sqrt{5}t + L_{12} \sin v\pi\sqrt{5}t\,)\,F_{12}.$$

$$u_{21} = (K_{21} \cos v\pi\,\sqrt{5}t + L_{21} \sin v\pi\sqrt{5}t\,)\,F_{21}.$$

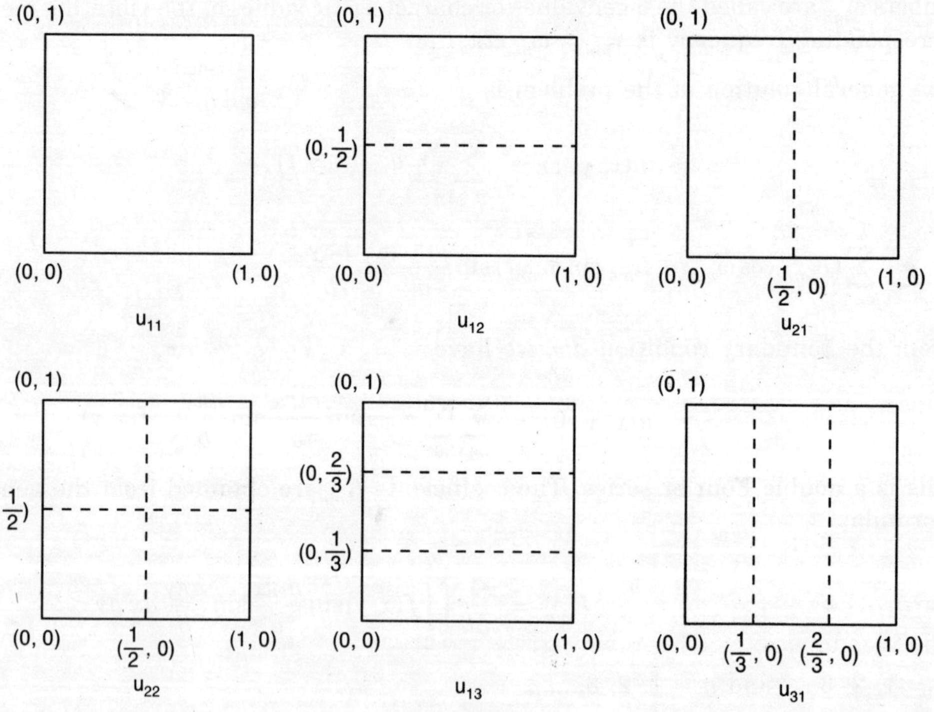

Fig. 8.12

have the nodal lines $y = \frac{1}{2}$ and $x = \frac{1}{2}$ respectively.

The nodal lines of the eigenfunctions $u_{11}, u_{12}, u_{21}, u_{22}, u_{13}$ and u_{31} of the square membrane are shown in Fig. 8.12.

Taking $K_{12} = 1$ and $L_{12} = L_{21} = 0$, we obtain

$$u_{12} + u_{21} = \cos v\pi\,\sqrt{5}t\,\,(F_{12} + K_{21}\,F_{21})$$

which also represents a vibrational mode with frequency $\nu_{12} = \sqrt{5}\,v/2$.

The nodal line of this mode of vibration is the solution of the equation

$$F_{12} + K_{21} F_{21} = \sin \pi x \sin 2\pi y + K_{21} \sin 2\pi x \sin \pi y = 0$$

or $$\cos \pi y + K_{21} \cos \pi x = 0.$$

The nature of the nodal lines depends on the value of K_{21}. Nodal lines of the solution of this equation for some values of K_{21} are shown in Fig. 8.13.

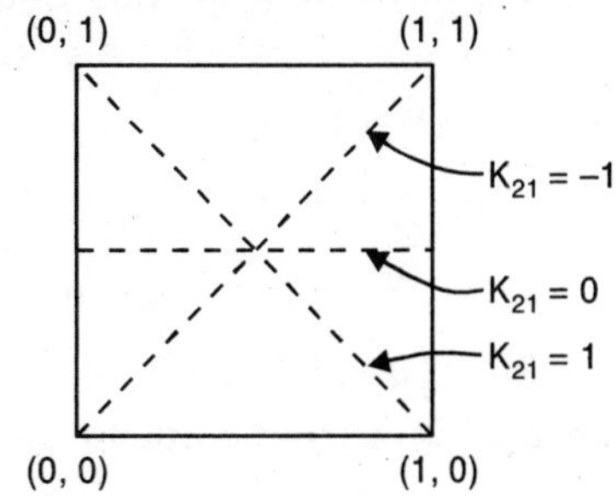

Fig. 8.13

22. *Find the deflection u(x, y, t) of the square membrane with a = b = 1 and v = 1, if the initial velocity is zero and the initial deflection is*

$$f(x, y) = B \sin \pi x \sin 2\pi y.$$

Solution

We have from Eqns. (8.28) and (8.29)

$$K_{mn} = 4B \int_0^1\int_0^1 \sin \pi x \sin 2\pi y \sin m\pi x \sin n\pi y \, dx\, dy$$

$$= \begin{cases} B & \text{when } m = 1 \text{ and } n = 2 \\ 0 & \text{otherwise} \end{cases}$$

and $L_{mn} = 0$.

since $\omega_{12} = \pi\sqrt{5}$,

$$u(x, y, t) = B \cos \pi\sqrt{5}t \sin \pi x \sin 2\pi y.$$

23. *Consider a stretched thin circular membrane of radius R, which is fixed along its entire boundary. Write the partial differential equation in polar coordinates for small transverse vibrations of the membrane. Find the deflection u(r, t) of the membrane when the initial deflection and initial velocity are given by*

$$u(r, 0) = f(r)$$

$$\left.\frac{\partial u}{\partial t}\right|_{t=0} = g(r)$$

Solution

The two-dimensional wave equation

$$\frac{\partial^2 u}{\partial t^2} = v^2 \nabla^2 u$$

takes the following form in polar coordinates.

$$\frac{\partial^2 u}{\partial t^2} = v^2 \left[\frac{\partial^2 u}{\partial r^2} + \frac{1}{r}\frac{\partial u}{\partial r} + \frac{1}{r^2}\frac{\partial^2 u}{\partial \theta^2}\right].$$

We shall consider only those solutions $u(r, t)$ of this equation which are radially symmetric *i.e.*, independent of θ. The wave equation then reduces to

$$\frac{\partial^2 u}{\partial t^2} = v^2 \left[\frac{\partial^2 u}{\partial r^2} + \frac{1}{r}\frac{\partial u}{\partial r}\right] \quad ...(8.30)$$

We have to solve this equation under the following boundary conditions:

(*i*) $u(R, t) = 0$ for all $t \geq 0$

(*ii*) $u(r, 0) = f(r)$

(*iii*) $\left.\dfrac{\partial u}{\partial t}\right|_{t=0} = g(r)$

(*iv*) $|u(r, t)| < M$

where M is a fixed number *i.e.*, the motion is bounded.

We apply the method of separation of variables to Eqn. (8.30) and assume

$$u(r, t) = H(r)\, G(t).$$

Substituting this equation into Eqn. (8.30) and dividing the resulting equation by $v^2 HG$, we get

$$\frac{1}{v^2 G}\frac{d^2 G}{dt^2} = \frac{1}{H}\left[\frac{d^2 H}{dr^2} + \frac{1}{r}\frac{dH}{dr}\right].$$

Since L.H.S. is a function of t only and the R.H.S. is a function of r only, the two sides must be equal to a constant. This constant should be negative in order to give a solution with proper boundary conditions. Thus, we have

$$\frac{1}{v^2 G}\frac{d^2 G}{dt^2} = \frac{1}{H}\left[\frac{d^2 H}{dr^2} + \frac{1}{r}\frac{dH}{dr}\right] = -p^2$$

The equation for H is

$$\frac{d^2 H}{dr^2} + \frac{1}{r}\frac{dH}{dr} + p^2 H = 0. \qquad \text{...(8.31)}$$

We introduce a new variable $s = pr$ so that

$$\frac{dH}{dr} = \frac{dH}{ds}\frac{ds}{dr} = p\frac{dH}{ds}.$$

and

$$\frac{d^2 H}{dr^2} = p^2 \frac{d^2 H}{ds^2}.$$

In the new variable s, Eqn. (8.31) becomes

$$\frac{d^2 H}{ds^2} + \frac{1}{s}\frac{dH}{ds} + H = 0.$$

This is Bessel's equation of order zero. The general solution of this equation is

$$H = c_1 J_0(s) + c_2 Y_0(s)$$

where c_1 and c_2 are two arbitrary constants and $J_0(s)$ and $Y_0(s)$ are the Bessel functions of the first and second kind of order zero. Since the deflection of the membrane is always finite, we cannot use $Y_0(s)$ as $Y_0(s)$ becomes infinite when s approaches zero. Hence, we put $c_2 = 0$. Clearly $c_1 \neq 0$ since otherwise $H = 0$. We may set $c_1 = 1$ and write

$$H(r) = J_0(pr). \qquad \text{...(8.32)}$$

On the boundary $r = R$ we must have boundary condition (i):

$$u(R, t) = H(R)G(t) = 0.$$

Since $G \equiv 0$ would imply $u \equiv 0$ for all r and t, we must have

$$H(R) = J_0(pR) = 0.$$

Denoting the positive zeros of $J_0(pR)$ by

$$pR = \alpha_1, \alpha_2, \alpha_3, \ldots$$

We obtain
$$p = \frac{\alpha_m}{R}, m = 1, 2, 3, \ldots.$$

The first four positive zeros of $J_0(pR)$ are given below:

$$\alpha_1 = 2.405, \alpha_2 = 5.520, \alpha_3 = 8.654, \alpha_4 = 11.792.$$

Hence, the functions

$$H_m(r) = J_0\left(\frac{\alpha_m}{R} r\right), \ m = 1, 2, 3, \ldots$$

are solutions of Eqn. (8.31) which vanish at $r = R$. The corresponding differential equation for $G(t)$ is

$$\frac{d^2 G_m}{dt^2} + \frac{v^2 \alpha_m^2}{R^2} G_m = 0$$

which has the general solution

$$G_m(t) = a_m \cos \omega_m t + b_m \sin \omega_m t,$$

where
$$\omega_m = \frac{v\alpha_m}{R}.$$

Therefore,

$$u_m(r, t) = H_m(r)\, G_m(t) = (a_m \cos \omega_m t + b_m \sin \omega_m t)\, J_0\left(\frac{\alpha_m}{R} r\right) \quad \ldots(8.33)$$

with $m = 1, 2, 3, \ldots$ are solutions of Eqn. (8.30) satisfying the boundary conditions (i), and (iv). These are the eigenfunctions of the problem and the corresponding eigenvalues are ω_m.

Nodal Lines: The nodal lines are obtained from the zeroes of $H_m(r)$.

For $m = 1$, $J_0\left(\frac{\alpha_1 r}{R}\right) = 0$ when $r = R$ *i.e.*, the circular membrane is fixed at its boundary.

There is no nodal line of the membrane. All the points of the membrane move upward (or downward) at the same time.

For $m = 2$, $J_0\left(\frac{\alpha_2 r}{R}\right) = 0$ when $\frac{\alpha_2 r}{R} = \alpha_1$

or
$$r = \frac{\alpha_1 R}{\alpha_2} = \frac{2.405}{5.520} R = 0.44\, R.$$

The circle $r = \alpha_1 R/\alpha_2$ is a nodal line. When the central part of the membrane ($r < \alpha_1 R/\alpha_2$) moves upward, the outer part ($r > \alpha_1 R/\alpha_2$) moves downward, and conversely.

For, $m = 3$, $J_0\left(\dfrac{\alpha_3 r}{R}\right) = 0$ when $\dfrac{\alpha_3 r}{R} = \alpha_1, \alpha_2$

or
$$r = \frac{\alpha_1 R}{\alpha_3}, \frac{\alpha_2 R}{\alpha_3}$$

or
$$r = \frac{2.405}{8.654} R = 0.28\,R$$

and
$$r = \frac{5.520}{8.654} R = 0.64\,R$$

The concentric circles $r = \alpha_1 R/\alpha_3$ and $r = \alpha_2\, R/\alpha_3$ are the nodal lines.

The nodal lines of the circular membrane for the normal modes $m = 1, 2, 3$ are shown in Fig. 8.14.

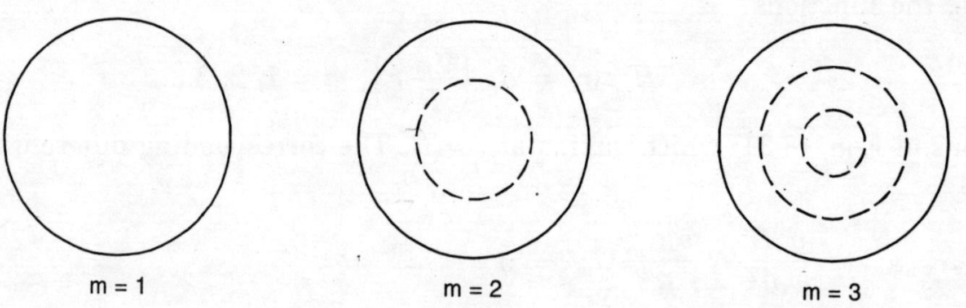

Fig. 8.14

In general u_{mn} (r, t) has $(m - 1)$ nodal lines which are concentric circles of radii $\alpha_1\, R/\alpha_m$, $\alpha_2\, R/\alpha_m$,....., $\alpha_{m-1}\, R/\alpha_m$.

Determination of the coefficients a_m and b_m of Eqn. (8.33).

We use the general solution of the problem:

$$u(r, t) = \sum_{m=1}^{\infty} u_m(r, t) = \sum_m (a_m \cos \omega_m t + b_m \sin \omega_m t) J_0(\alpha_m\, r/R) \quad ...(8.34)$$

From the boundary condition (*ii*), we have

$$u(r, 0) = \sum_m a_m J_0(\alpha_m\, r/R) = f(r).$$

Then, a_m is the coefficient of the Fourier-Bessel series which represents $f(r)$ in terms of J_0 $(\alpha_m\, r/R)$. Thus, we have

$$a_m = \frac{2}{R^2 J_1^2(\alpha_m)} \int_0^R r f(r) J_0(\alpha_m r / R) dr \quad ...(8.35)$$

with $m = 1, 2, 3,....$

From the boundary condition (*iii*), we have

$$\left.\frac{\partial u}{\partial t}\right|_{t=0} = \sum_m \omega_m b_m J_0(\alpha_m r / R) = g(r).$$

The coefficient b_m is obtained in a similar way

$$b_m = \frac{2}{R^2 J_1^2(\alpha_m)\omega_m}\int_0^R rg(r)J_0(\alpha_m r/R)dr \qquad ...(8.36)$$

with $m = 1, 2, 3,...$

The deflection $u(r, t)$ at any point at any time is given by Eqn. (8.34) where the coefficients a_m and b_m are obtained from Eqns. (8.35) and (8.36).

24. *Find the deflection u(r, t) of the circular membrane with R = 1 and v = 1, if the initial velocity is zero and the initial deflection is*

$$f(r) = k\,(1 - r^2).$$

Solution

We have

$$a_m = \frac{2}{J_1^2(\alpha_m)}\int_0^1 kr\left(1-r^2\right)J_0(\alpha_m r)dr$$

$$= \frac{2k}{J_1^2(\alpha_m)}\cdot\frac{2J_2(\alpha_m)}{\alpha_m^2}$$

$$= \frac{4kJ_2(\alpha_m)}{\alpha_m^2 J_1^2(\alpha_m)},$$

$$b_m = 0,$$

and

$$\omega_m = \alpha_m.$$

Thus,

$$u(r, t) = 4k\sum_{m=1}^{\infty}\frac{J_2(\alpha_m)}{\alpha_m^2 J_1^2(\alpha_m)}.\cos\alpha_m t\; J_0(\alpha_m r).$$

25. *A string tied between x = 0 and x = l vibrates in fundamental mode. The amplitude A, tension T and mass per unit length μ are given. Find the total energy of the string.*

(I.I.T. 2003)

Solution

The equation of the standing wave in the string in fundamental mode is given by

$$y = A\sin kx\cos\omega t$$

where $k = \dfrac{2\pi}{\lambda}$ and $\lambda = 2l$ for the fundamental mode,

$$\omega = 2\pi\nu = \frac{2\pi v}{\lambda} = \frac{\pi}{l}\sqrt{\frac{T}{\mu}}$$

and $\nu = \dfrac{1}{2l}\sqrt{\dfrac{T}{\mu}}$.

Total energy
$$E = \frac{\mu}{2}\int_0^l \left[\left(\frac{\partial y}{\partial t}\right)^2 + v^2\left(\frac{\partial y}{\partial x}\right)^2\right] dx$$
$$= \frac{\pi^2 A^2 T}{4l}$$

Again
$$(\text{P.E.})_{max} = \frac{\mu}{2}\int_0^l \left(\frac{\partial y}{\partial t}\right)^2 dx \Bigg|_{max}$$
$$= \frac{\pi^2 A^2 T}{4l} = \text{Total energy}$$

26. *The ends of a stretched wire of length L are fixed at x = 0 and x = L. In one experiment the displacement of the wire is*

$$y_1 = A \sin\left(\frac{\pi x}{L}\right) \sin \omega t$$

and energy is E_1. In another experiment its displacement is

$$y_2 = A \sin \frac{2\pi x}{L} \sin 2\omega t$$

and energy is E_2. Then

(a) $E_2 = E_1$ (b) $E_2 = 2E_1$ (c) $E_2 = 4E_1$ (d) $E_2 = 16E_1$

(I.I.T. 2001)

Solution

$$E_1 = (\text{P.E.})_{1\ max} = \frac{\mu}{2}\int_0^L \left(\frac{\partial y_1}{\partial t}\right)^2 dx\Bigg|_{max} = \frac{1}{4}\mu\omega^2 A^2 L$$

$$E_2 = (\text{P.E.})_{2\ max} = \frac{\mu}{2}\int_0^L \left(\frac{\partial y_2}{\partial t}\right)^2 dx\Bigg|_{max} = \mu\,\omega^2 A^2 L$$

Correct Choice: c.

27. *A sonometer wire resonates with a given tuning fork forming standing waves with five antinodes between the two bridges when a mass of 9 kg is suspended from the wire. When this mass is replaced by a mass M, the wire resonates with the same tuning fork forming three antinodes for the same positions of the bridges. The value of M is*

(a) 25 kg (b) 5 kg (c) 12.5 kg (d) 1/25 kg.

(I.I.T. 2002)

Solution

They are two nodes at the positions of the bridges. Thus the node-antinode structure of the string is like N A N A N A N A N A N. In this case $\frac{\lambda}{2} = \frac{l}{5}$

or,
$$\nu = \frac{v}{\lambda} = \frac{5}{2l}\sqrt{T/\mu} = \frac{5}{2l}\sqrt{9g/\mu}.$$

In the second case the node-antinode structure of the string is like N A N A N A N and

$$\nu = \frac{3}{2l}\sqrt{Mg/\mu}$$

Thus, $$\frac{5}{2l}\sqrt{9g/\mu} = \frac{3}{2l}\sqrt{Mg/\mu} \text{ or, } M = 25 \text{ kg}$$

Correct Choice : *a.*

28. *A massless rod is suspended by two identical strings AB and CD of equal length Fig. 8.15. A block of mass m is suspended from point O such that BO is equal to x. Further it is observed that the frequency of 1st harmonic (fundamental frequency) in AB is equal to 2nd harmonic frequency in CD. Then, length BO is*

(a) L/5 (b) 4L/5 (c) 3L/4 (d) L/4 (I.I.T. 2006)

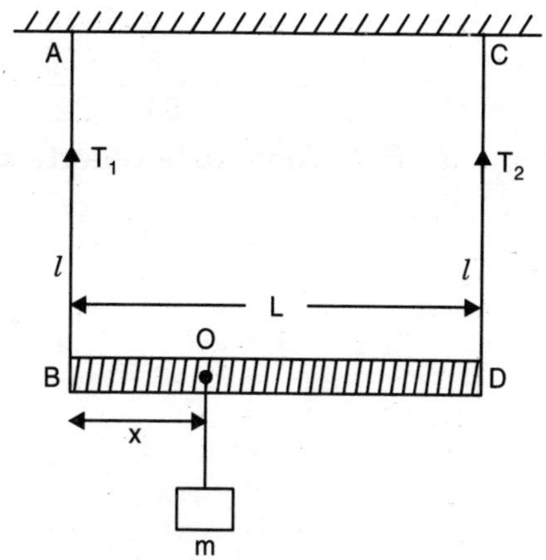

Fig. 8.15

Solution

Let T_1 and T_2 be tensions in two strings AB and CD respectively. Then

$$\frac{1}{2l}\sqrt{T_1/\mu} = \frac{2}{2l}\sqrt{T_2/\mu}$$

or $$T_2 = T_1/4$$

Taking torque about O for equilibrium, we have

$$T_1 x = T_2 (L - x)$$

or $$4T_2 x = T_2 (L - x)$$

or $$x = \frac{L}{5}$$

Correct Choice : *a.*

29. *A long wire PQR is made by joining two wires PQ and QR of equal radii. PQ has length 4.8 m and mass 0.06 kg. QR has length 2.56 m and mass 0.2 kg (Fig. 8.16)*

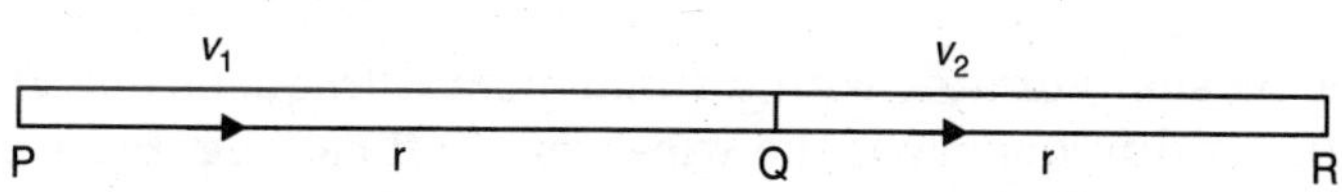

Fig. 8.16

The wire PQR is under a tension of 80 N. A sinusoidal wave-pulse of amplitude 3.5 cm is sent along the wire PQ from the end P. No power is dissipated during the propagation of the wave-pulse. Calculate

(a) The time taken by the wave-pulse to reach the other end R of the wire.

(b) The amplitude of the reflected and transmitted wave-pulse after the incident wave-pulse crosses the joint Q. (I.I.T. 1999)

Solution

Let $y = 3.5 \times 10^{-2} \sin(kx - \omega t)$

$$\mu_1 = \frac{0.06}{4.8} = \frac{1}{80} \text{ kg m}^{-1}$$

$$\mu_2 = \frac{0.2}{2.56} = \frac{1}{12.8} \text{ kg m}^{-1}$$

$$v_1 = \sqrt{\frac{T_1}{\mu_1}} = \sqrt{\frac{80}{1/80}} = 80 \text{ ms}^{-1}$$

$$v_2 = \sqrt{\frac{T_2}{\mu_2}} = \sqrt{\frac{80}{1/12.8}} = 32 \text{ ms}^{-1}$$

(*a*) Total time $= \frac{4.8}{80} + \frac{2.56}{32} = 0.14s$

(*b*) R_a = Amplitude reflection coefficient

$$= \frac{\text{Reflected amplitude}}{\text{Incident amplitude}} = \frac{\sqrt{\mu_1} - \sqrt{\mu_2}}{\sqrt{\mu_1} + \sqrt{\mu_2}}$$

$$= \frac{\sqrt{\frac{1}{80}} - \sqrt{\frac{1}{12.8}}}{\sqrt{\frac{1}{80}} + \sqrt{\frac{1}{12.8}}} = -\,0.40$$

T_a = Amplitude transmission coefficient

$$= \frac{\text{Transmitted amplitude}}{\text{Incident amplitude}} = \frac{2\sqrt{\mu_1}}{\sqrt{\mu_1} + \sqrt{\mu_2}}$$

$$= 0.60$$

Reflected amplitude $= 3.5 \times 10^{-2} \times 0.4 = 1.4 \times 10^{-2}$ m.

Transmitted amplitude $= 3.5 \times 10^{-2} \times 0.6 = 2.1 \times 10^{-2}$ m.

SUPPLEMENTARY PROBLEMS

1. A wire under tension vibrates with a fundamental frequency of 512 Hz. What would be the fundamental frequency if the wire were half as long, twice as thick and under one-fourth the tension?

2. Steel and silver wires of the same diameter and same length are stretched with equal tension. The densities of steel and silver are 7.8 and 10.6 g/cm^3 respectively. What is the fundamental frequency of the silver wire if that of steel is 256 Hz?

3. A string has mass 2 g and length 60 cm^3. What must be the tension so that when vibrating transversely its first overtone has frequency 200 Hz?

4. A wire having a linear density of 0.05 g/cm^3 is stretched between two rigid supports with a tension of 4.5×10^7 dynes. It is observed that the wire resonates at a frequency of 420 cycles/s. The next higher frequency at which the same wire resonates is 490 cycles/s. Find the length of the wire. (*I.I.T. 1971*)

$$\left[\textbf{Hints: } 420 = \frac{n}{2l}\sqrt{\frac{T}{\mu}};\ 490 = \frac{n+1}{2l}\sqrt{\frac{T}{\mu}}\right]$$

5. A wire of density 9 g/cm^3 is stretched between two clamps 1.00 m apart while subjected to an extension of 0.05 cm. What is the lowest frequency of transverse vibrations in the wire? Assume Young's modulus Y = 9 × 10^{10} N/m^2. *(I.I.T. 1975)*

$$\left[\textbf{Hints: } \nu = \frac{1}{2l}\sqrt{\frac{T}{A\rho}};\ Y = \frac{T/A}{dl/l}\right]$$

6. Three strings of equal length but stretched with different tensions are made to vibrate. If the masses per unit length are in the ratio 1 : 2 : 3 and frequencies are the same, calculate the ratio of the tensions.

7. Show that the dispersion relation for the normal modes of a homogeneous and flexible string is given by

$$\omega = \sqrt{\frac{T}{\mu}}k.$$

8. (*a*) A string of length l which is under tension T and is fixed at both ends has mass per unit length μ. The string is given initial deflection at its various points and is released at time $t = 0$. It executes small transverse vibrations. The initial deflection at any point x is denoted by $h(x)$.

The initial velocity is zero at any point x. Find the different normal modes of vibrations and the deflection $y(x, t)$ of the string at any point x and at any time $t > 0$.

(*b*) Show that $y\ (x, t)$ can be expressed as superposition of forward and backward waves.

9. Find the deflection $y(x, t)$ of the vibrating string of length $l = \pi$ and mass per unit length μ, fixed at both ends and under tension T, corresponding to zero initial velocity and initial deflection $h(x) = 0.01 \sin x$.

10. A long string of mass per unit length 0.1 kgm^{-1} is joined to another of mass per unit length 0.4 kgm^{-1}. They are under the same tension of 250 N. Find the characteristic impedances of the strings. What fraction of the power carried by the wave is transmitted from the first string to the second?

11. Consider the problem 16 (vibration of a plucked string) where the deflection of the string is given by

$$y(x, t) = \frac{2hl^2}{\pi^2 a(l-a)}\sum_{n=1}^{\infty}\frac{1}{n^2}\sin\frac{n\pi a}{l}\sin\frac{n\pi x}{l}\cos\frac{n\pi vt}{l}.$$

Hence show that

(*i*) $\displaystyle\sum_{n=1}^{\infty}\frac{1}{n^2}\sin^2\frac{n\pi a}{l} = \frac{\pi^2 a(l-a)}{2l^2}$

(*ii*) $\displaystyle\frac{1}{1^2}+\frac{1}{3^2}+\frac{1}{5^2}+\ldots=\frac{\pi^2}{8}.$

12. (*a*) If we write Eqn. (8.16) in the form

$$y(x, t) = \sum_{n=1}^{\infty} C_n \sin n\omega(t-\tau)\sin\frac{n\pi x}{l}$$

then show that

$$C_n \sin\frac{n\pi x}{l} = \frac{\tau(v_1+v_2)}{\pi^2 n^2}\sin\left(\frac{n\pi T_1}{\tau}\right)$$

(*b*) Assuming

$$C_n = \frac{\tau(v_1+v_2)}{\pi^2 n^2}$$

and $$\frac{n\pi x}{l} = \frac{n\pi T_1}{\tau}$$

show that $\tau = 2\,T_1$ at $x = \frac{l}{2}$ and the forward and backward velocities are equal in magnitude at $x = l/2$.

(*c*) Show that

$$\frac{vT_1}{2l} = \frac{x}{l}$$

and Eqn. (8.17) can be written in the form

$$y(x, t) = \frac{\tau(v_1+v_2)}{\pi^2}\sum_{n=1}^{\infty}\frac{1}{n^2}\sin\frac{n\pi x}{l}\sin\frac{2n\pi}{\tau}\left(t-\frac{\tau x}{2l}\right)$$

13. Consider the problem of transverse vibration of rectangular membrane of sides a and b. (*a*) How does the frequency change if the tension of the membrane is increased? (*b*) Find one eigenvalue of the rectangular membrane of sides $a = 2$, $b = 1$ such that two different eigenfunctions have the same eigenvalue.

14. Show that among all rectangular membranes of the same area $A = a \times b$ and the same velocity of propagation of wave on the membrane, the square membrane has the lowest frequency for the mode u_{11}.

$$\left[\textbf{Hints: } \omega_{11}^2 = v^2\pi^2\left[\frac{1}{a^2}+\frac{a^2}{A^2}\right] \text{and } \omega_{11}^2 \text{ is minimum for } a = \sqrt{A}\right]$$

15. Consider a square membrane with $a = b = 1$ and $v = 1$. If the initial velocity is zero and the initial deflection is $f(x, y) = x + y$, show that the deflection $u(x, y, t,)$ of the membrane is given by

$$u(x, y, t) = \frac{4}{\pi^2}\sum_m\sum_n\frac{1}{mn}\cos\pi\sqrt{m^2+n^2}\,t\ [(-1)^m\{(-1)^n-1\}$$
$$+ (-1)^n\{(-1)^m-1\}]\sin m\pi x \sin n\pi y.$$

16. A square membrane of side 10 cm is made of material of density 1 g/cm^2 and is under tension 32 dyne/cm. Find the lowest frequency of vibration of the membrane.

17. If the tension of the circular membrane is increased how does the frequency change?

18. Determine numerical values of the radii of the nodal lines of the 4th normal mode of vibration of a circular membrane of radius unity.

19. Find the deflection $u(r, t)$ of the circular membrane with $R = 1$ and $v = 1$, if the initial velocity is zero and the initial deflection is $f(r) = 0.1\, J_0(\alpha_2 r)$.

$$\left[\textbf{Hints: } \int_0^1 rJ_0(\alpha_2 r)J_0(\alpha_m r)dr = \begin{cases} \frac{1}{2}J_1^2(\alpha_2) & \text{if } \alpha_m = \alpha_2 \\ 0 & \text{if } \alpha_m \neq \alpha_2 \end{cases}\right]$$

20. A membrane having the form of a circular annulus of radii R_1 and R_2, is fixed along its entire boundary at $r = R_1$ and $r = R_2$. Show that the periods of the normal modes of vibration are of the form $2\pi/(vp)$ where p satisfies the equation

$$J_0(pR_1)Y_0(pR_2) - J_0(pR_2)Y_0(pR_1) = 0.$$

21. If $u(x, y, t)$ be the deflection of a stretched membrane at any point (x, y) at any instant t, show that the kinetic energy (K.E.) and potential energy (P.E.) of the membrane at that instant are given by

$$\text{K.E.} = \frac{\sigma}{2}\int\int\left(\frac{\partial u}{\partial t}\right)^2 dx\, dy,$$

$$\text{P.E.} = \frac{T}{2}\int\int\left[\left(\frac{\partial u}{\partial x}\right)^2 + \left(\frac{\partial u}{\partial y}\right)^2\right]dx\, dy$$

where σ = Mass per unit area of the membrane

and T = Tension per unit length of the membrane.

[**Hints:** Due to deflection of the membrane, the elementary area $dx\, dy$ becomes

$$dx\left[1 + \left(\frac{\partial u}{\partial x}\right)^2\right]^{1/2} dy\left[1 + \left(\frac{\partial u}{\partial y}\right)^2\right]^{1/2}$$

$$\approx dx\, dy + \frac{1}{2}\left[\left(\frac{\partial u}{\partial x}\right)^2 + \left(\frac{\partial u}{\partial y}\right)^2\right]dx\, dy\Bigg]$$

22. A circular membrane of radius 20 cm and density per unit area 1 g/cm^2 is stretched to a tension of 10^4 dynes/cm. Compute the four lowest frequencies of vibration of the membrane.

23. Two vibrating strings of the same material but lengths L and $2L$ have radii $2r$ and r respectively. They are stretched under the same tension. Both the strings vibrate in their fundamental modes, the one of length L with frequency ν_1 and the other with frequency ν_2.

The ratio ν_1/ν_2 is given by

(*a*) 2 (*b*) 4 (*c*) 8 (*d*) 1. *(I.I.T. 2000)*

24. The extension in a string obeying Hooke's law is x. The speed of sound in the stretched string is v. If the extension in the string is increased to $1.5x$, the speed of sound will be .... *(I.I.T. 1996)*

[**Hints:** $T = kx$ and $T' = k \times 1.5x$, k = Constant]

25. The (x, y) coordinates of the corners of a square plate are $(0, 0)$, $(L, 0)$, (L, L), $(0, L)$. The edges of the plate are clamped and transverse standing waves are set up in it. If $u(x, y)$ denotes the displacement of the plate at the point (x, y) at some instant of time, the possible expression (s) for u is (are) (a = positive constant)

(*a*) $a \cos \dfrac{\pi x}{2L} \cos \dfrac{\pi y}{2L}$ (*b*) $a \sin \dfrac{\pi x}{L} \sin \dfrac{\pi y}{L}$

(*c*) $a \sin \dfrac{\pi x}{L} \sin \dfrac{2\pi y}{L}$ (*d*) $a \cos \dfrac{2\pi x}{L} \sin \dfrac{\pi y}{L}$. *(I.I.T. 1998)*

[**Hints:** $u = 0$ at $x = 0, L$; $u = 0$ at $y = 0, L$]

9

The Doppler Effect

9.1 DOPPLER SHIFT

If the observer or the source or both are in motion then the observer notes an apparent change in frequency from the actual frequency of the wave emitted by the source. This phenomenon is called the Doppler effect and the difference between the actual and observed frequency or wavelength is known as Doppler shift. When both the source and the observer are in motion along the same straight line (Fig. 9.1), the moving observer will receive a wave whose apparent frequency f_{os} is given by

$$f_{os} = f\frac{v - v_o}{v - v_s} \qquad \text{...(9.1)}$$

Fig. 9.1

Here, f = Actual frequency of the wave emitted by the source through the medium which is at rest,

v = Velocity of the wave through the medium which is at rest,

v_o = Velocity of the observer (away from the source) with respect to the medium,

v_s = Velocity of the source (towards the observer) with respect to the medium.

We may consider the following special cases:

(*i*) When the source is at rest ($v_s = 0$) and the observer is moving away from the source, the apparent frequency is

$$f_o = f\frac{v - v_o}{v}, \; f_o < f$$

(*ii*) When the source is at rest ($v_s = 0$) and the observer is moving towards the source, the apparent frequency is

$$f_o = f\frac{v + v_o}{v}, \; f_o > f$$

(*iii*) When the observer is at rest ($v_o = 0$) and the source is moving towards the observer, the apparent frequency is

$$f_s = f\frac{v}{v - v_s}, \; f_s > f$$

(*iv*) When the observer is at rest ($v_o = 0$) and the source is moving away from the observer, the apparent frequency is

$$f_s = f\frac{v}{v + v_s}, \; f_s < f \text{ and } \lambda_s > \lambda,$$

where the apparent wavelength $\lambda_s = \dfrac{v}{f_s}$ and the actual wavelength $\lambda = \dfrac{v}{f}$.

When the light from a star is examined spectroscopically, it is found to contain the spectra of common terrestrial elements, but the spectral lines are shifted towards the red end of the spectrum. If the star is moving away from the earth, it is clear that $\lambda_s > \lambda$ and the observed 'red shift' can be explained.

In solving numerical problems Eqn. (9.1) should be remembered. The direction of sound is always taken as the direction from the source towards the observer. The velocity measured in the direction of sound is taken as positive while that in the opposite direction is taken as negative. When the observer moves toward the source, or, the source moves toward the observer or both move toward each other, the apparent frequency increases. In the other case when the observer moves away from the source, or the source moves away from the observer, or, both move away from each other, the apparent frequency decreases.

SOLVED PROBLEMS

1. *(a) When the source is at rest and the observer is moving towards the source, show that the moving observer will receive a wave whose apparent frequency f_o is given by*

$$f_o = f\frac{v + v_o}{v}$$

where f = actual frequency of the wave emitted by the source, v = velocity of the wave through the medium which is at rest, v_o = velocity of the observer (towards the source) with respect to the medium.

(b) What will happen when the observer is moving away from the source?

Solution

(*a*) The observer O is moving with velocity v_o towards the stationary source S (Fig. 9.2). The source S is emitting waves which are travelling through the medium with velocity v.

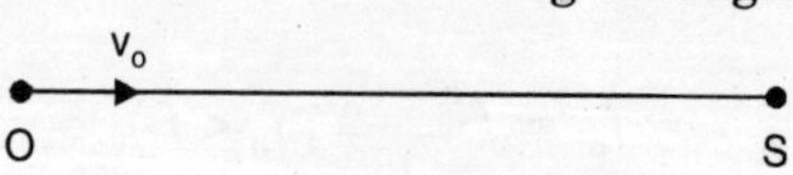

Fig. 9.2

Then, we have $v = f\lambda$, where λ is the wavelength of the wave, which is the distance between adjacent crests. If the observer were at rest he would have received f number of crests in unit time. Since the observer is moving towards the source he will receive more number of crests in unit time. The increase in the number of crests received by the observer in unit time is v_o/λ. Thus, the apparent frequency f_o is

$$f_o = f + \frac{v_o}{\lambda} = f + \frac{v_o f}{v} = f\frac{v + v_o}{v}. \qquad \text{...(9.2)}$$

(*b*) When the observer is moving away from the source, the apparent frequency as received by the observer is

$$f_o = f - \frac{v_o}{\lambda} = f\frac{v - v_o}{v}. \qquad ...(9.3)$$

2. (*a*) *When the observer is at rest and the source is in motion towards the observer show that the apparent frequency as received by the observer is*

$$f_s = f\frac{v}{v - v_s}.$$

(*b*) *What is the apparent frequency when the source is moving away from a stationary observer?*

Solution

(*a*) When the source is at rest, the successive wave crests emitted by it are one wavelength λ apart. When the source is moving towards the observer, the distance between adjacent crests is decreased by the distance the source travels in one cycle. The time taken in one cycle is $\frac{1}{f}$ and the distance traversed by the source in one cycle is v_s/f. Thus, the apparent wavelength λ_s is given by

$$\lambda_s = \lambda - \frac{v_s}{f} = \frac{v}{f} - \frac{v_s}{f} = \frac{1}{f}(v - v_s).$$

But, since the medium in which the wave is being propagated is at rest, the wave velocity remains unchanged, so that the frequency of the signal received by the stationary observer is

$$f_s = \frac{v}{\lambda_s} = \frac{vf}{v - v_s}. \qquad ...(9.4)$$

(*b*) The apparent frequency when the source is moving away from the observer is

$$f_s = \frac{vf}{v + v_s}. \qquad ...(9.5)$$

3. *When the source and the observer are in motion along the same straight line as shown in Fig. 9.1, show that the moving observer will receive a wave whose apparent frequency f_{os} is given by Eqn. (9.1).*

Solution

When the source is moving towards the stationary observer it gives rise to a wave whose apparent frequency f_s is given by Eqn. (9.4). If we take f_s to be the frequency of the wave itself, we need not concern ourselves further with the motion of the source. The moving observer will receive a wave whose apparent frequency f_{os} is obtained by substituting f_s for f in the right hand side of Eqn. (9.3).

$$f_{os} = f_s\frac{v - v_o}{v} = \frac{vf}{v - v_s}\cdot\frac{v - v_o}{v} = f\frac{v - v_o}{v - v_s}.$$

4. (*a*) *Show that if in the problem 3, v_s and v_o are both small compared with v, the fractional change in frequency $\Delta f/f$ is given by $(v_s - v_o)/v$. (b) Assuming that the result (a) is also applicable to the Doppler effect of light, show that $\Delta f/f = v_s/c$ or, $\Delta\lambda/\lambda = v_s/c$ when the observer is at rest and $v_s << c$, where c is the velocity of light.*

Solution

(*a*) We have from Eqn. (9.1)

$$f_{os} = f\frac{v - v_o}{v - v_s} = f\frac{1 - v_o / v}{1 - v_s / v}$$

$$\approx f(1 - v_o/v)(1 + v_s/v)$$

$$f_{os} \approx f(1 - v_o/v + v_s/v)$$

Thus, $$f_{os} - f = \Delta f = f(v_s/v - v_o/v)$$

Hence, $$\Delta f/f = (v_s - v_o)/v \quad ...(9.6)$$

(*b*) From Eqn. (9.6), we have

$$\frac{\Delta f}{f} = \frac{f_{os} - f}{f} = \frac{\frac{c}{\lambda_{os}} - \frac{c}{\lambda}}{\frac{c}{\lambda}} = \frac{\lambda}{\lambda_{os}} - 1 = \frac{\Delta\lambda}{\lambda_{os}} \approx \frac{\Delta\lambda}{\lambda}.$$

Thus, $$\frac{\Delta\lambda}{\lambda} = \frac{v_s}{c} \quad ...(9.7)$$

when the observer is at rest.

5. *Show that the Doppler effect is greater when the source approaches a stationary observer than when the observer approaches the stationary source with the same velocity.*

Solution

When the source approaches the stationary observer with velocity v_s, the apparent frequency is

$$f_s = f\frac{v}{v - v_s}.$$

When the observer approaches the stationary source with the velocity v_s, the apparent frequency is

$$f_o = f\frac{v + v_s}{v}.$$

Now, $$f_s - f_o = f\frac{v_s^2}{v(v - v_s)} > 0.$$

Hence, $$f_s > f_o.$$

6. *A police car, parked by the roadside, sounds its siren, which has a frequency f of 1000 Hz. What frequency f′ do you hear if (a) you are driving directly toward the police car at 30 m s*$^{-1}$*? (b) you are driving away from the police car at the same speed? (c) you are at rest and the police car is coming toward you at 30 m s*$^{-1}$*? (d) you are at rest and the police car is going away from you at the same speed? (e) both you and the police car are driving toward each other at 30 m s*$^{-1}$*? (f) both you and the police car are driving away from each other at 30 m s*$^{-1}$*?*

[Velocity of sound in air is 340 m s^{-1}*]*

Solution

(*a*) $f' = f\dfrac{v+v_o}{v} = 1000 \times \dfrac{340+30}{340} = 1088.2$ Hz

(*b*) $f' = f\dfrac{v-v_o}{v} = 1000 \times \dfrac{340-30}{340} = 911.8$ Hz

(*c*) $f' = f\dfrac{v}{v-v_s} = 1000 \times \dfrac{340}{340-30} = 1096.8$ Hz.

(*d*) $f' = f\dfrac{v}{v+v_s} = 1000 \times \dfrac{340}{340+30} = 918.9$ Hz.

(*e*) $f' = f\dfrac{v+v_o}{v-v_s} = 1000 \times \dfrac{340+30}{340-30} = 1193.5$ Hz

(*f*) $f' = f\dfrac{v-v_o}{v+v_s} = 1000 \times \dfrac{340-30}{340+30} = 837.8$ Hz.

7. *Two aeroplanes A and B are approaching each other and their velocities are 108 km/h and 144 km/h respectively. The frequency of a note emitted by A as heard by the passengers in B is 1170 Hz. Calculate the frequency of the note heard by the passengers in A. Velocity of sound = 350 m s*$^{-1}$.

Solution

We have

$$f_{os} = \frac{v+v_o}{v-v_s}f,$$

where $v_s = 108 \text{ km/h} = \dfrac{108\times1000}{60\times60} = 30 \text{ m s}^{-1}$

$$v_o = 144 \text{ km/h} = \frac{144\times1000}{60\times60} = 40 \text{ m s}^{-1}$$

Thus, $$f = f_{os}\frac{v-v_s}{v+v_o} = 1170 \times \frac{350-30}{350+40} = 960 \text{ Hz.}$$

8. *An observer on a railway platform observed that as a train passed through the station at 108 km/h, the frequency of the whistle appeared to drop by 300 Hz. Find the frequency of the whistle. Velocity of sound in air = 350 m s*$^{-1}$.

Solution

The apparent frequency when the source is approaching the observer, is

$$f_s = f\frac{v}{v-v_s}.$$

The apparent frequency when the source is going away from the observer, is

$$f_s' = f\frac{v}{v+v_s}.$$

Thus, $$f_s - f_s' = \frac{2fvv_s}{v^2 - v_s^2}$$

or
$$f = \frac{(f_s - f_s')(v^2 - v_s^2)}{2vv_s}$$

$$= 300 \times \frac{(350)^2 - (30)^2}{2 \times 350 \times 30} = 1737.1 \text{ Hz.}$$

9. *A source emitting a sinusoidal sound wave of frequency 480 Hz travels towards a wall at a speed 15 m s^{-1}. Find the beat frequency perceived by an observer moving at a speed 10 ms^{-1} under the following situations, if the speed of sound in air is 330 m s^{-1}:*

(i) the observer is going away from the source and the wall (Fig. 9.3).

(ii) the observer is travelling towards the source and the wall.

(iii) the observer is in between the wall and the source, and moving away from the wall.

(iv) the observer is in between the wall and the source, and moving towards the wall.

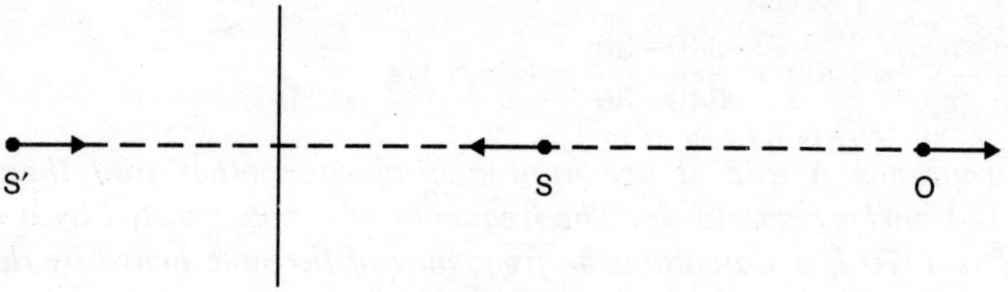

Fig. 9.3

Solution

(*i*) First we consider the sound travelling directly from the source (S) to the observer (O) [Fig. 9.3].

The frequency received by O is

$$f_{os} = f\frac{v - v_o}{v + v_s} = 480 \times \frac{320}{345} = 445.2 \text{ Hz.}$$

The observer will also receive the sound reflected from the wall. Suppose at any instant, S' is the image of S behind the wall (Fig. 9.3). The image S' is moving at a speed 15 m s^{-1} towards the observer. We may consider S' as the source of the reflected sound wave. Thus, the frequency perceived by O due to the reflected wave is

$$f'_{os} = 480 \times \frac{330 - 10}{330 - 15} = 487.6 \text{ Hz.}$$

The beat frequency = 487.6 – 445.2 = 42.4 Hz.

(*ii*) The frequencies perceived by O due to the waves from the source and the reflected waves are

$$f_{os} = f\frac{v + v_o}{v + v_s} = 480 \times \frac{330 + 10}{330 + 15} = 473.0 \text{ Hz}$$

$$f'_{os} = f\frac{v + v_o}{v - v_s} = 480 \times \frac{330 + 10}{330 - 15} = 518.1 \text{ Hz}$$

The beat frequency = 518.1 – 473.0 = 45.1 Hz.

(*iii*) The apparent frequencies are

$$f_{os} = f\frac{v+v_o}{v-v_s} = 518.1 \text{ Hz}$$

$$f'_{os} = f\frac{v-v_o}{v-v_s} = 487.6 \text{ Hz}$$

The beat frequency = 518.1 – 487.6 = 30.5 Hz.

(*iv*) The apparent frequencies are

$$f_{os} = \frac{v-v_o}{v-v_s} = 487.6 \text{ Hz.}$$

$$f'_{os} = \frac{v+v_o}{v-v_s} = 518.1 \text{ Hz.}$$

The beat frequency = 518.1 – 487.6 = 30.5 Hz.

10. *A car is travelling along a road. A stationary policeman observes that the frequency ratio of the siren of the car is 5/4 as it passes. What is the speed of the car? [Velocity of sound in air = 333 m s^{-1}]*

Solution

The apparent frequency when the car is approaching the policeman is

$$f_s = f\frac{v}{v-v_s}$$

The apparent frequency when the car is going away from the policeman is

$$f'_s = f\frac{v}{v+v_s}.$$

Thus,
$$f_s/f'_s = \frac{v+v_s}{v-v_s} = \frac{5}{4}$$

or
$$v_s = \frac{v}{9} = \frac{333}{9} = 37 \text{ m s}^{-1}.$$

11. *A whistle of frequency 540 Hz moves in a circle of radius 2.0 ft at an angular speed of 15 rad/s. What are (a) lowest and (b) the highest frequencies heard by a listener a long distance away at rest with respect to the centre of the circle? Velocity of sound in air = 1125 ft/s.*

Solution

The linear speed of the whistle = $v_s = r\omega = 2 \times 15 = 30$ ft/s.

(*a*) The minimum frequency is heard when the source moves away from the listener. The lowest frequency heard by the listener is

$$f' = f\frac{v}{v+v_s} = 540 \times \frac{1125}{1125+30} = 526 \text{ Hz.}$$

(*b*) The maximum frequency is heard when the source approaches the listener. The highest frequency heard by the listener is

$$f' = f\frac{v}{v-v_s} = 540 \times \frac{1125}{1125-30} = 554.8 \text{ Hz.}$$

12. *Figure 9.4 shows a transmitter and receiver of waves contained in a single instrument. It is used to measure the speed V of a target object (idealized as a flat plate) that is moving directly toward the unit, by analyzing the waves reflected from it. (a) Apply Doppler equations twice, first with the target as observer and then with the target as a source, and show that the frequency f_r of the reflected waves at the receiver is related to their source frequency f_s by*

$$f_r = f_s \frac{v+V}{v-V}$$

where v is the speed of the waves. (b) In many practical situations, V << v, show that, then the above equation becomes

$$f_r = f_s\left(1+\frac{2V}{v}\right).$$

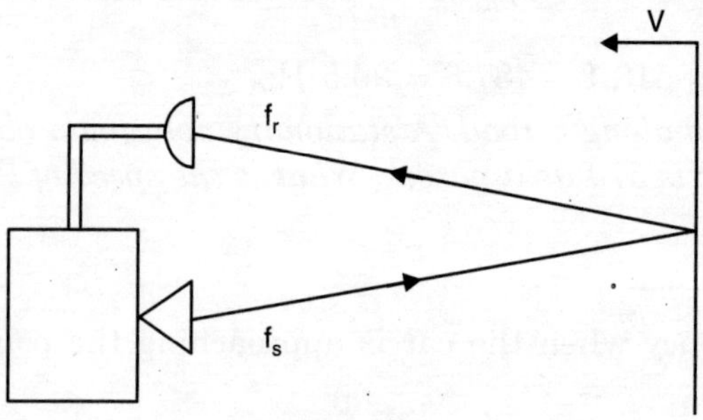

Fig. 9.4

Solution

(*a*) The frequency as received by the target (as observer) is

$$f' = f_s \frac{v+V}{v}.$$

With the target as the source, the frequency received by the receiver is

$$f_r = f' \frac{v}{v-V} = f_s \frac{v+V}{v-V}$$

(*b*) When $v >> V$, we have

$$f_r = f_s \frac{v\left(1+\frac{V}{v}\right)}{v\left(1-\frac{V}{v}\right)} \approx f_s\left(1+\frac{V}{v}\right)\left(1+\frac{V}{v}\right) \approx f_s\left(1+\frac{2V}{v}\right).$$

13. *A sonometer wire under tension of 64 Newton vibrating in its fundamental mode is in resonance with a vibrating tuning fork. The vibrating portion of the sonometer wire has a length of 10 cm and a mass of one gm. The vibrating tuning fork is now moved away from the vibrating wire with a constant speed and an observer standing near sonometer hears one beat per second. Calculate the speed with which the tuning fork is moved, if the velocity of sound in air is 300 m s^{-1}.* (I.I.T. 1983)

Solution

The fundamental frequency of the sonometer wire is

$$f = \frac{1}{2l}\sqrt{\frac{T}{\mu}} = \frac{1}{2 \times 0.1}\sqrt{\frac{65}{10^{-3}/0.1}} = 400 \text{ Hz.}$$

Actual frequency of the vibrating tuning fork = f = 400 Hz. The apparent frequency of the tuning fork is

$$f_s = f\frac{v}{v+v_s} = 399 \text{ Hz.}$$

where v = velocity of sound in air = 300 m s^{-1}.

v_s = velocity of the tuning fork, which is moving away from the vibrating wire.

Thus, $$v_s = \frac{v}{399} = \frac{300}{399} = 0.75 \text{ m s}^{-1}.$$

14. *A girl is sitting near the open window of a train that is moving at a speed of 10 m/s^{-1} to the east. The girl's uncle stands near the tracks and watches the train move away. The locomotive whistle vibrates at 500 Hz. (a) The air is still. What frequency does the uncle hear? What frequency does the girl hear? (b) A wind begins to blow from the east at 10 m/s. What frequency does the uncle now hear? What frequency does the girl now hear?*

[Velocity of sound in still air = 343 m s^{-1}]

Solution

(*a*) The uncle hears the frequency

$$f' = f\frac{v}{v+v_s} = 500\times\frac{343}{343+10} = 485.84 \text{ Hz.}$$

The girl and the engine move together with the same speed: $v_0 = v_s$ = 10 m s^{-1}.

(*i*) If the engine is in front of the compartment carrying the girl, she hears the frequency

$$f' = f\frac{v+v_o}{v+v_s} = f = 500 \text{ Hz.}$$

(*ii*) If the girl's compartment is in front of the engine,

$$f' = f\frac{v-v_o}{v-v_s} = f = 500 \text{ Hz.}$$

(*b*) If the wind blows from the source towards the listener with a velocity v_ω, then the effective velocity of sound = $v + v_\omega$. Thus, the uncle hears the frequency

$$f' = f\frac{(v+v_\omega)}{(v+v_\omega)+v_s} = 500\times\frac{343+10}{(343+10)+10} = 486.23 \text{ Hz.}$$

(*i*) If the wind blows from the locomotive whistle towards the girl, the effective velocity of sound = $v + v_\omega$, and the girl hears the frequency.

$$f' = f\frac{(v+v_\omega)+v_o}{(v+v_\omega)+v_s} = f = 500 \text{ Hz.}$$

(*ii*) If the wind blows from the girl to the whistle (the girl is in front of the engine), the effective velocity of sound = $v - v_\omega$. In this case also the girl hears the frequency.

$$f' = f\frac{(v-v_\omega)-v_o}{(v-v_\omega)-v_s} = f = 500 \text{ Hz.}$$

15. *Two tuning forks of frequencies 350 Hz each move relative to a stationary observer. One fork moves away from the observer, while the other moves toward him at the same speed. The observer hears beats of frequency 4 Hz. Find the speed of each tuning fork relative to the stationary observer. Velocity of sound in air = 340 m s^{-1}.*

Solution

The apparent frequency of the tuning fork coming towards the stationary observer is

$$f_1 = f\frac{v}{v - v_s}.$$

where v_s = speed of the tuning fork.

The apparent frequency of the tuning fork going away from the observer is

$$f_2 = f\frac{v}{v + v_s}.$$

Thus, the beat frequency is

$$4 = f_1 - f_2 = \frac{2fvv_s}{v^2 - v_s^2}.$$

or

$$2v_s^2 + fvv_s - 2v^2 = 0.$$

Since v_s is positive, the positive root of the above equation is

$$v_s = \frac{-fv + [f^2v^2 + 16v^2]^{1/2}}{4}$$

$$= 1.94 \text{ m s}^{-1}.$$

16. *A train approaching a hill at a speed of 40 km/h sounds a whistle of frequency 580 Hz when it is at a distance of 1 km from the hill. A wind with a speed of 40 km/h is blowing in the direction of motion of the train. Find (i) the frequency of the whistle as heard by an observer on the hill, (ii) the distance from the hill at which the echo from the hill is heard by the driver and its frequency. [Velocity of sound in air is 1200 km/h]* (I.I.T 1988)

Solution

(*i*) The apparent frequency is

$$f' = f\frac{(v + v_\omega)}{(v + v_\omega) - v_s}$$

$$= 580 \times \frac{1200 + 40}{(1200 + 40) - 40} = 599.33 \text{ Hz}.$$

(*ii*) Let the driver be at O' at a distance of x km from the hill when he hears the echo (Fig. 9.5). Time taken by the train to reach the point O' from O is

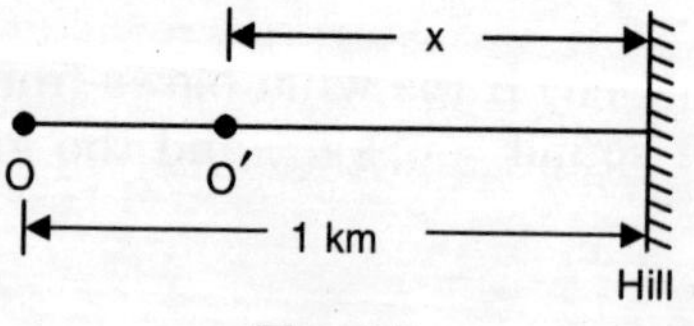

Fig. 9.5

$$t = \frac{1 - x}{40} \text{ h}.$$

Time taken by the sound to reach the point O' from O after reflection at the hill is

$$t = \frac{1}{v + v_\omega} + \frac{x}{v - v_\omega}$$

Thus, we have

$$\frac{1 - x}{40} = \frac{1}{1200 + 40} + \frac{x}{1200 - 40}$$

or
$$\frac{1}{40} - \frac{1}{1240} = \frac{x}{40} + \frac{x}{1160}$$

or
$$x = \frac{29}{31}\text{ km} = 0.935 \text{ km.}$$

For finding the frequency f'' as received by the driver we apply Doppler equation twice, first with the hill as observer (f') and then the hill as the source (see problem 12). Thus, we have

$$f'' = f'\left[\frac{(v - v_\omega) + v_o}{(v - v_\omega)}\right]$$

$$= 599.33 \times \frac{1200}{1160} = 620 \text{ Hz.}$$

17. *A source of sound is moving along a circular orbit of radius 3 m with an angular velocity of 10 rad/s. A sound detector located far away from the source is executing simple harmonic motion along the line BD (Fig. 9.6) with an amplitude BC = CD = 6 m. The frequency of oscillation of the detector is 5/π per second. The source is at the point A when the detector is at the point B. If the source emits a continuous sound wave of the frequency 340 Hz find the maximum and the minimum frequencies recorded by the detector.* (I.I.T. 1990)

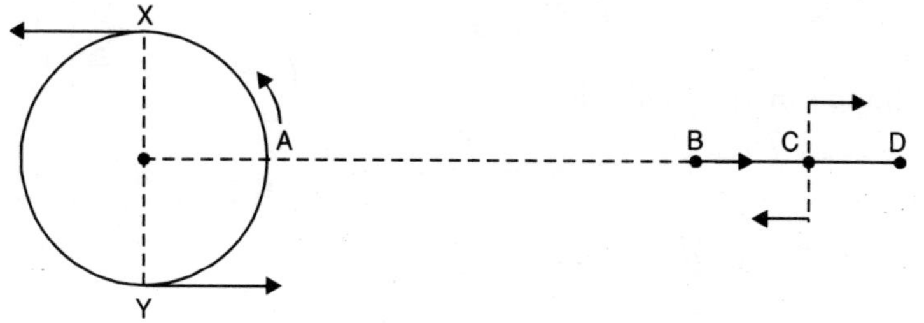

Fig. 9.6

Solution

The speed of the source = $v_s = r\omega = 30 \text{ m s}^{-1}$.

Time period of the source = $T = \frac{2\pi}{\omega} = \frac{\pi}{5}$ s.

The time periods for the circular motion and S.H.M are the same. Thus, when the source is at X, the detector is at C [they are moving away from each other] and the apparent frequency is minimum. When the source is at Y, the detector is at C [they are moving toward each other] and the apparent frequency is maximum. The velocity of the detector at C is

$$v_o = 6 \times 2\pi\nu = 6 \times 2\pi \times \frac{5}{\pi} = 60 \text{ m s}^{-1}.$$

Minimum frequency recorded is

$$f_{\min} = f\frac{v - v_o}{v + v_s} = 340 \times \frac{330 - 60}{330 + 30} = 255 \text{ Hz.}$$

Maximum frequency recorded is

$$f_{\max} = f\frac{v + v_o}{v - v_s} = 340 \times \frac{330 + 60}{330 - 30} = 442 \text{ Hz.}$$

18. *An earth satellite, transmitting on a frequency f passes directly over a radio receiving station at an altitude of 400 km and at a speed of 3.0 × 10⁴ km/h. Find the change in* *frequency, attributable to the Doppler effect, as a function of time, counting t = 0 as the instant the satellite is over the station. [Neglect the curvatures of the earth and of the satellite orbit]*

Solution

Altitude = 400 × 10^3 m = 4 × 10^5 m.

Velocity of the satellite = $v = \dfrac{3\times10^7}{60\times60} = \dfrac{1}{12}$ × 10^5 m s^{-1}.

Velocity of radio signal = c = 3 × 10^8 m s^{-1}.

Suppose at time t = 0 the satellite is at point A just over the radio receiving station E. The satellite is moving with velocity v and in time t it reaches the point B (Fig. 9.7). The component of the velocity v along EB is v sin θ moving away from E where ∠AEB = θ. The apparent frequency f' is given by

$$f' = f\frac{c}{c+v\sin\theta}$$

Since $\dfrac{v}{c} \ll 1$, we may write

$$f' = \frac{f}{1+\dfrac{v}{c}\sin\theta} \approx f\left(1-\frac{v}{c}\sin\theta\right)$$

Hence, the change in frequency is

$$\Delta f = f - f' = \frac{fv}{c}\sin\theta$$

Since
$$\sin\theta = \frac{vt}{[(4\times10^5)^2 + v^2t^2]^{1/2}},$$

$$\Delta f = \frac{ft}{36\times10^3}\frac{1}{[2304+t^2]^{1/2}} \text{ Hz.}$$

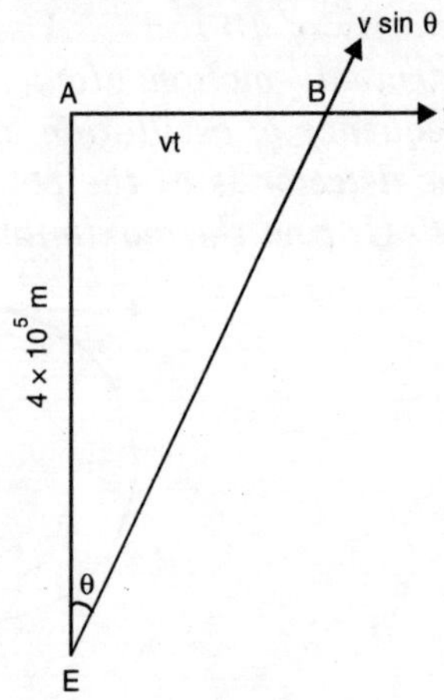

Fig. 9.7

19. *A spectral line of wavelength 6000 Å from a star is found to be shifted 1 Å towards the red. Find the velocity at which the star is receding from the earth.*

Solution

We have $\dfrac{\Delta\lambda}{\lambda} = \dfrac{v_s}{c}$ [see problem 4(*b*)]

Thus, $v_s = \dfrac{1}{6000}\times3\times10^8 = 5\times10^4$ m s^{-1}.

20. *A motor cyclist is moving towards a stationary car which is emitting sound of 165 Hz. and a police car is chasing the motor cyclist blowing siren at frequency 176 Hz. If the speed of police car is 22 m s^{-1}, then the speed of motor cyclist for which the motor cyclist hears no beats is*

(a) zero (b) 11 m s^{-1} (c) 22 m s^{-1} (d) 33 m s^{-1} (I.I.T. 2003)

Solution

The frequency recorded by motorcyclist from the sound of the stationary car is

$$f' = \frac{v+v_o}{v}f = \frac{330+v_o}{330}\times 165$$

The frequency recorded by motorcyclist from the sound of the moving police car is

$$f'' = \frac{v - v_o}{v - v_s} f = \frac{330 - v_o}{330 - 22} \times 176$$

For no beats, $f' = f''$

Thus,
$$\frac{330 + v_o}{330} \times 165 = \frac{330 - v_o}{308} \times 176$$

or
$$v_o = 22 \text{ m s}^{-1}.$$

Correct Choice: *c.*

21. *A siren placed at a railway platform is emitting sound of frequency 5 kHz. A passenger sitting in a moving train A records a frequency of 5.5 kHz while the train approaches the siren.*

During his return journey in a different train B he records a frequency of 6.0 kHz while approaching the same siren. The ratio of the velocity of train B to that of train A is

(a) 242/252 (b) 2 (c) 5/6 (d) 11/6. (I.I.T. 2002)

Solution

$$f_A = f\frac{v + v_A}{v} = f\left(1 + \frac{v_A}{v}\right)$$

$$f_A - f = f\,\frac{v_A}{v} = 0.5 \text{ kHz.}$$

$$f_B = f\frac{v + v_B}{v} = f\left(1 + \frac{v_B}{v}\right)$$

$$f_B - f = f\,\frac{v_B}{v} = 1 \text{ kHz}$$

Thus,
$$\frac{v_B}{v_A} = \frac{1}{0.5} = 2$$

Correct Choice: *b.*

22. *A boat is travelling in a river with a speed 10 m s^{-1} along the stream flowing with a speed 2 m s^{-1}. From this boat a sound transmitter is lowered into the river through a rigid support. The wavelength of the sound emitted from the transmitter inside the water is 14.45 mm.*

Assume that the attenuation of sound in water and air is negligible.

(a) What will be the frequency detected by a receiver kept inside the river downstream?

(b) The transmitter and the receiver are now pulled up to air. The air is blowing with a speed 5 m s^{-1} in the direction opposite the river stream. Determine the frequency of the sound detected by the receiver.

[Temperature of the air and water = 20°C,

Density of river water = 10^3 kg/m^3

Bulk modulus of the water = 2.088×10^9 Pa

Gas constant R = 8.31 J/mol-K

Mean molecular mass of water = 28.8×10^{-3} kg/mol

C_p/C_v for air = 1.4] (I.I.T. 2001)

Solution

(*a*) Velocity of sound in still water

$$V_{sw} = \sqrt{\frac{K}{\rho}} = \left[\frac{2.088 \times 10^9}{10^3}\right]^{\frac{1}{2}} = 1.445 \times 10^3 \text{ m s}^{-1}$$

where K = Bulk modulus of the fluid

ρ = Equilibrium density of the fluid

The frequency of sound emitted from the transmitter

$$= f = \frac{V_{sw}}{\lambda_w} = \frac{1.445 \times 10^3}{14.45 \times 10^{-3}} = 10^5 \text{ Hz.}$$

Here the source of sound is moving with velocity v_s = 10 m s^{-1} towards the receiver. The receiver is at rest v_o = 0. The water is flowing along the direction of sound with velocity v_w = 2 m s^{-1}.

$$f' = f\frac{v_{sw} + v_w}{(v_{sw} + v_w) - v_s} = 10^5 \times \frac{1.445 \times 10^3 + 2}{1.445 \times 10^3 + 2 - 10}$$

$$= 1.00696 \times 10^5 \text{ Hz}$$

(*b*) Velocity of sound in still air

$$v_{sa} = \sqrt{\frac{\gamma RT}{M}} = \left[\frac{1.4 \times 8.31 \times 293}{28.8 \times 10^{-3}}\right]^{\frac{1}{2}} = 344.03 \text{ m s}^{-1}$$

Here the source of sound moves towards stationary receiver with velocity v_s=10 m s^{-1}. The air is blowing opposite to the direction of sound. Hence the effective velocity of sound in air is

$$v_{sa} - v_a = 344.05 - 5 = 339.05 \text{ m s}^{-1}$$

$$f' = f\frac{v_{sa} - v_a}{(v_{sa} - v_a) - v_s} = 10^5 \times \frac{339.05}{329.03}$$

$$= 1.03045 \times 10^5 \text{ Hz.}$$

23. *A band playing music at frequency f is moving towards a wall at a speed v_b. A motorist is following the band with a speed v_m. If v is the speed of sound, obtain an expression for the beat frequency heard by the motorist.* (I.I.T. 1997)

Solution

The frequency directly received by the observer

$$f' = f\frac{v + v_o}{v + v_s}$$

The frequency received by the observer after reflection

$$f'' = f\frac{v + v_o}{v - v_s}$$

$$\text{Beat frequency} = n = f'' - f' = 2f\, v_b \frac{v + v_m}{v^2 - v_b^2}$$

where we put $v_o = v_m$, $v_s = v_b$.

24. *Two trains A and B with speeds 20 m s^{-1} and 30 m s^{-1} respectively in the same direction on the straight track, with B ahead of A, the engines are at the front ends. The engine of train A blows a long whistle.*

Assume that the sound of whistle is composed of components varying in frequency from f_1 = 800 Hz to f_2 = 1120 Hz, as shown in Fig. 9.8. The spread in the frequency (highest frequency – lowest frequency) is thus 320 Hz. The speed of sound in still air is 340 m s^{-1}.

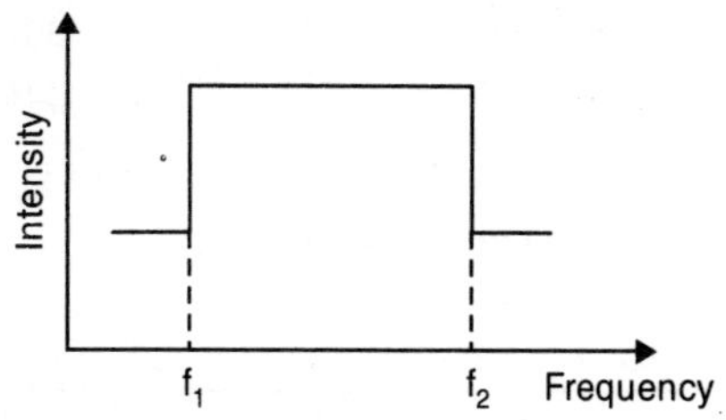

Fig. 9.8

(*i*) *The speed of sound of the whistle is*

(A) 340 m s^{-1} for passengers in A and 310 m s^{-1} for passengers in B

(B) 360 m s^{-1} for passengers in A and 310 m s^{-1} for passengers in B

(C) 310 m s^{-1} for passengers in A and 360 m s^{-1} for passengers in B

(D) 340 m s^{-1} for passengers in both the trains.

(*ii*) *The distribution of the sound intensity of the whistle as observed by the passengers in train A is best represented by*

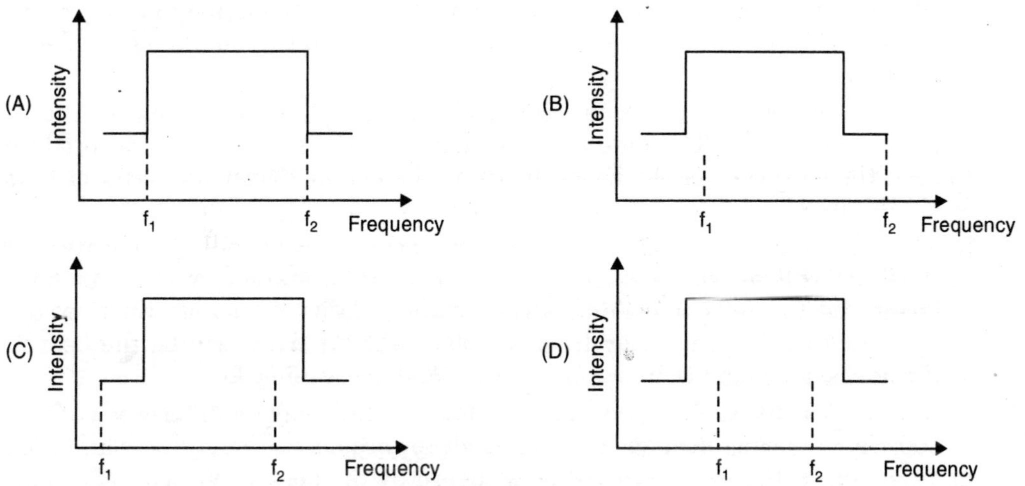

Fig. 9.9

(*iii*) *The spread of frequency as observed by the passengers in train B is*

(A) 310 Hz (B) 330 Hz (C) 350 Hz (D) 290 Hz. (I.I.T. 2007)

Solution

(*i*) The speed of sound does not depend on speed of source or observer.

Correct Choice: *D.*

(*ii*) Since there is no relative motion between the source and the observer, the frequency of sound heard by the passengers in train *A* will be same as the original frequency of sound emitted by the whistle of train *A*.

Correct Choice: *A.*

(*iii*) $f_B = f_A \dfrac{v - v_O}{v - v_S}$

$$f_B \text{ (for } f_A = 800 \text{ Hz)} = \frac{340 - 30}{340 - 20} \times 800 = 775 \text{ Hz}$$

$$f_B \text{ (for } f_A = 1120 \text{ Hz)} = \frac{340-30}{340-20} \times 1120 = 1085 \text{ Hz}$$

So the spread of frequency = 1085 – 775 = 310 Hz.

Correct Choice: *A.*

SUPPLEMENTARY PROBLEMS

1. Show that it is possible for zero frequency to be perceived if the observer is in motion, but not if the source is in motion and the observer is stationary.

2. A train approaches a railway station with a speed of 30 miles/h, continuously blowing a whistle of frequency 300 Hz. What is the frequency apparent to a person waiting on the platform of the station?
[Velocity of sound in air at that time = 1110 ft/s.]

3. An engine moves towards you with a speed of 33 m s^{-1} blowing a whistle of frequency 900 Hz. At the same time, you are moving towards the engine in a car with an equal speed. Calculate the apparent frequency as observed by you. [Velocity of sound in air = 333 m s^{-1}]

4. A train approaches a station with a velocity of 22 ft/s whistling all the time with frequency 250 Hz. The note on reflection from the station building produces beats. Find the frequency of the beats heard by the engine driver. [Velocity of sound in air = 1120 ft/s]

5. A tuning fork of frequency 500 Hz approaches a wall with a velocity of 5 m s^{-1}. (*a*) What will be the number of beats heard by a stationary observer between the direct and the reflected sound waves, if the velocity of sound is 330 m s^{-1} and the tuning fork is moving away from the observer? (*b*) What will be the beat frequency if the observer stands between the wall and the tuning fork?

6. A motor car fitted with two sounding horns, which have a difference in frequency by 300 Hz, is moving at a speed of 48 km/h towards a stationary person. Calculate the difference in the frequencies of notes heard by the person. Velocity of sound in air is 330 m s^{-1}.

7. Two cars pass each other in opposite directions, one of them blowing its horn, the frequency of the note emitted being 480 Hz. Calculate the frequencies heard on the other car before and after they have passed each other. The velocity of either car is 72 km/h and the velocity of sound is 320 m s^{-1}.

8. A train approaches a railway station with a speed of 90 miles/h. A sharp blast is blown with the whistle of the engine at intervals of one second. Find the interval between the successive blasts as heard by a person on the platform of the station. [Velocity of sound in air = 1120 ft/s.]

9. Calculate the percentage difference between the frequency of a note emitted by the whistle of a train approaching an observer with a velocity of 60 miles/h and that heard by the observer. [Velocity of sound in air = 1120 ft/s.]

10. The 15000 Hz whine of the turbines in the jet engines of an aircraft moving with speed 200 m s^{-1} is heard at what frequency by the pilot of a second craft trying to overtake the first at a speed of 250 m s^{-1}? Velocity of sound in air = 340 m s^{-1}.

11. An ambulance emitting a whine at 1500 Hz overtakes and passes a cyclist pedaling a bike at 10 ft/s. After being passed, the cyclist hears a frequency of 1490 Hz. How fast is the ambulance moving? [Velocity of sound in air = 1120 ft/s.]

12. A person in a train and another person near the rail track blow trumpets of same frequency 440 Hz. If there are 4 beats/s as they approach each other, what is the speed of the train? Velocity of sound in air = 1120 ft/s.

13. A submarine moving north with a speed of 75 km/h with respect to the ocean floor emits a sonar signal (sound wave in water) of frequency 1000 Hz. If the ocean at the point has a current moving north at 15 km/h relative to the land, what frequency is observed by a ship north to the submarine that does not have its engine running? Sonar waves travel at 5470 km/h.

14. An acoustic burglar alarm consists of a source emitting waves of frequency 20 kHz. What will be the beat frequency of waves reflected from an intruder walking at 0.9 m s^{-1} directly away from the alarm? Velocity of sound = 340 m s^{-1}.

15. Two trains are travelling toward each other at 100 ft/s relative to the ground. One train is blowing a whistle at 480 Hz. (*a*) What frequency will be heard on the other train in still air? (*b*) What frequency will be heard on the other train if the wind is blowing at 100 ft/s toward the whistle and away from the listener? (*c*) What frequency will be heard if the wind direction is reversed? [Velocity of sound in still air = 1125 ft/s]

16. A source of sound of frequency 256 Hz is moving rapidly towards a wall with a velocity of 5 m s^{-1}. How may beats per second will be heard if sound travels at a speed of 330 m s^{-1}? (*I.I.T. 1981*)

17. The calcium lines in the spectrum of light from a distant galaxy are found to occur at longer wavelengths than those for terrestrial light sources containing calcium. The measurements indicate that this galaxy is receding from us at 2.2×10^4 km/s. Calculate the fractional shift in wavelengths ($\Delta\lambda/\lambda$) of the calcium lines.

18. Could you go through a red light fast enough to have it appear green? Take 630 nm as the wavelength of red light and 540 nm as the wavelength of green light.

19. A whistle emitting a sound of frequency 440 Hz is tied to a string of 1.5 m length and rotated with an angular velocity of 20 rad.s^{-1} in the horizontal plane. Calculate the range of frequencies heard by an observer stationed at a large distance from the whistle. Speed of sound in air = 330 m s^{-1}. (*I.I.T. 1996*)

20. Certain characteristic wavelengths in the light from a galaxy are observed to be increased in wavelength, as compared with terrestrial sources, by about 0.4%. What is the radial speed of this galaxy with respect to the earth? Is it approaching or receding?

21. A train moves towards a stationary observer with speed 34 m s^{-1}. The train sounds a whistle and its frequency registered by the observer is f_1. If the train's speed is reduced to 17 m s^{-1}, the frequency registered is f_2. If the speed of sound is 340 m s^{-1}, then the ratio f_1/f_2 is

(*a*) 2 (*b*) $\frac{1}{2}$ (*c*) $\frac{18}{19}$ (*d*) $\frac{19}{18}$. (*I.I.T. 2000*)

22. A bus is moving towards a huge wall with a velocity of 5 m s^{-1}. The driver sounds a horn of frequency 200 Hz. The frequency of the beats heard by a passenger of the bus will be .... Hz. (*I.I.T. 1994*)

23. A whistling train approaches a junction. An observer standing at the junction observes the frequency to be 2.2 kHz. and 1.8 kHz of the approaching and the receding train. Find the speed of the train (speed of sound = 300 m s^{-1}). (*I.I.T. 2005*)

10 Acoustics of Buildings

10.1 REVERBERATION

If a loud sound wave is produced in an ordinary room with good reflecting walls, the wave undergoes a large number of reflections at the walls. The repeated reflections produce persistence of sound—this phenomenon is called reverberation. In an auditorium or classroom excessive reverberation is not desirable. However, some reverberation is necessary in a concert hall.

10.2 TIME OF REVERBERATION

It is the time required by the energy density to fall to the minimum audibility value (E) from an initial steady value 10^6E (*i.e.* million times minimum audibility) when the source of the sound wave is removed. The optimum time of reverberation is about 0.5 s for a medium sized room, 0.8 s to 1.5 s for an auditorium, 1 s to 2 s for a music room and greater than 2 s for a temple.

10.3 SABINE'S LAW

From a large number of experiments Sabine has given the following equation for the time of reverberation (T):

$$T = K\frac{V}{a}$$

where K is a constant, V is the volume of the enclosure and a is the total absorption power of all surfaces. The latter is measured in unit Sabin. Again, $a = S\overline{a}$, where S is the total surface area in sq. ft. and $\overline{a}$ is the mean absorption coefficient. The absorption coefficient is the fraction of the energy absorbed to the energy incident on the surface. For an open window, absorption coefficient = 1. For marble, the absorption coefficient is found to be 0.01 *i.e.* it absorbs only 1% of the sound energy at each incidence.

10.4 DECIBEL (dB) UNIT OF SOUND LEVEL

Instead of speaking of the intensity I of a sound wave, it is found to be more convenient to speak of a *sound level* β, which is defined as

$$\beta = 10 \log \left(\frac{I}{I_o}\right) \text{ dB},$$

where I_o is the standard reference intensity = 10^{-12} W/m^2 that is near the lower limit of human audibility. When $I = I_o$, $\beta = 0$. Thus, the threshold audibility corresponds to zero decibel.

SOLVED PROBLEMS

1. *Discuss the theory of growth and decay of sound in a 'live room' and find an expression for the reverberation time.*

Solution

In a 'live room', the absorption coefficient of the material of the walls of the room is small (less than 0.4) so that there is increase in loudness of sound due to reverberation.

Growth of sound energy—Suppose that there is a source of sound of constant output in a live room. Due to reverberation, initially the energy density of the sound increases and after sometime it attains a maximum value. We shall find an expression of the intensity of sound energy in the enclosure at any time under the following assumptions:

(*i*) the source emits sound energy at a constant rate.

(*ii*) the sound propagates in all possible directions.

(*iii*) there is uniform distribution of energy inside the room.

Consider an elementary area dS on the wall and an elementary volume dV inside the room at a distance r from the surface dS (Fig. 10.1). The normal to the surface dS makes an angle θ with the direction of r. In the spherical polar coordinate system, we have

$$dV = r^2 \sin\theta \, d\theta \, d\phi \, dr.$$

Let the solid angle subtended by dS at the elementary vloume be $d\Omega$,

$$d\Omega = \frac{dS \cos\theta}{r^2}.$$

Let the energy density inside the room be E at any time. Then the energy contained in the volume dV is EdV which propagates in all directions.

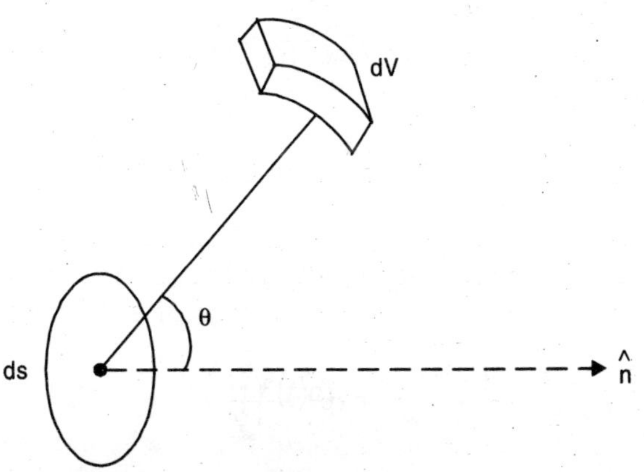

Fig. 10.1

Out of this energy, the amount that is directed towards the soild angle $d\Omega$ is

$$\frac{d\Omega}{4\pi}EdV = \frac{E}{4\pi}\frac{dS\cos\theta}{r^2}.r^2\sin\theta\, d\theta\, d\phi\, dr.$$

Total energy incident on one side of the surface dS of the wall in unit time interval is

$$\int_{r=0}^{v}\int_{\theta=0}^{\pi/2}\int_{\phi=0}^{2\pi}\frac{EdS}{4\pi}\sin\theta\cos\theta\, d\theta\, d\phi\, dr = \frac{EvdS}{4},$$

where v is the velocity of sound wave in air. Thus, the energy of the wave falling on unit area of the wall per unit time, *i.e.*, the intensity of the sound wave, is

$$I = \frac{Ev}{4} \qquad ...(10.1)$$

If the interior of the walls is made of different materials of areas, ds_1, ds_2,..., with absorption coefficients a_1, a_2,... respectively, then the total absorptive power of the walls may be written as

$$a = \sum_{r=1}^{n} a_r ds_r.$$

The total rate of absorption by the wall is $\frac{Eva}{4}$.

Thus, the total rate of increase of energy of the wave inside the room of total volume V is

$$V\frac{dE}{dt} = P - \frac{Eva}{4} \qquad ...(10.2)$$

where P is the constant output power of the source of the sound wave. From Eqn. (10.2), we have

$$\frac{VdE}{P-\frac{Eva}{4}} = dt.$$

Solving this equation, we get

$$-\frac{4V}{av}\ln\left(P-\frac{Eva}{4}\right) = t + C.$$

Now suppose that the source starts operating at time $t = 0$ so that at $t = 0$, $E = 0$. With this initial condition, we obtain

$$t = \frac{4V}{av}\left[\ln P - \ln\left(P-\frac{Eva}{4}\right)\right]$$

or

$$\frac{P}{P-\frac{Eva}{4}} = \exp\left(\frac{avt}{4V}\right)$$

or

$$1 - \frac{Eva}{4P} = \exp\left(-\frac{avt}{4V}\right)$$

or
$$E = \frac{4P}{av}\left[1 - \exp\left(-\frac{avt}{4V}\right)\right]. \quad ...(10.3)$$

Eqn. (10.3) gives the energy density of the sound wave in the live room at any time t.

When $t \rightarrow \propto$, $\quad E = E_{max} = \frac{4P}{av}.$

Thus,
$$E = E_{max}\left[1 - \exp\left(-\frac{avt}{4V}\right)\right] \quad ...(10.4)$$

and
$$I = I_{max}\left[1 - \exp\left(-\frac{avt}{4V}\right)\right] \quad ...(10.5)$$

where I_{max} is the maximum intensity of the sound wave and it is seen form Eqn. (10.1) that
$$I_{max} = \frac{E_{max}v}{4} = \frac{P}{a}.$$

Decay of sound energy: When the energy density E attains its maximum value E_{max}, the source of the sound wave is cut off. The energy density then begins to fall. In Eqn. (10.2), we put $P = 0$:
$$V\frac{dE}{dt} = -\frac{Eva}{4}. \quad ...(10.6)$$

Solving this equation, we get
$$E = C\exp\left(-\frac{avt}{4V}\right).$$

When $t = 0$, $E = E_{max} = \frac{4P}{av}$. Thus, we get
$$E = E_{max}\exp\left(-\frac{avt}{4V}\right) \quad ...(10.7)$$

and
$$I = I_{max}\exp\left(-\frac{avt}{4V}\right). \quad ...(10.8)$$

Eqns. (10.7) and (10.8) show that, when the source is removed, the energy density does not go to zero immediately, but falls exponentially.

Time of reverberation: We take in Eqn. (10.7) the initial steady value of the energy density $E_{max} = 10^6E$, where E is the energy density at the threshold audibility. If T is the time of reverberation, we have
$$E = 10^6E\ \exp\left(-\frac{avT}{4V}\right)$$

or
$$\frac{avT}{4V} = 6\ln 10 = 6 \times 2.303$$

or
$$T = \frac{55.27V}{av} \quad ...(10.9)$$

Thus $$T = K\frac{V}{a} \quad ...(10.10)$$

where $K = \dfrac{55.27}{v}$ is a constant.

Eqn. (10.10) is nothing but Sabine's law of time of reverberation. Note that a has the dimension of (length)^2.

When v = 1100 ft/s, K = 0.05

When v = 344 m s^{-1}, K = 0.16.

2. *Derive the expression of the time of reverberation in a 'dead room'.*

Solution

In a 'dead room' the absorption coefficient of the material of the walls of the room is large (greater than 0.4) so that the increase in loudness of sound due to reverberation is very small. In this case higher order reflections should be taken into consideration. If $\bar{a}$ is the mean absorption coefficient of the materials of the walls, everytime the wave strikes the wall, a fraction $\bar{a}$ of its energy is absorbed and $(1-\bar{a})$ fraction is reflected. Thus, after removing the source of sound, the intensity falls to $I = I_0(1-\bar{a})^n$ after n successive reflections, where I_0 is the initial intensity. If $I_0 = 10^6 I$, where I is the threshold audibility, then we have

$$\frac{I}{I_0} = 10^{-6} = (1-\bar{a})^n$$

or $$n = \frac{-6}{\log_{10}(1-\bar{a})} = \frac{-6\times 2.303}{\ln(1-\bar{a})}.$$

If l is the average distance traversed by the wave between successive reflections on the walls, then $nl = vT$, where v is the velocity of the sound wave and T is the reverberation time. We also have $l = 4V/S$ where V = total volume and S = total surface area.

Thus, we have

$$l = \frac{4V}{S} = \frac{vT}{n}$$

or $$T = \frac{4nV}{vS} = \frac{4V}{vS}\times\frac{(-6)\times 2.303}{\ln(1-\bar{a})}$$

or $$T = \frac{-55.27V}{vS\ln(1-\bar{a})} \quad ...(10.11)$$

which is known as Eyring's formula.

Note: When $\bar{a}$ is small, $\ln(1-\bar{a}) = -\bar{a}$ and $S\bar{a} = a$ = total absorption power of the walls, Eqn. (10.11) reduces to

$$T = \frac{55.27V}{va}$$

which is same as Sabine's law [Eqn. (10.9)].

3. *In an auditorium a source of sound of power P_1 is switched on. After sufficiently long time, the source is switched off and the time t_1 during which the energy density falls to threshold audibility is noted. The same experiment is performed with a different source of* *power P_2 and the corresponding time t_2 is noted. Show that the average absorption coefficient $\bar{a}$ of the materials of the walls is given by*

$$\bar{a} = \frac{4V\ln(P_1/P_2)}{vS(t_1 - t_2)}$$

where V = *Volume of the auditorium*
S = *Surface area of the walls of the auditorium*
v = *Velocity of sound in air.*

Solution

If E is the energy density of the threshold audibility, then from Eqn. (10.7), we have

$$E = \frac{4P_1}{av}\exp\left(-\frac{avt_1}{4V}\right) = \frac{4P_2}{av}\exp\left(-\frac{avt_2}{4V}\right)$$

or

$$\frac{P_1}{P_2} = \exp\left[\frac{av}{4V}(t_1 - t_2)\right]$$

or

$$a = \frac{4V\ln(P_1/P_2)}{v(t_1 - t_2)}.$$

Since $a = \bar{a}S$, we have

$$\bar{a} = \frac{4V\ln(P_1/P_2)}{vS(t_1 - t_2)}.$$

4. *Average absorption coefficient of a room of height 15 ft, breadth 20 ft and length 30 ft, is 0.2. Find the reverberation time of the room.*

Solution

Volume of the room = $V = 15 \times 20 \times 30 = 9000$ cu. ft.

Surface area of the room = $2 \times [15 \times 20 + 15 \times 30 + 20 \times 30] = 2700$ sq. ft.

From Eqn. (10.10), we have

$$T = \frac{0.05V}{\bar{a}S} = \frac{0.05 \times 9000}{0.2 \times 2700} = 0.83 \text{ s.}$$

5. *A hall has a volume of 2000 m^3. Its total absorption is equivalent to 90 m^2 of open window. What will be the effect on the reverberation time if an audience fills the hall and thereby increases the absorption by another 90 m^2?*

Solution

$$T = \frac{0.16V}{a} = \frac{0.16 \times 2000}{90} = 3.56 \text{ s.}$$

When the hall is filled with the audience, the total absorption a' becomes $90 + 90 = 180$ m^2 of open window. The new reverberation time T' is

$$T' = \frac{0.16 \times 2000}{180} = \frac{T}{2} = 1.72 \text{ s.}$$

6. *A studio measuring 20 m × 12 m × 8 m, has a reverberation time of 1.0 s when empty. What will be the reverberation time when an audience of 200 are present inside the studio? Assume that each person is equivalent to 0.4 m² of absorption.*

Solution

Total absorption power of the empty studio is

$$a = \frac{0.16V}{T} = \frac{0.16 \times 20 \times 12 \times 8}{1.0} = 307.2 \text{ units.}$$

Due to presence of the audience, the value of a increases by 200 × 0.4 = 80 m². Thus, the new reverberation time is

$$T' = \frac{0.16V}{387.2} = 0.79 \text{ s.}$$

7. *An auditorium is 20 ft high, 50 ft wide and 100 ft long and contains 500 wooden seats. Each seat has total absorption power 0.2 unit. The walls, floor and ceiling have an average absorption coefficient 0.03. What is the reverberation time when the auditorium is empty?*

Solution

Volume of the auditorium = V = 20 × 50 × 100 = 10^5 cu. ft.

Total surface area of the walls, ceiling and floor

= 2 × (20 × 50 + 20 × 100 + 50 × 100) = 16000 sq. ft.

Absorption units for the walls, ceiling and floor = 0.03 × 16000 = 480 units.

Absorption units for the empty wooden seats = 0.2 × 500 = 100 units

Total absorption power = a = 580 units.

$$\text{Reverberation time} = T = \frac{0.05V}{a} = 8.62 \text{ s.}$$

8. *In problem 7, how much acoustic material of absorption coefficient 0.4 must be placed in the room so that the reverberation time becomes 2 s when the room is empty?*

Solution

When T = 2 s, the value of a is

$$\frac{0.05V}{T} = \frac{0.05 \times 10^5}{2} = 2500 \text{ units.}$$

Thus, the acoustic material of 2500 – 580 = 1920 units of absorption is required. The area of the required acoustic material $S = \frac{1920}{0.4} = 4800$ sq. ft.

9. *In problem 7 what would be the reverberation time when the auditorium is full? Assume that each person has total absorption power 4.7 units.*

Solution

Total absorption units of 500 persons = 500 × 4.7 = 2350 units

Hence, the total absorption power of full auditorium

= a = 2350 + 580 = 2930 Sabin.

$$\text{Reverberation time} = \frac{0.05V}{a} = \frac{0.05 \times 10^5}{2930} = 1.71 \text{ s.}$$

10. *It is found that one source of sound is 20 dB louder than another source. What is the ratio of their intensities?*

Solution

We have, $10 \log (I_1/I_0) - 10 \log (I_2/I_0) = 20$

or $\log (I_1/I_2) = 2$

or $I_1/I_2 = 10^2 = 100.$

The intensity of sound coming from the first source is 100 times greater than that of the second one.

11. *The sound level in a classroom is 50 dB. How much is the intensity of sound wave (W/m²) in that classroom? What is the corresponding value of energy density of sound wave in the room?*

[Velocity of sound in air = 330 m s⁻¹]

Solution

We know, $\beta = 10 \log (I_1/I_0) = 50$

or $I_1 = I_0 \times 10^5 = 10^{-12} \times 10^5 = 10^{-7}$ W/m².

From Eqn. (10.1) we have for the energy density

$$E = \frac{4I}{v} = \frac{4 \times 10^{-7}}{330} = 1.2 \times 10^{-9} \text{ J/m}^3.$$

12. *Spherical sound waves are produced uniformly in all directions from a point source, the radiated power P being 40 W. What is the intensity of the sound waves at a distance 2 m from the source. What is the corresponding sound level?*

Solution

All the radiated power must pass through a sphere of radius r centered at the source. Hence,

$$I = \frac{P}{4\pi r^2} = \frac{40}{4\pi \times 4} = 0.80 \text{ W/m}^2.$$

The corresponding value of the sound level β is

$$\beta = 10 \log (I/I_0) = 10 \log\left(\frac{0.80}{10^{-12}}\right) = 119 \text{ dB}.$$

Note: This value is very close to the threshold of pain (120 dB) of sound level for the human beings.

13. *An office room of size 10 ft high, 20 ft wide and 30 ft long has walls made of plaster, wood and glass with mean absorption coefficient 0.03. The floor is covered with a carpet of absorption coefficient 0.15 and the ceiling with acoustic plaster of absorption coefficient 0.40. What is the reverberation time if five persons are present in the room? Each person has total absorption power of 4.6 Sabin. If four persons are typing on four type writers, each producing 1 erg of sound per second, what will be the intensity level in the room after sufficiently long time?*

Solution

$$\text{Volume of the room} = V = 10 \times 20 \times 30 = 6000 \text{ cu. ft.}$$

$$\text{Absorption units for four walls} = 0.03 \times 1000 = 30 \text{ units.}$$

$$\text{Absorption units for the floor} = 0.15 \times 600 = 90 \text{ units}$$

$$\text{Absorption units for the ceiling} = 0.40 \times 600 = 240 \text{ units}$$

$$\text{Absorption units for 5 persons} = 4.6 \times 5 = 23 \text{ units}$$

$$\text{Total absorption power} = a = 383 \text{ units.}$$

$$\text{Hence, reverberation time} = \frac{0.05V}{a} = \frac{0.05 \times 6000}{383} = 0.78 \text{ s.}$$

Maximum energy density after sufficiently long time is

$$E_m = \frac{4P}{av}$$

$$\text{Maximum intensity} = I_m = \frac{E_m v}{4} = \frac{P}{a}$$

Hence,

$$I_m = \frac{4 \times 1 \text{ erg/s}}{383 \text{ sq.ft}} = \frac{4 \times 10^{-7} \text{ J/s}}{383 \times 9.29 \times 10^{-2} \text{ m}^2}$$

$$= 1.12 \times 10^{-8} \text{ W/m}^2$$

In terms of decibel (dB) unit of sound level it is

$$\beta = 10 \log \left(\frac{1.12 \times 10^{-8}}{10^{-12}} \right) = 40.5 \text{ dB.}$$

SUPPLEMENTARY PROBLEMS

1. In an empty room of size 10 ft high, 20 ft wide and 30 ft long a source of sound of power 10 W is switched on. After sufficiently long time the source is switched off and the time during which the energy density falls to threshold audibility is found to be 1.2 s. This same experiment is performed with a different source of power 20 W and the corresponding time of decay to threshold audibility is found to be 1.3 s. Find the average absorption coefficient $\bar{a}$ of the materials of the walls. What is the reverberation time of the empty room?
2. A hall of volume 48000 cu. ft. is found to have a reverberation time of 2 s. If the area of the sound absorbing surface be 6000 sq. ft., calculate the mean absorption coefficient.
3. In a dining room of rectangular shape 11 ft high, 20 ft wide and 30 ft long has 20 seats. Each seat has total absorption power of 0.15 units. The walls, floor and ceiling have an average absorption coefficient 0.03. What is the reverberation time when the room is empty? What is the reverberation time when the dining room is full? Assume that each person has total absorption power of 4.7 units.
4. A hall has a volume of 2300 m^3 and its total absorption is equivalent to 93 m^2 of an open window. How many persons must sit in the hall so that the reverberation time

becomes 2 s, given that the absorption area of one person is equivalent to 0.379 m^2 of open window. Calculate also the reverberation time of the empty hall.

5. The threshold of pain for sound waves for human being is 120 dB. What is the corresponding value of intensity of sound wave?
6. Two sound waves have intensities I_1 and I_2. What is the difference in their sound levels?
7. How much more intense is an 80 dB shout than a 20 dB whisper?
8. A certain sound level is increased by an additional 30 dB. Show that its intensity increases by a factor of 1000.
9. A single violin produces an intensity level of 50 dB at a particular seat. How many decibels will be produced at that position by 10 such violins playing together.
10. The source of a sound wave delivers 2 μW of power. If it is a point source (*a*) What is the intensity 2 m away and (*b*) What is the sound level in decibels at that distance?
11. (*a*) Show that the intensity I is the product of the energy per unit volume E and the speed of propagation v of a wave disturbance in *free space.* (*b*) Radio waves travel at a speed of 3×10^8 m/s. Find the energy density in a radiowave 500 km from a 50,000 W source, assuming the waves to be spherical and the propagation to be isotropic.
12. You are standing at a distance x from an isotropic source of sound waves. You walk 5 m towards the source and observe that the intensity of these waves has doubled. Calculate the distance x.
13. In a test a subsonic jet flies overhead at an altitude of 100 m. The sound intensity on the ground as the jet passes overhead is 160 dB. At what altitude should the plane fly so that the ground noise is no greater than 120 dB, the threshold of pain?

11 Electromagnetic Waves

11.1 MAXWELL'S EQUATIONS

The four basic principles of electromagnetic theory are stated in mathematical form as the four equations of Maxwell:

I. Gauss's law for electricity

$$\vec{\nabla}.\vec{D} = \rho \qquad ...(11.1)$$

II. Gauss's law for magnetism

$$\vec{\nabla}.\vec{B} = 0 \qquad ...(11.2)$$

III. Faraday's law of induction

$$\vec{\nabla}\times\vec{E} = -\frac{\partial \vec{B}}{\partial t} \qquad ...(11.3)$$

IV. Ampere-Maxwell's law

$$\vec{\nabla}\times\vec{H} = \frac{\partial \vec{D}}{\partial t}+\vec{J} \qquad ...(11.4)$$

Here,

$\vec{D}$ = Electric displacement in C/m^2

$\vec{B}$ = Magnetic induction in Wb/m^2 or Tesla (T)

$\vec{E}$ = Electric field intensity in V/m

$\vec{H}$ = Magnetic field intensity in A/m

ρ = Free charge density in C/m^3

$\vec{J}$ = Current density in A/m^2.

11.2 PROPAGATION OF PLANE ELECTROMAGNETIC WAVES IN MATTER

For propagation of plane electromagnetic waves in homogeneous, isotropic, linear and stationary media, the following relations hold:

$$\vec{D} = K_e\varepsilon_0\vec{E} = \varepsilon\vec{E} \qquad ...(11.5)$$

$$\vec{B} = K_m\mu_0\vec{H} = \mu\vec{H} \qquad ...(11.6)$$

$$\vec{J} = \sigma\vec{E}. \quad ...(11.7)$$

Here,

ε = Permittivity of the medium

μ = Magnetic permeability of the medium

ε_0 = Permittivity constant = 8.85×10^{-12} $C^2/N.m^2$ or F/m

μ_0 = Permeability constant

= $4\pi \times 10^{-7}$ H/m or T.m/A = 1.26×10^{-6} H/m or T.m/A

K_e = Dielectric constant or relative permittivity = $\varepsilon/\varepsilon_0$

K_m = Relative permeability = μ/μ_0

σ = Conductivity of the medium in mho/m or A/V.m.

The electromagnetic waves are transverse in nature and the electric vector $\vec{E}$ and the magnetic vector $\vec{H}$ are always mutually perpendicular. These vectors $\vec{E}$ and $\vec{H}$ are so oriented that their vector product $\vec{E} \times \vec{H}$ points in the direction of propagation of the electromagnetic wave. In non-conductors ($\sigma = 0$), the phase velocity of the electromagnetic wave is

$$u = \frac{c}{(K_e K_m)^{1/2}}$$

[c is the velocity of light in free space = 3×10^8 m s^{-1}]

In non-magnetic media ($K_m = 1$), the index of refraction n is related to the dielectric coefficient by the relation

$$n = K_e^{1/2} \quad ...(11.8)$$

In non-conductors, the vectors $\vec{E}$ and $\vec{H}$ are in phase, and the electric and magnetic energy densities are equal:

$$\frac{1}{2}\varepsilon E^2 = \frac{1}{2}\mu H^2. \quad ...(11.9)$$

11.3 ENERGY FLOW AND POYNTING VECTOR

The Poynting vector, defined by

$$\vec{S} = \vec{E}\times\vec{H} \quad ...(11.10)$$

gives the energy flux (W/m^2) in an electromagnetic wave. For plane waves in non-conductors, the intensity of the waves, *i.e.*, the average of S, is

$$I = \overline{S} = \left(\frac{\varepsilon}{\mu}\right)^{1/2} E_{\text{rms}}^2 = 2.65\times10^{-3}\left(\frac{K_e}{K_m}\right)^{1/2} E_{\text{rms}}^2 \text{ W/m}^2 \quad ...(11.11)$$

In free space $K_e = K_m = 1$, and the intensity of the wave is

$$I = \overline{S} = \left(\frac{\varepsilon_0}{\mu_0}\right)^{1/2} E_{\text{rms}}^2 = 2.65\times10^{-3} E_{\text{rms}}^2 \text{ W/m}^2. \quad ...(11.12)$$

11.4 RADIATION PRESSURE

Electromagnetic radiation falling on a surface exerts a pressure on it. The average force on a unit area of a plane mirror due to radiation falling normally on it in free space or the radiation pressure is given by

$$p = \frac{2\overline{S}}{c} = 1.77 \times 10^{-11} E_{rms}^2 \text{ N/m}^2 \qquad ...(11.13)$$

We can ascribe this pressure to a change in momentum $2\overline{S}/c$ per unit time and per unit area in the incident wave, the factor 2 being required because the wave is reflected with a momentum equal to its initial momentum but of opposite sign. The factor 2 will not be required if the radiation falls on a perfect absorber.

11.5 POLARIZATION OF ELECTROMAGNETIC WAVE

The transverse electromagnetic wave is said to be polarized (more specifically, *plane polarized*) if the electric field vectors are parallel to a particular direction for all points in the wave. The direction of the electric field vector $\vec{E}$ is called the *direction of polarization;* the plane containing $\vec{E}$ and the direction of propagation is called the *plane of vibration*.

In a sheet of polarizing material called *polaroid,* there exists a certain characteristic polarizing direction. The sheet will transmit only those wave train components whose electric vectors vibrate parallel to the polarizing direction and will absorb those that vibrate at right angles to this direction. The intensity of polarized light passing through a sheet of polaroid is reduced from I_0 to I where

$$I = I_0 \cos^2 \theta \text{ (Law of Malus)} \qquad ...(11.14)$$

Here, θ is the angle between the plane of vibration and the polarizing direction of the sheet.

We consider the case of plane waves travelling in the z-direction. If E_x is different from zero, but E_y is equal to zero for all z and t, the waves are said to be plane polarized along x-direction. We can also have any combination of E_x and E_y (in the case of a single frequency) with an arbitrary relative phase between E_x and E_y. Then we have a general state of polarization called *elliptical polarization* [See chapter 2].

SOLVED PROBLEMS

1. *Set up Maxwell's electromagnetic field equation in differential forms and explain the physical significance of each.*

Solution

I. First Equation

Gauss' law for a distribution of changes of volume V in a dielectric medium is given by

$$\oint_S \vec{D}.d\vec{s} = \int_V \rho \, dV$$

where ρ is the volume charge density and V is the volume enclosed by the closed surface S.

Using the divergence theorem, we have

$$\oint_s \vec{D}\cdot d\vec{s} = \int_V \vec{\nabla}\cdot\vec{D}\, dV$$

Thus, $$\int_V (\vec{\nabla}\cdot\vec{D} - \rho)\, dV = 0$$

which is true for any arbitrary small volume dV.

Thus, $$\vec{V}\cdot\vec{D} = \rho$$

It is just the differential form of Gauss's law in electrostatics.

II. Second Equation

Gauss's theorem for magnetostatics states that the total flux of magnetic induction through any closed surface is zero. *i.e.*,

$$\oint_s \vec{B}\cdot d\vec{s} = 0$$

Using divergence theorem, we get

$$\int_V \vec{\nabla}.\vec{B}\, dV = 0$$

It is true for any arbitrary volume V.

Thus $$\vec{\nabla}.\vec{B} = 0$$

It is just the differential form of Gauss' law in magnetostatics. It also signifies the non existence of magnetic monopoles.

III. Third Equation

The emf induced in a closed loop is given by Faraday's law of electromagnetic induction:

$$\varepsilon = -\frac{d\phi}{dt} = -\int_{\substack{\text{Enclosed}\\ \text{surface}}} \frac{\partial\vec{B}}{\partial t}.d\vec{s}$$

where ϕ is the magnetic flux linked with the closed surface. The negative sign indicates that the induced emf opposes the change in flux.

Again this emf is also given by

$$\varepsilon = \int_{\substack{\text{Enclosed}\\ \text{curve}}} \vec{E}.d\vec{l}$$

Thus, we have

$$\int_{\substack{\text{Enclosed}\\ \text{curve}}} \vec{E}.d\vec{l} = -\int_{\substack{\text{Enclosed}\\ \text{surface}}} \frac{\partial\vec{B}}{\partial t}.d\vec{s}$$

Using Stokes' theorem, we have

$$\int_{\substack{\text{Enclosed}\\ \text{curve}}} \vec{E}.d\vec{l} = \int_s (\vec{\nabla}\times\vec{E}).d\vec{s}$$

Thus,

$$\int_s \left[\left(\vec{\nabla}\times\vec{E}\right)+\frac{\partial\vec{B}}{\partial t}\right].d\vec{s}=0$$

The vanishing of the integral for any arbitrary closed surface requires that the integrand itself is zero.

$$\vec{\nabla}\times\vec{E} + \frac{\partial\vec{B}}{\partial t} = 0$$

It is just the differential form of Faraday's law of electromagnetic induction.

IV. Fourth Equation

The differential form of Ampere's Circuital law is

$$\vec{\nabla}\times\vec{H} = \vec{J}$$

where $\vec{J}$ is the conduction current density in A/m^2.

Since the divergence of any curl of a vector is zero, we have

$$\vec{\nabla}\cdot\left(\vec{\nabla}\times\vec{H}\right) = \vec{\nabla}\times\vec{J} = 0$$

This is true only when the charge density ρ is a constant or, the electric field is steady. When the charge density changes with time it satisfies the equation of continuity

$$\vec{\nabla}\cdot\vec{J} + \frac{\partial\rho}{\partial t} = 0$$

Thus in general the equation $\vec{\nabla}\cdot\vec{J} = 0$ is not in conformity with the equation of continuity. To resolve this problem Maxwell introduced the concept of **displacement current**.

We know from Gauss's law for electricity

$$\vec{\nabla}\cdot\vec{D} = \rho$$

From equation of continuity we have

$$\vec{\nabla}.\vec{J}+\frac{\partial}{\partial t}(\vec{\nabla}.\vec{D})=\vec{\nabla}.\vec{J}+\vec{\nabla}.\left(\frac{\partial\vec{D}}{\partial t}\right)$$

$$= \vec{\nabla}\cdot\left(\vec{J}+\frac{\partial\vec{D}}{\partial t}\right) = 0$$

If we add $\frac{\partial\vec{D}}{\partial t}$ to $\vec{J}$ then the divergence of total current density $\left(\vec{J}+\frac{\partial\vec{D}}{\partial t}\right)$ is zero. Thus, we have the generalised Ampere's law

$$\vec{\nabla}\times\vec{H} = \vec{J}+\frac{\partial\vec{D}}{\partial t}$$

which is known as Ampere-Maxwell law. It incorporates the equation of continuity. The quantity $\frac{\partial\vec{D}}{\partial t}$ introduced by Maxwell is called the **displacement current density** . When the electric field does not change with time, the displacement current is zero and Ampere's circuital law is applicable.

2. *Show that the displacement current in the dielectric of a parallel-plate capacitor is equal to the conduction current in the connecting leads.*

Solution

The capacitance of a parallel-plate capacitor is given by

$$C = \frac{\varepsilon A}{d}$$

where A is the area of the plate and d is the plate separation.

At any instant if V is the voltage across the capacitor then the charge on the plate of the capacitor is

$$q = CV$$

Now, the electric field in the dielectric is (neglecting the fringe effect)

$$E = \frac{V}{d}.$$

We also have

$$\vec{D} = \varepsilon \vec{E}$$

where $\vec{D}$ is normal to the plates.

The displacement current is

$$i_D = \int_s \frac{\partial \vec{D}}{\partial t} \cdot d\vec{s} = \int_s \varepsilon \frac{\partial \vec{E}}{\partial t} \cdot d\vec{s}$$

$$= \frac{\varepsilon A}{d} \frac{dV}{dt} = C \frac{dV}{dt} = \frac{d}{dt}(CV)$$

$$= \frac{dq}{dt} = i_C$$

where i_C is the conduction current.

3. *(a) In a medium which is neither a good conductor nor a perfect dielectric has electrical conductivity σ mho/m and dielectric constant $K_e = \varepsilon/\varepsilon_0$. The electric field is given by*

$$\vec{E} = \vec{E}_0 \, exp\left[j\left(\omega t - \vec{k} \cdot \vec{r}\right)\right].$$

Find the conduction and displacement current densities and the frequency at which they have equal magnitudes.

(b) Find the frequency at which the conduction current density and displacement current density are equal in magnitude in distilled water where $\sigma = 2.0 \times 10^{-4}$ mho/m and $K_e = 80$.

Solution

(*a*) The conduction current density is given by

$$\vec{J}_C = \sigma \vec{E} = \sigma \vec{E}_0 \; \exp\left[j\left(\omega t - \vec{k} \cdot \vec{r}\right)\right] \text{A/m}^2.$$

The displacement current density is

$$\vec{J}_D = \varepsilon \frac{\partial \vec{E}}{\partial t} = K_e \varepsilon_0 j\omega \, \vec{E}_0 \; \exp\left[j\left(\omega t - \vec{k} \cdot \vec{r}\right)\right] \text{A/m}^2.$$

Thus, $\left|\vec{J}_C\right| = \left|\vec{J}_D\right|$ when $\sigma = K_e\, \varepsilon_0\, \omega$

or
$$\omega = 2\pi f = \frac{\sigma}{K_e \varepsilon_0}$$

Thus, the frequency at which $\left|\vec{J}_C\right| = \left|\vec{J}_D\right|$ is

$$f = \frac{1}{2\pi}\frac{\sigma}{K_e \varepsilon_0}\ \text{Hz}.$$

(*b*) $f = \dfrac{1}{2\pi}\dfrac{2\times10^{-4}}{80\times 8.85\times10^{-12}} = 4.50 \times 10^4$ Hz.

4. *Write the complete set of Maxwell's equations in integral form, assuming that a charge-density distribution and current-density distribution exist in the region of interest and that* $\mu = \mu_0$, $\varepsilon = \varepsilon_0$ *for the medium under consideration.*

Solution

Integrating Eqn. (11.1) over a volume, and using the divergence theorem, we get

$$\int_{\substack{\text{Enclosed}\\ \text{surface}}} \vec{E}\cdot d\vec{s} = \frac{1}{\varepsilon_0}\int_{\text{Volume}} \rho dV.$$

This equation is just Gauss's law for the electric field $\vec{E}$. Similarly, Eqn. (11.2) gives

$$\int_{\substack{\text{Enclosed}\\ \text{surface}}} \vec{B}\cdot d\vec{s} = 0$$

This is the mathematical form of Gauss's law for the magnetic field $\vec{B}$, it shows that there are no magnetic monopoles.

Integrating Eqn. (11.3) over a surface enclosing the region under consideration, we obtain

$$\int (\vec{\nabla}\times\vec{E})\cdot d\vec{s} = -\int \frac{\partial \vec{J}}{\partial t}.d\vec{s}$$

or
$$\int_{\substack{\text{Enclosed}\\ \text{curve}}} \vec{E}.d\vec{l} = -\int_{\substack{\text{Enclosed}\\ \text{surface}}} \frac{\partial \vec{B}}{\partial t}.d\vec{s}$$

This is just the integral form of Faraday's law of induction.

Finally, Eqn. (11.4) yields the result

$$\int_{\substack{\text{Enclosed}\\ \text{curve}}} \vec{B}.d\vec{l} = \mu_0\left[\varepsilon_0 \int_{\substack{\text{Enclosed}\\ \text{surface}}} \frac{\partial \vec{E}}{\partial t}.d\vec{s} + \int_{\substack{\text{Enclosed}\\ \text{surface}}} \vec{J}.d\vec{s}\right].$$

This is just the mathematical form of Ampere-Maxwell law, with conduction and displacement currents on an equal footing. The first term on the right within the bracket is the displacement current.

5. *Starting from Maxwell's equations of electromagnetism in vacuum, obtain the classical wave equations for the four field vectors $\vec{E}, \vec{D}, \vec{B}$ and $\vec{H}$. Show that the field vectors can be propagated as waves in free space with velocity of propagation*

$$c = \frac{1}{\sqrt{\varepsilon_0 \mu_0}}.$$

Solution

In free space, $\rho = 0$, $\vec{J} = 0$, $\varepsilon = \varepsilon_0$ and $\mu = \mu_0$. From Eqn. (11.3), we have

$$\vec{\nabla} \times \left(\vec{\nabla} \times \vec{E}\right) + \frac{\partial}{\partial t}\left(\vec{\nabla} \times \vec{B}\right) = 0$$

or
$$\vec{\nabla}(\vec{\nabla}.\vec{E}) - \nabla^2 \vec{E} + \frac{\partial}{\partial t}(\vec{\nabla} \times \mu_0 \vec{H}) = 0$$

or
$$-\nabla^2 \vec{E} + \mu_0 \frac{\partial^2 \vec{D}}{dt^2} = 0 \quad \text{[Using Eqn. 11.4]}$$

or
$$\nabla^2 \vec{E} = \mu_0 \varepsilon_0 \frac{\partial^2 \vec{E}}{\partial t^2}. \qquad ...(11.15)$$

Similarly, we get

$$\nabla^2 \vec{H} = \mu_0 \varepsilon_0 \frac{\partial^2 \vec{H}}{dt^2} \qquad ...(11.16)$$

Multiplying Eqn. (11.15) by ε_0 and Eqn. (11.16) by μ_0, we get identical equations for $\vec{D}$ and $\vec{B}$. The vector equation (11.15) consists of three separate partial differential equations:

$$\nabla^2 E_x = \mu_0 \varepsilon_0 \frac{\partial^2 E_x}{\partial t^2}; \quad \nabla^2 E_y = \mu_0 \varepsilon_0 \frac{\partial^2 E_y}{\partial t^2}; \quad \nabla^2 E_z = \mu_0 \varepsilon_0 \frac{\partial^2 E_z}{\partial t^2}.$$

There are four such families of equations for the four field vectors $\vec{E}, \vec{D}, \vec{H}$ and $\vec{B}$. They satisfy the classical wave equation with velocity of propagation.

$$c = \frac{1}{\sqrt{\varepsilon_0 \mu_0}} = \left[\frac{1}{8.85 \times 10^{-12} \times 1.26 \times 10^{-6}}\right]^{1/2} = 3 \times 10^8 \text{ m s}^{-1}.$$

6. *From Maxwell's equations of electromagnetism in free space show that electromagnetic plane waves are transverse.*

For a plane electromagnetic wave travelling along the z-axis in free space, show that

(i) $$\frac{\partial H_x}{\partial t} = \frac{1}{\mu_0} \frac{\partial E_y}{\partial z} \qquad \frac{\partial H_x}{\partial z} = \varepsilon_0 \frac{\partial E_y}{\partial t}$$

(ii) $$\frac{\partial H_y}{\partial t} = -\frac{1}{\mu_0} \frac{\partial E_x}{\partial z} \qquad \frac{\partial H_y}{\partial t} = -\varepsilon_0 \frac{\partial E_x}{\partial t}.$$

Solution

We consider the case of plane waves travelling in the z-direction (along $\hat{k}$, the direction of propagation of the wave) so that the wave fronts are planes parallel to the xy-plane. If the vibrations are to be represented by variations of $\vec{E}$ and $\vec{H}$, we see that in any wavefront they must be constant over the whole plane at any instant, and their partial derivatives with respect to x and y must vanish. None of the components of $\vec{E}$ and $\vec{H}$ depends on either of the transverse coordinates x and y. Now,

$$\vec{E} = \hat{i}E_x(z,t) + \hat{j}E_y(z,t) + \hat{k}E_z(z,t)$$

$$\vec{H} = \hat{i}H_x(z,t) + \hat{j}H_y(z,t) + \hat{k}H_z(z,t).$$

Then,
$$\vec{\nabla}.\vec{E} = \frac{\partial E_z}{\partial z} = 0$$

which says that E_z is independent of z. That E_z is also independent of t can be seen by considering Maxwell's Eqn. (11.4) in free space:

$$\vec{\nabla} \times \vec{H} = \varepsilon_0 \frac{\partial \vec{E}}{\partial t}.$$

We take the z-component of this equation:

$$\varepsilon_0 \frac{\partial E_z}{\partial t} = \frac{\partial H_y}{\partial x} - \frac{\partial H_x}{\partial y} = 0$$

Thus, E_z is a constant. For simplicity, we take this constant to be zero. Similarly, we can show that H_z is a constant and we again take H_z to be zero. Thus, we conclude that apart from the nonwave like constant fields, the electromagnetic plane waves are transverse waves. Thus, the electric and magnetic fields are perpendicular to the direction of propagation $\vec{k}$ (z-direction):

$$\vec{E} = \vec{i}E_x(z,t) + \vec{j}E_y(z,t)$$

$$\vec{H} = \vec{i}H_x(z,t) + \vec{j}H_y(z,t)$$

and
$$\vec{E} \cdot \hat{k} = 0,\ \vec{H}.\hat{k} = 0, \vec{E} \times \vec{H} = (E_x H_y - E_y H_x)\hat{k}.$$

(*i*) We take the x-component of Maxwell's equation (11.3) and the y-component of Maxwell's equation (11.4):

$$\frac{\partial B_x}{\partial t} = \frac{\partial E_y}{\partial z} \text{ and } \frac{\partial D_y}{\partial t} = \frac{\partial H_x}{\partial z}$$

or
$$\frac{\partial H_x}{\partial t} = \frac{1}{\mu_0}\frac{\partial E_y}{\partial z} \qquad ...(11.17)$$

$$\frac{\partial H_x}{\partial z} = \varepsilon_0 \frac{\partial E_y}{\partial t} \qquad ...(11.18)$$

Thus, E_y and H_x are coupled.

(*ii*) Similarly if we take y-component of Eqn. (11.3) and the x-component of Eqn. (11.4), we shall find that E_x and H_y are coupled:

$$\frac{\partial H_y}{\partial t} = -\frac{1}{\mu_0}\frac{\partial E_x}{\partial z} \qquad ...(11.19)$$

$$\frac{\partial H_y}{\partial z} = -\varepsilon_0 \frac{\partial E_x}{\partial t} \qquad ...(11.20)$$

7. *In problem 3 we choose our system of axes such that the x-axis is parallel to the vector $\vec{E}$, i.e.,*

$$\vec{E} = \hat{i}E_x(z,t)$$

and consider harmonic waves travelling in the positive direction of the z-axis, i.e.,

$$E_x = E_{xo}\, exp\left[i\omega\left(t-\frac{z}{c}\right)+\theta\right],$$

θ being the phase angle at $t = 0$ and $z = 0$.

Show that (i) $\vec{E}\,.\,\vec{H} = 0$

(ii) $\vec{E}\times\vec{H}$ is along $\hat{k}$ (iii) $\dfrac{E_x}{H_y} = \mu_0 c = \left(\dfrac{\mu_0}{\varepsilon_0}\right)^{1/2} = 377$ ohms or, $\dfrac{E_x}{B_y} = c$

(iv) the electric and magnetic energy densities are equal, i.e., $\frac{1}{2}\varepsilon_0 E^2 = \frac{1}{2}\mu_0 H^2$.

Solution

We have $\vec{E} = \hat{i}E_x(z,t), E_y = 0$ and $E_z = 0$. Then from Eqns. (11.17) and (11.18), we get $H_x = 0$. We also have $H_z = 0$, and the magnetic field $\vec{H}$ is given by

$$\vec{H} = \hat{j}\vec{H}_y(z,t).$$

Thus, (*i*) $\vec{E}\,.\,\vec{H} = 0$ and (*ii*) $\vec{E}\times\vec{H}$ is along $\hat{k}$. (*iii*) Using Eqns. (11.19) and (11.20), we find

$$\frac{\partial H_y}{\partial t} = \frac{i\omega}{\mu_0 c}E_x = \frac{1}{\mu_0 c}\frac{\partial E_x}{\partial t} \qquad ...(11.21)$$

and

$$\frac{\partial H_y}{\partial z} = -i\omega\varepsilon_0 E_x = \varepsilon_0 c \frac{\partial E_x}{\partial z} \qquad ...(11.22)$$

Since $\dfrac{1}{\mu_0 c} = \varepsilon_0 c$, we find from Eqns. (11.21) and (11.22) that in a travelling plane wave H_y and $\dfrac{E_x}{\mu_0 c}$ are equal aside from uninteresting additive constant which we put equal to zero.

Thus,

$$\frac{E_x}{H_y} = \mu_0 c \quad \text{or,} \quad \frac{E_x}{B_y} = c. \qquad ...(11.23)$$

Since $\vec{E}$ is along the x-axis and $\vec{H}$ is along the y-axis we can write

$$\frac{E}{H} = \mu_0 c \quad \text{or,} \quad \frac{E}{B} = c.$$

Putting the values of μ_0 and ε_0 we obtain

$$\frac{E_x}{H_y} = \left(\frac{\mu_0}{\varepsilon_0}\right)^{1/2} = 377 \text{ ohms} \qquad ...(11.24)$$

E/H has the dimension of impedance.

(*iv*)

$$\frac{\frac{1}{2}\varepsilon_0 E^2}{\frac{1}{2}\mu_0 H^2} = \frac{\varepsilon_0}{\mu_0}\frac{E^2}{H^2} = 1 \qquad ...(11.25)$$

8. *Imagine an electromagnetic plane wave in vacuum whose $\vec{E}$-field (in SI units) is given by*

$$E_x = 10^2 \sin \pi(3\times10^6 z - 9\times10^{14} t), E_y = 0, E_z = 0.$$

Determine (i) the speed, frequency, wavelength, period, initial phase and $\vec{E}$-field amplitude and polarization, (ii) the magnetic field $\vec{B}$.

Solution

(*i*) The wavefunction has the form

$$E_x(z, t) = E_{x_0} \sin k(z - vt).$$

Here, $E_x = 10^2 \sin [3 \times 10^6\pi(z - 3 \times 10^8 t)]$.

We see that $k = 3 \times 10^6 \pi\, m^{-1}$ and $v = 3 \times 10^8$ m s^{-1}.

Hence, $\lambda = \frac{2\pi}{k} = 666.7$ nm,

$$\nu = \frac{v}{\lambda} = \frac{3\times10^8}{\frac{2}{3}\times10^{-6}} = 4.5\times10^{14} \text{ Hz}.$$

The period $T = \frac{1}{\nu} = 2.2 \times 10^{-15}$ s, and the initial phase is zero. The electric field amplitude is $E_{x_0} = 10^2$ Vm^{-1}. The wave is linearly polarized in the *x*-direction and propagates along the *z*-axis.

(*ii*) The wave is propagating in the *z*-direction whereas the electric field $\vec{E}$ oscillates along the *x*-axis, *i.e.*, the field $\vec{E}$ resides in the *xz*-plane. Now, $\vec{B}$ is normal to both $\vec{E}$ and the *z*-axis, so it resides in the *yz*-plane. Thus, $B_x = 0$, $B_z = 0$ and $\vec{B} = \hat{j}\, B_y\,(z, t)$. Since, $E = cB$, we see that

$$B_y(z,t) = 0.33\times10^{-6} \sin \pi(3\times10^6 z - 9\times10^{14} t)\, T.$$

9. *Show from Maxwell's equations that the differential equations for the field vectors $\vec{E}$ and $\vec{H}$ in homogeneous, isotropic, linear and stationary media are given by*

$$\nabla^2 \vec{E} = \varepsilon\mu \frac{\partial^2 \vec{E}}{\partial t^2} + \sigma\mu \frac{\partial \vec{E}}{\partial t} + \vec{\nabla}(\rho/\varepsilon)$$

and

$$\nabla^2 \vec{H} = \varepsilon\mu \frac{\partial^2 \vec{H}}{\partial t^2} + \sigma\mu \frac{\partial \vec{H}}{\partial t}.$$

Solution

From Maxwell's equation (11.3), we have

$$\vec{\nabla}\times(\vec{\nabla}\times\vec{E})+\frac{\partial}{\partial t}(\vec{\nabla}\times\mu\vec{H}) = 0$$

or
$$\vec{\nabla}\times(\vec{\nabla}\times\vec{E})+\frac{\partial}{\partial t}\left(\mu\vec{J}+\varepsilon\mu\frac{\partial^2\vec{E}}{\partial t}\right) = 0$$

or
$$\vec{\nabla}(\vec{\nabla}.\vec{E})-\nabla^2\vec{E}+\sigma\mu\frac{\partial\vec{E}}{\partial t}+\varepsilon\mu\frac{\partial^2\vec{E}}{\partial t} = 0$$

or
$$\nabla^2\vec{E} = \varepsilon\mu\frac{\partial^2\vec{E}}{\partial t^2}+\sigma\mu\frac{\partial\vec{E}}{\partial t}+\vec{\nabla}(\rho/\varepsilon) \quad ...(11.26)$$

Similarly,
$$\vec{\nabla}\times(\vec{\nabla}\times\vec{H}) = \frac{\partial}{\partial t}(\vec{\nabla}\times\vec{D})+\vec{\nabla}\times\vec{J}$$

or
$$\vec{\nabla}(\vec{\nabla}.\vec{H})-\nabla^2\vec{H} = -\varepsilon\mu\frac{\partial^2\vec{H}}{\partial t^2}+\sigma\vec{\nabla}\times\vec{E}$$

or
$$-\nabla^2\vec{H} = -\varepsilon\mu\frac{\partial^2\vec{H}}{\partial t^2}-\sigma\mu\frac{\partial\vec{H}}{\partial t}$$

or
$$\nabla^2\vec{H} = \varepsilon\mu\frac{\partial^2\vec{H}}{\partial t^2}+\sigma\mu\frac{\partial\vec{H}}{\partial t} \quad ...(11.27)$$

In vacuum $\rho = \sigma = 0$ and $\varepsilon = \varepsilon_0$, $\mu = \mu_0$. Then Eqns. (11.26) and (11.27) reduce to free space wave Eqns. (11.15) and (11.16).

10. *Consider a plane electromagnetic wave travelling along the z-axis in a homogeneous, isotropic, linear and stationary medium. Show that both $\vec{E}$ and $\vec{H}$ are transverse.*

Solution

We consider a plane wave propagating in the positive direction along the z-axis such that all derivatives with respect to x and y are zero. From Maxwell's equation (11.1), we have

$$\vec{\nabla}.\vec{E} = \frac{\partial}{\partial z}E_z = \rho/\varepsilon \quad ...(11.28)$$

and
$$\frac{\partial}{\partial z}(\rho/\varepsilon)\hat{k} = \frac{\partial^2}{\partial z^2}E_z\hat{k}.$$

Eqn. (11.26) can be written as

$$\frac{\partial^2}{\partial z^2}(\hat{i}E_x+\hat{j}E_y+\hat{k}E_z) = \mu\left(\varepsilon\frac{\partial^2}{\partial t^2}+\sigma\frac{\partial}{\partial t}\right)(\hat{i}E_x+\hat{j}E_y+\hat{k}E_z)+\frac{\partial}{\partial z}(\rho/\varepsilon)\hat{k}.$$

The z-component of this equation gives

$$\varepsilon\frac{\partial^2 E_z}{\partial t^2}+\sigma\frac{\partial E_z}{\partial t} = 0 \quad ...(11.29)$$

Thus, the longitudinal component E_z of the electric field, if it exists, must be of the form

$$E_z = a + be^{-\sigma t/\varepsilon}$$

where a and b are constants. Thus E_z must decrease exponentially with time and we may set $E_z = 0$. Since we are concerned solely with wave propagation, we may put $\rho = 0$ as obtained from Eqn. (11.28). The wave has no longitudinal component of $\vec{E}$.

It is easy to show that the $\vec{H}$ vector is also transverse. The divergence of $\vec{E}$ being equal to zero, we must have $\frac{\partial H_z}{\partial z} = 0$ because the derivatives with respect to x and y are both zero. Since H_z is not a function of z, we may set $H_z = 0$ for propagation of waves. Thus $\vec{H}$ vector is also transverse.

11. *Consider the propagation of a plane electromagnetic wave along the z-axis in a homogeneous, isotropic, linear and stationary medium. Show that if* $\vec{E}$ *is along the x-axis, then* $\vec{H}$ *is along the y-axis and the vector product* $\vec{E} \times \vec{H}$ *points in the direction of propagation. Show further that* $\frac{E_x}{H_y} = \frac{\omega\mu}{k}$ *where* ω *is the angular frequency of the wave.*

Solution

Since both $\vec{E}$ and $\vec{H}$ are transverse, we can write

$$\vec{E} = \hat{i}E_x(z,t).$$

$$\vec{H} = \hat{i}H_x(z,t) + \hat{j}H_y(z,t).$$

From Maxwell's Eqn. (11.3), we get

$$-\frac{\partial E_y}{\partial z} = -\mu\frac{\partial H_x}{\partial t}$$

and

$$\frac{\partial E_x}{\partial z} = -\mu\frac{\partial H_y}{\partial t}$$

Any wave of angular frequency ω propagating in the positive direction of the z-axis must involve the exponential function

$$exp\{i(\omega t - kz)\}$$

where k is the wave number. Thus, we obtain

$$ikE_y = -\mu i\omega H_x$$

and

$$-ikE_x = -\mu i\omega H_y$$

or

$$-\frac{E_y}{H_x} = \frac{\omega\mu}{k} = \frac{E_x}{H_y} \qquad \text{...(11.30)}$$

When $E_y = 0$, $H_x = 0$. If $\vec{E}$ is along the x-axis, then $\vec{H}$ is along the y-axis and $\vec{E} \times \vec{H}$ is along the z-axis.

12. *Light from a laser is propagating in the z-direction. If the amplitude of the electric field in the light wave is* 6.3×10^3 *V/m, and if the electric field points in the x-direction, what are the direction and amplitude of the magnetic field?*

Solution

The magnetic field is along the y-direction.

$$B_0 = \frac{E_0}{c} = \frac{6.3 \times 10^3}{3 \times 10^8} = 2.1 \times 10^{-5}\text{ T}$$

and

$$H_0 = \frac{E_0}{377} = \frac{6.3 \times 10^3}{377} = 16.71\text{ A/m}$$

[H_0 may be obtained from the equation $H_0 = B_0/\mu_0$]

13. *Show that when integrated over a closed surface, the normal component of the Poynting vector ($\vec{S}$) in free space gives the total outward flow of energy per unit time. Show further that for a plane wave, the Poynting vector is the product of the energy density and the wave velocity c, and its average value in free space is given by*

$$\bar{S} = 2.65 \times 10^{-3} E_{rms}^2 \text{ W/m}^2.$$

Solution

The quantity $\vec{S} = \vec{E} \times \vec{H} = \frac{1}{\mu_0} \vec{E} \times \vec{B}$ is called the Poynting vector and it is in the direction of propagation of a plane electromagnetic wave in free space. Let us calculate the divergence of this vector for any electromagnetic field in free space:

$$\vec{\nabla}.(\vec{E} \times \vec{H}) = -\vec{E} \cdot (\vec{\nabla} \times \vec{H}) + \vec{H}.(\vec{\nabla} \times \vec{E})$$

$$= -\vec{E}\frac{\partial \vec{D}}{\partial t} - \vec{H}.\frac{\partial \vec{B}}{\partial t}$$

$$= -\frac{\partial}{\partial t}\left[\frac{1}{2}\varepsilon_0 E^2 + \frac{1}{2}\mu_0 H^2\right] \qquad ...(11.31)$$

Integrating over a volume V bounded by a surface A, and using the divergence theorem, we have

$$\int_A (\vec{E} \times \vec{H}) \cdot d\vec{s} = -\frac{\partial}{\partial t}\int_V \left(\frac{1}{2}\varepsilon_0 E^2 + \frac{1}{2}\mu_0 H^2\right) dV \qquad ...(11.32)$$

The integral on the right-hand side is the sum of the electric and magnetic energies. The right-hand side is the energy lost per unit time by the volume V, and the left-hand side must be the total outward flux of energy in Watts over the surface A bounding the volume V. Thus, when integrated over a closed surface the normal component of the Poynting vector gives the total outward flow of the energy per unit time. This is known as **Poynting's Theorem.**

For a plane wave with the electric field $\vec{E}$ along the x-direction, we have, from problem 4,

$$\vec{S} = \vec{E} \times \vec{H} = E_x H_y \hat{k} = \left(\frac{\varepsilon_0}{\mu_0}\right)^{1/2} E^2 \hat{k} = \varepsilon_0 c E^2 \hat{k}$$

$$= \left(\frac{\mu_0}{\varepsilon_0}\right)^{\frac{1}{2}} H^2 \hat{k} = \mu_0 c H^2 \hat{k}.$$

Thus for a plane wave the average value of $\vec{S}$ is given by

$$\vec{S} = \varepsilon_0 c E_{rms}^2 = \frac{1}{2}\varepsilon_0 c E_0^2$$

$$= 2.65 \times 10^{-3} E_{rms}^2 \text{ W/m}^2$$

where E_0 is the amplitude of the wave of the electric field.

Average energy density of the electromagnetic wave in free space

$$= \frac{1}{2}\varepsilon_0 E_{rms}^2 + \frac{1}{2}\mu_0 H_{rms}^2$$

$$= \varepsilon_0 E_{rms}^2$$

and $$\overline{S} = (\varepsilon_0 E_{rms}^2) \times c$$

Thus the energy can be considered to travel with an average density $\varepsilon_0 E_{rms}^2$ at the velocity of propagation c in the direction of propagation $\hat{k}$.

14. *Find the rms values of the electric and magnetic fields in air at a distance x metres from a radiating point source of power W watts.*

Solution

Average value of the Poynting vector at a distance x from the source is

$$\overline{S} = \frac{W}{4\pi x^2} = 2.65 \times 10^{-3} E_{rms}^2$$

Thus, $$E_{rms} = \frac{1}{x}\left[\frac{1000W}{4\pi \times 2.65}\right]^{1/2} \text{ V/m}$$

$$H_{rms} = \frac{E_{rms}}{377} \text{ A/m.}$$

15. *An observer is 2 m from a point light source whose power output is 100 W. Calculate the rms values of the electric and magnetic fields and the radiation pressure at the position of the observer. Assume that the source radiates uniformly in all directions.*

Solution

Average value of the Poynting vector at a distance 2 m from the source is

$$\overline{S} = \frac{100}{4\pi . 2^2} = 2.65 \times 10^{-3} E_{rms}^2.$$

Thus, $$E_{rms} = \left[\frac{25 \times 10^3}{4\pi \times 2.65}\right]^{1/2} = 27.40 \text{ V/m.}$$

$$B_{rms} = \frac{E_{rms}}{c} = 9.13 \times 10^{-8} \text{ T}$$

$$H_{rms} = \frac{B_{rms}}{\mu_0} = 7.25 \times 10^{-2} \text{ A/m.}$$

The radiation pressure on a perfect absorber is

$$p = \frac{\overline{S}}{c} = \frac{25 \times 10^{-8}}{4\pi \times 3} = 6.63 \times 10^{-9} \text{ N/m}^2.$$

16. *A beam of light with an energy flux S of 10 W/cm² falls normally on a perfectly reflecting plane mirror of 2 cm² area. What force acts on the mirror?*

Solution

$S = 10 \times 10^4 \text{ W/m}^2 = 10^5 \text{ W/m}^2$.

Radiation pressure, $p = \dfrac{2S}{c} = \dfrac{2 \times 10^5}{3 \times 10^8} = \dfrac{2}{3} \times 10^{-3} \text{ N/m}^2$

Force on the mirror $= \dfrac{2}{3} \times 10^{-3} \times 2 \times 10^{-4} = 1.33 \times 10^{-7}$ N.

17. *The power radiated by the sun is 3.8 × 10²⁶ W; the average distance between the sun and the earth is 1.5 × 10¹¹ m, (a) What is the average value of the Poynting vector on the surface of the earth? (b) Calculate the rms values of the electric and magnetic fields on the surface of the earth due to solar radiation. (c) Show that the average solar energy incident on the earth is ~ 2 calorie/(cm² minute) and the radiation pressure on a perfect absorber is 4.47 × 10⁻⁶ N/m².*

[Assume that no solar radiation is absorbed by the atmosphere]

Solution

(*a*) Average value of the Poynting vector on the surface of the earth is

$$\overline{S}_E = \frac{3.8 \times 10^{26}}{4\pi \times (1.5)^2 \times 10^{22}} = 1.34 \times 10^3 \text{ W/m}^2$$

(*b*) $\overline{S}_E = 2.65 \times 10^{-3} E_{\text{rms}}^2$

Thus,

$$E_{\text{rms}} = \left[\frac{1.34 \times 10^3}{2.65 \times 10^{-3}}\right]^{\frac{1}{2}} = 711.1 \text{ V/m}$$

$$H_{\text{rms}} = \frac{E_{\text{rms}}}{377} = 1.89 \text{ A/m}$$

(*c*) Average solar energy incident on the earth per cm² per minute is

$$\frac{\overline{S}_E \times 60}{10^4 \times 4.19} = \frac{1.34 \times 10^3 \times 60}{10^4 \times 4.19} = 1.92 \text{ cal/(cm}^2\text{.minute)}$$

$$\sim 2 \text{ cal/(cm}^2\text{.minute)}$$

Radiation pressure on a perfect absorber is

$$p = \frac{\overline{S}_E}{c} = \frac{1.34 \times 10^3}{3 \times 10^8} = 4.47 \times 10^{-6} \text{ N/m}^2.$$

18. *A particle in the solar system is under the combined influence of the sun's gravitational attraction and the radiation force due to the sun's rays. Assume that the particle is a sphere of density 1.0 × 10³ kg/m³ and that all of the incident light is absorbed. (a) Show that all particles with radius less than some critical radius R_0, will be blown out of the solar system. (b) Calculate R_0.*

Solution

When the gravitational attraction and the radiation force are equal we have

$$G\frac{Mm}{r^2} = \frac{W}{4\pi r^2}\cdot\frac{1}{c}\pi R_0^2$$

where, M = Mass of the sun = 1.99×10^{30} kg

m = Mass of the particle of radius R_0

r = Distance of the particle from the sun

W = Power radiated by the sun = 3.9×10^{26} W

c = Velocity of light = 3×10^8 m s^{-1}.

Thus, $$GM \times \frac{4}{3}\pi R_0^3 \times 1.0 \times 10^3 = \frac{WR_0^2}{4c}$$

Note that R_0 does not depend on the distance from the particle to the sun.

Hence,

$$R_0 = \frac{3W}{16\pi G\, cM \times 10^3}$$

$$= \frac{3 \times 3.9 \times 10^{26}}{16\pi \times 6.67 \times 10^{-11} \times 3 \times 10^8 \times 1.99 \times 10^{30} \times 10^3}$$

$$= 5.85 \times 10^{-7} \text{ m} = 585 \text{ nm}.$$

19. *A plane electromagnetic harmonic wave of frequency 600 × 10^{12} Hz, propagating in the positive x-direction in vacuum, has an electric field amplitude of 42.42 Vm^{-1}. The wave is linearly polarized such that the plane of vibration of the electric field is at 45° to the xz-plane. Obtain the vector $\vec{E}$ and $\vec{B}$.*

Solution

The electric vector $\vec{E}$ is given by

$$\vec{E} = \vec{E}_0 \sin\left[2\pi \times 600 \times 10^{12}\left(t - \frac{x}{3 \times 10^8}\right)\right]$$

Here, $$E_x = 0, E_0 = \left(E_{oy}^2 + E_{oz}^2\right)^{1/2}, \quad \text{and } E_{oy} = E_{oz}$$

since $$E_{oy} = E_{oz} = \frac{1}{\sqrt{2}}E_0 = 30 \text{ Vm}^{-1}$$

We get, $$E_x = 0, E_y = E_z = 30\sin\left[2\pi \times 600 \times 10^{12}\left(t - \frac{x}{3 \times 10^8}\right)\right]$$

Also, $$E = cB, \text{ hence}$$

$$B_x = 0,\ B_z = -B_y = 10^{-7}\sin\left[2\pi \times 600 \times 10^{12}\left(t - \frac{x}{3 \times 10^8}\right)\right]$$

So,

$$\vec{E} = E_y\hat{j} + E_z\hat{k} = E_y(\hat{j} + \hat{k})$$

$$\vec{B} = B_y\hat{j} + B_z\hat{k} = B_z(-\hat{j}+\hat{k})$$

Then, $\vec{E}.\vec{B}=0$, or $\vec{E}$ is normal to $\vec{B}$

and $\vec{E}\times\vec{B} = E_yB_z(\hat{i}+\hat{i}) = \frac{2E_y^2}{c}\hat{i}$, as required.

20. *Derive the Malus law [Eqn. 11.14].*

Solution

Let an incident plane-polarized wave $\vec{E} = \vec{E}_0 \sin(\omega t + \alpha)$ make an angle θ with the transmission axis. Decomposing the field $\vec{E}$ into two plane-polarized waves, we obtain.

$\vec{E}_1 = (\vec{E}_0\cos\theta)\sin(\omega t+\alpha)$, $\vec{E}_1 = (\vec{E}_0\sin\theta)\sin(\omega t+\alpha)$.

Since intensity is proportional to the square of the amplitude, and since only $\vec{E}_{\|}$ is transmitted,

$$\frac{I}{I_0} = \frac{(\vec{E}_0\cos\theta)\cdot(\vec{E}_0\cos\theta)}{\vec{E}_0\cdot\vec{E}_0} = \cos^2\theta.$$

21. *Polarized light of initial intensity I_0 passes through two analyzers—the first with its axis at 45° to the amplitude of the initial beam and the second with its axis at 90° to the initial amplitude. What is the intensity of the light that emerges from this system and what is the direction of its amplitude?*

Solution

The angle between the axis of the first analyzer and the initial amplitude $\vec{A}_0$ is 45°, hence the intensity I' after passing through the first analyzer is equal to $I_0 \cos^2 45° = 0.5\, I_0$. The transmitted amplitude $\vec{A}'$ is at an angle 45° with respect to the axis of the second analyzer, so the final intensity $I = I' \cos^2 45° = 0.5 \times 0.5\, I_0 = 0.25\, I_0$. Also, the final amplitude $\vec{A}$ is at 90° with respect to initial amplitude $\vec{A}_0$.

Note that if only the second analyzer were in place, no light would pass through, since $\vec{A}_0$ is normal to its transmission axis.

22. *Two polarizing sheets have their polarizing direction parallel so that the intensity I_0 of the transmitted light is a maximum. Through what angle must either sheet be rotated if the intensity is to drop by one-fourth?*

Solution

We have from law of Malus

$$\frac{1}{4}I_0 = I_0\cos^2\theta$$

or $$\cos\theta = \pm\frac{1}{2} \text{ or, } \theta = 60° \text{ or } 120°.$$

23. Describe the state of polarization of the wave represented by the following equations:

$$E_x = E_0 \cos (kz - \omega t)$$
$$E_y = \mp E_0 \cos (kz - \omega t)$$

Solution

The wave propagates in the positive z-direction.

$$E^2 = E_x^2 + E_y^2 = E_0^2.$$

The electric vector is constant is magnitude but sweeps around a circle at a frequency ω. For the upper sign the rotation is counter clockwise and the wave is called left circularly polarized. For the lower sign the rotation is clockwise and the wave is called right circularly polarized.

24. *Let (ε_0) denote the dimensional formula of the permittivity of the vacuum, and (μ_0) that of the permeability of the vacuum. If M = mass,*

L = length, T = time and I = electric current, then

(a) $[\varepsilon_0] = M^{-1} L^{-3} T^2 I$ *(c)* $[\mu_0] = M L T^{-2} I^{-2}$

(b) $[\varepsilon_0] = M^{-1} L^{-3} T^4 I^2$ *(d)* $[\mu_0] = M L^2 T^{-1} I$ (I.I.T. 1999)

Solution

We know

$$F = \frac{q_1 q_2}{4\pi\varepsilon_0 r^2}$$

or

$$\varepsilon_0 = \frac{q_1 q_2}{4\pi r^2 F}$$

Charge = current × time

Force = mass × acceleration

Thus,

$$(\varepsilon_0) = \frac{I\,T\,I\,T}{L^2 M\,L\,T^{-2}} = M^{-1} L^{-3} T^4 I^2$$

Since $c = \dfrac{1}{\sqrt{\varepsilon_0 \mu_0}}$ = velocity of light in vacuum.

$$[\mu_0] = \frac{1}{[c^2 \varepsilon_0]} = \frac{1}{L^2 T^{-2} M^{-1} L^{-3} T^4 I^2} = M L T^{-2} I^{-2}$$

Correct Choice: *b, c.*

25. *If ε_0 and μ_0 are respectively the electric permittivity and magnetic permeability of free space, ε and μ the corresponding quantities in a medium, the index of refraction of the medium in terms of the above parameters is ......* (I.I.T. 1992)

Solution

$$n = \frac{\text{Speed of light in vacuum}}{\text{Speed of light in the medium}} = \frac{1/\sqrt{\varepsilon_0 \mu_0}}{1/\sqrt{\varepsilon\,\mu}} = \sqrt{\frac{\varepsilon\,\mu}{\varepsilon_0 \mu_0}}.$$

26. *A light of wavelength 6000 Å in air, enters a medium with refractive index 1.5. Inside the medium its frequency is .... Hz and its wavelength is ...... Å.* (I.I.T. 1997)

Solution

Frequency, $\nu = \dfrac{c}{\lambda} = \dfrac{3 \times 10^8}{6000 \times 10^{-10}} = 5 \times 10^{14}$ Hz.

The frequency does not change. Inside the medium the frequency is 5×10^{14} Hz.

Inside the medium, velocity $= \dfrac{c}{n}$

Wavelength $\lambda' = \dfrac{c}{n\nu} = \dfrac{\lambda}{n} = \dfrac{6000 \text{ Å}}{1.5} = 4000$ Å.

27. *Earth receives 1400 W/m² of solar power. If all the solar energy falling on a lens of area 0.2 m² is focussed on a block of ice of mass 280 g, the time taken to melt the ice will be .... minutes.*

[Latent heat of fusion of ice = 3.3×10^5 J/kg] (I.I.T. 1997)

Solution

Heat required to melt 280 g of ice is

$$280 \times 10^{-3} \times 3.3 \times 10^5 \text{ J}$$

Solar energy received by ice per second = 1400×0.2 J

Time taken by the ice to melt is

$$\frac{280 \times 3.3 \times 10^2}{1400 \times 0.2} \text{s} = 330 \text{ s} = 5.5 \text{ minutes.}$$

28. *In a wave motion $y = a \sin(kx - \omega t)$, y can represent*

(a) electric field (b) magnetic field (c) displacement (d) pressure. (I.I.T. 1999)

Solution

Displacement and pressure of the sound wave can be sinusoidal. In case of electromagnetic wave the electric and magnetic field can be sinusoidal.

Correct Choice: *a, b, c, d.*

SUPPLEMENTARY PROBLEMS

1. Show that the wave equation in free space for the field vector $\vec{E}$ can be written in the form

$$\left(\nabla^2 + k^2\right)\vec{E} = 0$$

where k is the wave number.

2. Starting from Maxwell's equations of electromagnetism in a non-conducting medium ($\sigma = 0$) obtain the classical wave equations for the field vectors $\vec{E}$ and $\vec{H}$ and show that the waves propagate with phase velocity

$$u = \frac{1}{(\varepsilon\mu)^{1/2}} = \frac{c}{(K_e K_m)^{1/2}}.$$

Find the relation between the index of refraction n and the dielectric coefficient K_e of a non-magnetic non-conductor.

3. In the previous problem choose the x-axis to be parallel to $\vec{E}$ and then show that

(*i*) $H_x = 0$ and $\dfrac{E_x}{H_y} = \left(\dfrac{\mu}{\varepsilon}\right)^{1/2}$,

(*ii*) the electric and magnetic energy densities are equal, *i.e.*,

$$\frac{1}{2}\varepsilon E^2 = \frac{1}{2}\mu H^2$$

(*iii*) the average value of the Poynting vector is

$$\overline{S} = \left(\frac{\varepsilon}{\mu}\right)^{1/2} E_{\text{rms}}^2 = 2.65\times10^{-3}\left(\frac{K_e}{K_m}\right)^{1/2} E_{\text{rms}}^2 \text{ W/m}^2.$$

4. Consider a plane harmonic electromagnetic wave moving along the positive x-direction with the electric vector $\vec{E}$ along the z-axis, *i.e.*, $\vec{E} = \hat{k}\, E_z(x, t)$. Show that $\vec{E} \times \vec{B} = \hat{i}\dfrac{1}{c}E_z^2$.

5. At a particular point, the instantaneous electric field of an electromagnetic wave points in the $+y$-direction, while the magnetic field points in the $-z$-direction. In what direction is the wave propagating?

6. A plane electromagnetic wave with wavelength 2.0 m travels in free space in the $+x$-direction with its electric vector $\vec{E}$ of amplitude 300 V/m, directed along the y-axis. (*a*) What is the frequency ν of the wave? (*b*) What is the direction and amplitude of the magnetic field $\vec{B}$ associated with wave? (*c*) If $E = E_0 \sin(kx - \omega t)$ what are the values of k and ω? (*d*) What is the time-averaged rate of energy flow in W/m^2 associated with this wave? (*e*) If the wave falls upon a perfectly absorbing sheet of area 3.0 m^2, at what rate would momentum be delivered to the sheet and what is the radiation pressure exerted on the sheet?

7. Calculate the electric field intensity and the radiation pressure due to solar radiation at the surface of the sun from the following data: power radiated by the sun $= 3.9 \times 10^{26}$ W; radius of the sun $= 7.0 \times 10^8$ m.

8. The electric field associated with a plane electromagnetic wave is given by $E_y = 0$, $E_z = 0$, $E_x = 1.5 \cos\left[10^{15}\pi\left(t - \dfrac{z}{c}\right)\right]$. Write expressions for the components of the magnetic field of the wave.

9. An electromagnetic wave in which the rms value of E is 25 V/m falls normally on an absorbing mass of 10^{-3} g/cm^2 and of specific heat 0.1 cal/(g.°C). Assuming that no heat is lost, calculate the rate at which the temperature of the absorber rises.

10. Radiation from the sun striking the earth has an intensity of 1.4 kW/m^2. (*a*) Assuming that the earth behaves like a flat disk at right angles to the sun's rays and that all the incident energy is absorbed, calculate the force on the earth due to radiation pressure. (*b*) Compare it with the force due to the sun's gravitational attraction. [Given, $G = 6.67 \times 10^{-11}$ N.m²/kg², mass of the sun = 1.99×10^{30} kg, mass of the earth = 5.98×10^{24} kg, the mean radius of the earth = 6.37×10^{6} m]

11. The earth's mean radius is 6.37×10^{6} m and the mean earth-sun distance is 1.5×10^{8} km. What fraction of the radiation emitted by the sun is intercepted by the disc of the earth?

12. The intensity of direct solar radiation that was unabsorbed by the atmosphere is found to be 80 W/m². How close would you have to stand to a 1.0 kW electric heater to feel the same intensity? Assume that the heater radiates uniformly in all directions.

13. Sunlight strikes the earth outside its atmosphere, with an intensity of 1.4 kW/m². Calculate E_0 and B_0 for sunlight, assuming it to be plane.

14. An airplane flying at a distance of 10 km from a radio transmitter receives a signal of power 20 μW/m². Calculate (*a*) the amplitude of the electric field at the airplane due to this signal, (*b*) the amplitude of the magnetic field at the airplane, (*c*) the total power radiated by the transmitter, assuming the transmitter to radiate uniformly in all directions.

15. Show that in a plane electromagnetic wave the average intensity is given by

$$\overline{S} = \frac{E_0^2}{2\mu_0 c} = \frac{cB_0^2}{2\mu_0} = \frac{c\mu_0}{2} H_0^2.$$

16. If the maximum value of the magnetic field component is $B_0 = 1.0 \times 10^{-4}$ T, what is the average intensity of a plane travelling electromagnetic wave?

17. You walk 100 m directly toward a street lamp and find that the intensity increases 2 times the intensity at your original position. How far from the lamp were you first standing?

18. High-power lasers are used to compress gas plasmas by radiation pressure. The reflectivity of a plasma is unity if the electron density is high enough. A laser generating pulses of radiation of peak power 1.65×10^{3} MW is focussed onto 1.0 mm² of high-electron density plasma. Find the pressure exerted on the plasma.

19. Describe the state of polarization of the wave represented by the following equations:

(*a*) $E_y = A \cos \omega(t - x/c)$
$E_z = A \sin \omega(t - x/c)$

(*b*) $E_y = A \cos \omega(t - x/c)$
$E_z = -A \cos \omega(t - x/c)$

(*c*) $E_y = A \cos \omega(t - x/c)$
$E_z = A \cos[\omega(t - x/c) + \pi/4]$

20. The magnetic field of an electromagnetic wave is given by

$$\vec{B} = B_0[\sin(kx - \omega t)\hat{k} - \cos(kx - \omega t)\hat{j}]$$

(*a*) Find the electric field of the electromagnetic wave.

(*b*) Show that the wave is circularly polarized.

12 Interference

The phenomenon of interference is due to superposition of two wave trains with a constant phase difference between them.

12.1 YOUNG'S EXPERIMENT

Sunlight is allowed to pass through a pin hole S and then through two pin holes S_1 and S_2 (Fig. 12.1). The pin holes S_1 and S_2 act as point sources. According to Huygens' principle, secondary waves are emitted from S_1 and S_2 and they are superimposed at the screen with a phase difference to produce interference fringes. Fringes are easily observed if monochromatic light is used and the pin holes are replaced by parallel slits. The distance between any two consecutive bright or dark fringes is the fringe width (β),

$$\beta = \frac{\lambda D}{d} \qquad \text{...(12.1)}$$

where λ = wavelength of the monochromatic light

d = the separation between the coherent sources S_1 and S_2

and D = distance of the screen from the coherent sources (S_1S_2).

It may be noted that interference (constructive or destructive) can occur only between overlapping waves from coherent sources. Further, two sources S_1 and S_2 are coherent provided there is a fixed difference between the phases of the wave emitted by S_1 at S_1 and the wave from S_2 at S_2.

12.2 DISPLACEMENT OF FRINGES DUE TO INTERPOSITION OF THIN FILM

If a thin strip of a transparent body of uniform thickness t is introduced in the path of one of the light rays coming from S_1 and S_2, the optical path of that beam is changed and the central fringe is shifted through a distance x. Then,

$$t = \frac{x\lambda}{(n-1)\beta} \qquad \text{...(12.2)}$$

where n is the refractive index of the transparent body.

12.3 FRESNEL'S BIPRISM

Two virtual sources S_1 and S_2 are produced by using Fresnel's biprism. The distance d between the virtual sources is usually measured by the displacement method. A convex lens is placed between the biprism and the eyepiece. The eyepiece is fixed at a distance from the slit which is greater than four times the focal length of the lens. For two positions of the lens, distinct images of S_1 and S_2 are formed in the plane of the cross wires of the eyepiece. If the separation between the images of S_1 and S_2 at the two positions of the lens are d_1 and d_2, then

$$d = \sqrt{d_1 d_2}. \quad ...(12.3)$$

12.4 CHANGE OF PHASE DUE TO REFLECTION

A phase change of π occurs when a ray of light is reflected from the surface of an optically denser medium on to a rarer medium.

12.5 LLYOD'S MIRROR

The interference fringes are formed by two narrow pencils of light, one proceeding directly from the source and the other being reflected by a mirror. In this case, the reflected ray suffers a phase change of π and the first fringe close to the mirror is dark.

12.6 THIN-FILM INTERFERENCE

When light falls on a thin transparent film, the light waves reflected from the upper and the lower surfaces of the film produce an interference pattern.

12.7 THE MICHELSON INTERFEROMETER

In Michelson interferometer, a light beam is split into two sub-beams which after traversing different optical paths are recombined to form an interference fringe pattern. By varying the path length of one of the sub-beams, the distances can be measured accurately in terms of wavelengths of light.

SOLVED PROBLEMS

1. *Obtain the conditions for maximum and minimum intensity of light in Young's double slit experiment. Find the average intensity of the interference pattern and show that it is exactly that which would exist in the absence of interference.*

Solution

Monochromatic light of wavelength λ is allowed to pass through a slit S and then at a considerable distance away, through two parallel slits S_1 and S_2 (Fig. 12.1).

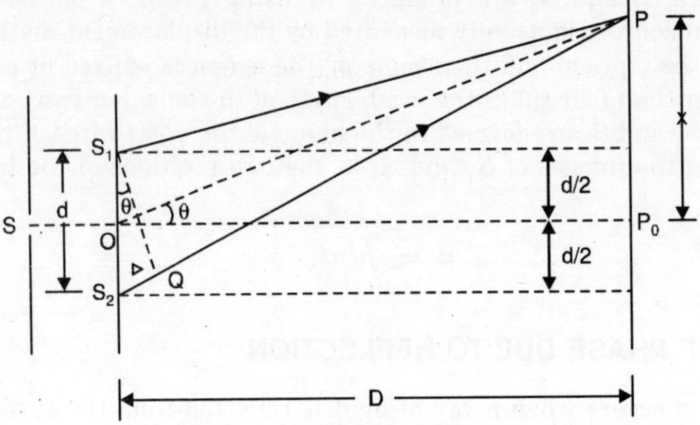

Fig. 12.1

The slits S_1 and S_2 are equidistant from S so that S_1 and S_2 act as coherent sources in the same phase; if they are of same width, they will emit disturbances of equal amplitudes. These disturbances are superimposed at P on the screen with a phase difference

$$\delta = \frac{2\pi}{\lambda}(S_2P - S_1P) = \frac{2\pi}{\lambda}\Delta$$

where $\Delta = S_2P - S_1P$ is the path difference. If y_1 and y_2 are displacements at P due to the waves coming from S_1 and S_2, we have

$$y_1 = a \sin \omega t$$

$$y_2 = a \sin (\omega t + \delta)$$

where a is the amplitude of both the waves. According to the principle of superposition, the resultant displacement y is

$$y = y_1 + y_2 = A \sin (\omega t + \phi)$$

where $A \cos \phi = a\,(1 + \cos \delta)$

$A \sin \phi = a \sin \delta$

and A is the amplitude of the resultant displacement. Thus,

$$A^2 = 4a^2 \cos^2\left(\frac{\delta}{2}\right).$$

The intensity I of the light at P is proportional to the square of the resultant amplitude.

$$I \propto A^2 = 4a^2 \cos^2\left(\frac{\delta}{2}\right). \qquad \text{...(12.4}a\text{)}$$

Thus, we have

$$I = 4I_0 \cos^2\left(\frac{\delta}{2}\right) \qquad \text{...(12.4}b\text{)}$$

where I_0 is the intensity on the screen associated with light from one of the two slits, the other slit being temporarily covered. For maximum intensity, $\delta = 0, 2\pi, 4\pi,...$ or, path difference $\Delta = 0, \lambda, 2\lambda...$; and for minimum intensity $\delta = \pi, 3\pi, 5\pi,...$ or, $\Delta = \frac{\lambda}{2}, \frac{3\lambda}{2}, \frac{5\lambda}{2}, ...$

The point P_0 on the screen is equidistant from S_1 and S_2. At P_0, $\Delta = 0$ and $\delta = 0$. We have maximum intensity at P_0. From S_1 we drop a perpendicular S_1Q on S_2P and suppose $\angle S_2S_1Q = \theta$, then $\Delta = d \sin \theta$ and $\delta = \frac{2\pi}{\lambda} d \sin \theta$, where $d = S_1S_2$ = separation between the coherent sources. Thus, the conditions for maxima and minima are

maxima $$d \sin \theta = m\lambda, \; m = 0, \pm 1, \pm 2, \ldots \qquad \ldots(12.5a)$$

minima $$d \sin \theta = \left(m + \frac{1}{2}\right)\lambda, \; m = 0, \pm 1, \pm 2, \ldots \qquad \ldots(12.5b)$$

Suppose the point P is at a distance x from P_0 and D is the distance of the screen from the coherent sources. Then

$$(S_1P)^2 = D^2 + \left(x - \frac{d}{2}\right)^2$$

$$(S_2P)^2 = D^2 + \left(x + \frac{d}{2}\right)^2$$

Usually D is very large compared to d or x. Then, we can write

$$S_1P \approx D\left[1 + \frac{1}{2}\left(x - \frac{d}{2}\right)^2 \Big/ D^2\right]$$

and $$S_2P \approx D\left[1 + \frac{1}{2}\left(x + \frac{d}{2}\right)^2 \Big/ D^2\right]$$

The path difference Δ becomes

$$\Delta = S_2P - S_1P = \frac{xd}{D} = d \tan \theta \approx d \sin \theta$$

where $\angle POP_0 = \theta$ and $\frac{x}{D} = \tan \theta$.

The conditions for bright and dark fringes are

bright fringes $$x = m\frac{\lambda D}{d}, \; m = 0, \pm 1, \pm 2, \ldots \qquad \ldots(12.6a)$$

dark fringes $$x = \left(m + \frac{1}{2}\right)\frac{\lambda D}{d}, \; m = 0, \pm 1, \pm 2, \ldots \qquad \ldots(12.6b)$$

At P_0, $m = 0$, *i.e.*, zeroth order bright fringe is formed at P_0.

The distance between any two consecutive bright or dark fringes is known as fringe width and it is given by

$$\beta = (m + 1)\frac{\lambda D}{d} - m\frac{\lambda D}{d} = \frac{\lambda D}{d}.$$

Thus, alternately dark and bright parallel fringes are formed on both sides of P_0. The width of the bright fringe is equal to the width of the dark fringe.

The intensity at P is given by Eqn. (12.4a). Since the amplitude a decreases with the increasing distance S_1P, the intensity decreases with increasing x, and the intensity decreases with increasing D.

The intensity at bright points is $4I_0$ and at dark point it is zero. Here, the energy is transferred from the points of minimum intensity to the points of maximum intensity. The average value of $\cos^2(\delta/2)$ is 1/2. Therefore, the average intensity of the interference pattern is $<I> = 2I_0$. Now, with two independent sources, each beam acts separately and contributes I_0 and so without interference we would have a uniform intensity of $2I_0$.

2. *Two straight narrow slits 0.25 mm apart are illuminated by a monochromatic source of wavelength 589 nm. Fringes are obtained at a distance of 50 cm from the slit. Find the width of the fringes.*

Solution

Fringe width
$$\beta = \frac{\lambda D}{d} = \frac{589\times10^{-9}\times50\times10^{-2}}{0.25\times10^{-3}}\text{ m}$$
$$= 1.178\times10^{-3}\text{ m.}$$

3. *Two coherent sources are 0.2 mm apart and the fringes are observed on a screen 80 cm away. It is found that with a certain monochromatic light the fifth bright fringe is situated at a distance of 12 mm away from the central fringe. Find the wavelength of light.*

Solution

We know
$$x = \frac{m\lambda D}{d}$$
or
$$\lambda = \frac{xd}{mD} = \frac{12\times0.2}{5\times80\times10}\text{ mm} = 6\times10^{-4}\text{ mm}$$
$$= 6000\text{ Å}$$

4. *Find the resultant E (t) of the following disturbances:*

$$E_1 = E_0 \sin \omega t$$
$$E_2 = E_0 \sin (\omega t + 15°)$$
$$E_3 = E_0 \sin (\omega t + 30°)$$
$$E_4 = E_0 \sin (\omega t + 45°)$$

Solution

$E(t) = E_1 + E_2 + E_3 + E_4 = A \sin(\omega t + \varphi)$ say. Equating the coefficients of $\sin \omega t$ and $\cos \omega t$ from both sides we get

$$A\cos\phi = E_0(1 + \cos 15° + \cos 30° + \cos 45°) = 3.539E_0$$
$$A\sin\phi = E_0(\sin 15° + \sin 30° + \sin 45°) = 1.4659E_0$$

which give
$$A = 3.83E_0 \text{ and } \phi = 22.5°$$

Thus,
$$E(t) = 3.83E_0 \sin(\omega t + 22.5°).$$

5. *What is the phase difference between the waves from the two slits at the mth dark fringe in a Young's double-slit experiment?*

Solution

$$\delta = \frac{2\pi}{\lambda} d\sin\theta = \frac{2\pi}{\lambda}\left(m + \frac{1}{2}\right)\lambda = (2m+1)\pi.$$

6. *Monochromatic light of wavelength 600 nm illuminates two parallel slits 6 mm apart. Calculate the angular deviation of the third order bright fringe (a) in radians, and (b) in degrees.*

Solution

For mth order bright fringe we have $d \sin \theta = m\lambda$. Now since θ is very small, $\sin \theta \approx \theta = m\lambda/d$.

$$(a)\ \theta = \frac{3 \times 600 \times 10^{-9}}{6 \times 10^{-3}} = 3 \times 10^{-4} \text{ radian}$$

$$(b)\ \theta = \frac{180^\circ}{\pi} \times 3 \times 10^{-4} = 0.017^\circ.$$

7. *A double-slit arrangement produces interference fringes for sodium light ($\lambda = 589$ nm) that are 0.004 rad apart. For what wavelength would the angular separation be 10% greater?*

Solution

Angular separation of the interference fringe $= \Delta\theta = \dfrac{\lambda}{d}$.

Thus, $$d = \frac{589 \times 10^{-9}}{0.004} \text{ m} = 14725 \times 10^{-8} \text{ m}$$

$$\text{New angular separation} = \frac{110}{100} \times 0.004 \text{ rad} = 0.0044 \text{ rad}$$

$$\text{The required wavelength} = 0.0044 \times 14725 \times 10^{-8} \text{ m} = 647.9 \text{ nm}.$$

8. *In a double-slit experiment $\lambda = 546$ nm, $d = 0.10$ mm and $D = 20$ cm. What is the linear distance between the fifth maximum and seventh minimum from the central maximum?*

Solution

Linear distance of the fifth maximum from the central maximum $= x_1 = 5\lambda\, D/d$.

Linear distance of the seventh minimum from the central maximum $= x_2 = \left(6 + \frac{1}{2}\right) \lambda D/d$.

Thus, the required linear distance $= x_2 - x_1 = 1.638$ mm.

9. *As shown in Fig. 12.2, O and Y are two identical radiators of waves that are in phase and of the same wavelength λ. The radiators are separated by distance 3λ. Find the largest distance from O along OX for which destructive interference occurs. Express this in terms of λ.*

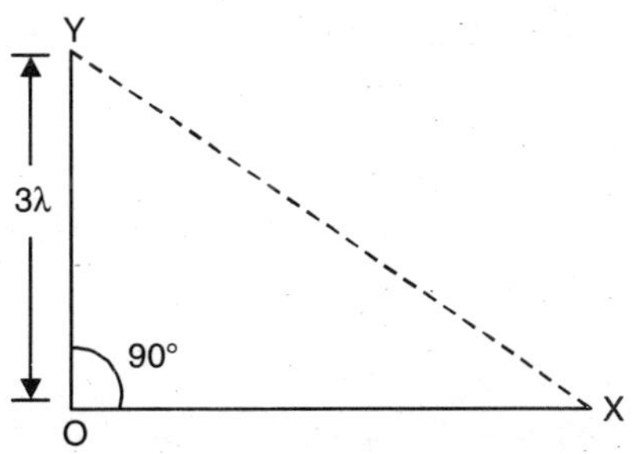

Fig. 12.2

Solution

Suppose X is the point where destructive interference is occurring. So, we have

$$YX - OX = \left(m + \frac{1}{2}\right)\lambda \quad m = 0, 1, 2,...$$

and $$YX^2 - OX^2 = 9\lambda^2$$

From these two equations, we find

$$OX = \left[\frac{36-(2m+1)^2}{4(2m+1)}\right]\lambda$$

which is largest when $m = 0$ and its value is $\frac{35}{4}\lambda$.

10. *The two elements of the double slit are moving apart symmetrically with relative velocity v. Calculate the rate at which the fringes pass a point x cm from the centre of the fringe system formed on a screen D cm from the double slit.*

Solution

Let N be the number of fringes within the length x cm from the central maximum. Then, we have

$$x = \beta N = \frac{\lambda D N}{d}$$

or

$$N = \frac{xd}{\lambda D}.$$

Thus,

$$\frac{\partial N}{\partial t} = \frac{\partial N}{\partial d}\frac{\partial d}{\partial t} = \frac{x}{\lambda D}v.$$

If v is positive, $\frac{\partial N}{\partial t}$ is also positive. As d increases N increases and the fringes move towards the centre of the pattern.

11. *In a double-slit experiment, the slits are 2 mm apart and are illuminated with a mixture of two wavelengths, λ = 750 nm and λ′ = 900 nm. At what minimum distance from the common central bright fringe on a screen 2 m from the slits will a bright fringe from one interference pattern coincide with a bright fringe from the other?*

Solution

From Eqn. (12.6*a*) we see that the *m*th bright fringe of the λ-pattern and the *m*-th bright fringe of the λ′ pattern are located at

$$x_m = \frac{m\lambda D}{d} \text{ and } x'_{m'} = \frac{m'\lambda' D}{d}.$$

Equating these distances, we get

$$\frac{m}{m'} = \frac{\lambda'}{\lambda} = \frac{900}{750} = \frac{6}{5}.$$

Hence, the first position at which overlapping of bright fringes occurs is

$$x_6 = x'_5 = \frac{6\times 750\times 10^{-9}\times 2}{2\times 10^{-3}} \text{ m} = 4.5 \text{ mm}.$$

12. *A double-slit arrangement produces interference fringes for sodium (λ = 589 nm) that are 0.2° apart. What is the angular fringe separation if the entire arrangement is immersed in water (n = 1.33)?*

Solution

Angular fringe separation in water

$$\Delta\theta = \frac{\lambda}{\ldots} = \frac{0.2°}{\ldots} = 0.15°$$

[If the double slit apparatus is immersed in a liquid of refractive index n, the optical path difference between S_2P and S_1P of Fig. 12.1 becomes $n(S_2P - S_1P) \approx (nd)\frac{x}{D} \approx (nd)$ sin θ. In Eqns. (12.5) and (12.6) d is replaced by (nd) so that the fringe width becomes $\lambda D\,(nd)$ *i.e.*, the fringes are shrunk].

13. *In Fresnel's biprism experiment show that the distance d between the two virtual sources S_1 and S_2 produced by it is given by*

$$d = 2l\,(n - 1)\alpha$$

where l = *distance between the slit and the biprism*

n = *refractive index of the material of the prism*

α = *refracting angle of the biprism.*

Solution

The biprism ABC is highly obtuse (*i.e.*, angle B is 178° and the refracting angles A and C are very small (Fig. 12.3). When light falls from the source S on the upper portion of the biprism it is bent downwards and appears to come from the virtual source S_1. Similarly when light falls on the lower portion of the biprism it is bent upwards and appears to come from the virtual source S_2. Thus, S_1 and S_2 act as two coherent sources distant d apart. The conditions for bright and dark fringes on the screen are given by Eqns. (12.5) and (12.6) of Young's double slit experiment.

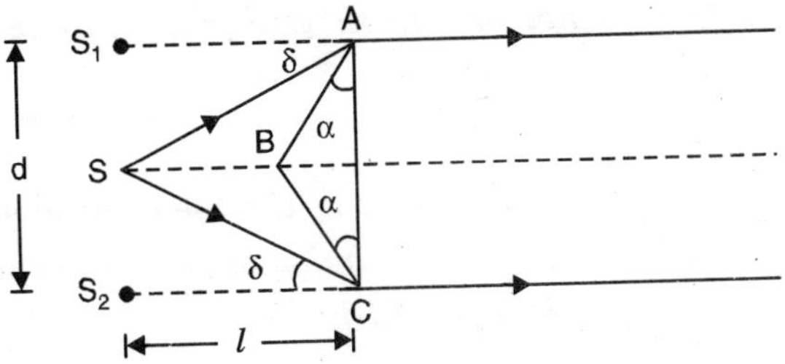

Fig. 12.3

Since each of the constituent prisms of the biprism is thin, the deviation produced in a ray is given by

$$\delta = (n - 1)\alpha$$

where n is the refractive index of the material of the prism. Since $d = S_1S_2$ is small, we may write

$$\delta = \frac{d/2}{l} = (n - 1)\,\alpha.$$

Thus, we have

$$d = 2l\,(n - 1)\,\alpha. \qquad ...(12.7)$$

14. *A source of wavelength 6000 Å illuminates a Fresnel's biprism of refracting angle 1° placed at a distance of 10 cm from it. Find the fringe width on a screen 100 cm from the biprism, the refractive index of the material being 1.5.*

Solution

$$\text{Fringe width} = \frac{\lambda D}{d} = \frac{\lambda D}{2l(n-1)\alpha}$$

$$= \frac{6000 \times 10^{-8} \times (100 + 10)}{2 \times 10 \times 0.5 \times \frac{\pi}{180}}\,\text{cm}$$

$$= 0.38 \text{ mm.}$$

15. *In an experiment with a biprism the distance between the focal plane of the eyepiece and the plane of the interfering sources is 100 cm. A lens inserted between the biprism and the eyepiece gives two images of the slit in two positions. In one case the two images of slit are 0.3 mm and in the other are 1.2 mm apart. If sodium light (λ = 589.3 nm) is used find the distance between the interfering bands.*

Solution

Here, $d = \sqrt{0.3 \times 1.2}$ mm = 0.6 mm

$$\text{Fringe width} = \frac{\lambda D}{d} = \frac{589.3 \times 10^{-9} \times 100 \times 10^{-2}}{0.6 \times 10^{-3}} \text{ m} = 0.98 \text{ mm}$$

16. *When a thin strip of a transparent body of uniform thickness t is introduced at right angles to the path of one of the light rays coming from two slits of Young's experiment, the central fringe on the screen is found to be shifted through a distance x. Show that*

$$t = \frac{x\lambda}{(n-1)\beta}$$

where λ = Wavelength of the monochromatic light used

n = Refractive index of the transparent body

β = Fringe width.

Solution

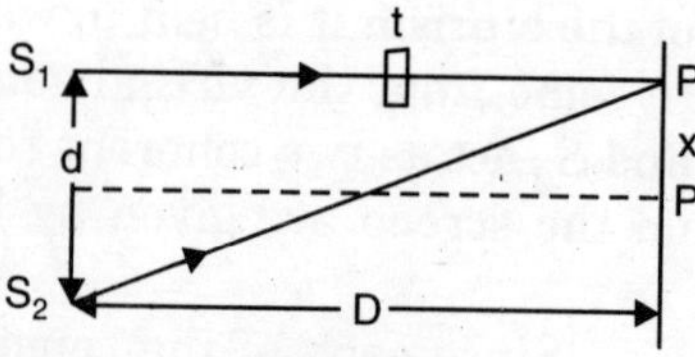

Fig. 12.4

S_1 and S_2 are two coherent monochromatic sources. The disturbances emitted by them are superimposed on the screen P_0P_0' to produce interference fringes (Fig. 12.4). P_0 is the position of the central band. If a thin transparent film of uniform thickness t is introduced in the path of one of the light rays the optical path of that ray changes and the position of the central fringe will be displaced. After the introduction of the transparent strip of thickness t (refractive index n) at right angles to the beam S_1P_0' the central fringe is displaced to P_0' from P_0 on the screen. Thus the optical path difference between the rays = $S_2P_0' - (S_1P_0' - t) - nt = 0$ which gives

$$S_2P_0' - S_1P_0' = (n-1)t.$$

The path difference between the rays at P_0' in the absence of the thin film = xd/D [see problem 1].

Thus, $$\frac{xd}{D} = (n-1)t$$

or $$x = (n-1)t\, D/d.$$

Now, fringe width $\beta = \lambda D/d$.

Thus, $$t = \frac{x\lambda}{(n-1)\beta}.$$

17. *A thin flake of mica (n = 1.58) is used to cover one slit of a double-slit arrangement. The central point on the screen is occupied by what used to be the sixth bright fringe. If λ = 580 nm, what is the thickness of the mica?*

Solution

$$t = \frac{x\lambda}{(n-1)\beta} = \frac{6\beta\lambda}{(n-1)\beta} = \frac{6 \times 580}{0.58} \text{ nm} = 6 \text{ µm}.$$

18. *One slit of a double-slit arrangement is covered by a thin glass plate of refractive index 1.45 and the other by a thin glass plate of refractive index 1.65. The point on the screen where the central maximum fall before the glass plates were inserted is now occupied by what had been the m = 4 bright fringe before. Assume λ = 550 nm and that the plates have the same thickness t. Find the value of t.*

Solution

In this case,

$$\frac{xd}{D} = (n_2 - n_1)t$$

and

$$t = \frac{x\lambda}{(n_2 - n_1)\beta} = \frac{4\beta\lambda}{(n_2 - n_1)\beta} = \frac{4 \times 550}{0.2}\text{ nm} = 11\ \mu\text{m}.$$

19. *Prove that the increase in optical path produced by rotating a plane-parallel plate of thickness t and refractive index n through an angle ϕ from the perpendicular position is given by*

$$\delta\Delta = (\sqrt{n^2 - \sin^2\phi} - \cos\phi - n + 1)t$$

which reduces to the form

$$\delta\Delta \approx \frac{t\phi^2}{2n}(n - 1)$$

when ϕ is small.

Solution

The optical path before the rotation of the plate (Fig. 12.5) (When the ray OA is normal to the plate)

$$= OA + nt + BC.$$

The optical path after rotation through an angle ϕ

$$= OA + nAD + DF + FE$$

$$= OA + \frac{nt}{\cos r} + [t - AD\cos(\phi - r)] + FE$$

Thus, the increase in the optical path is

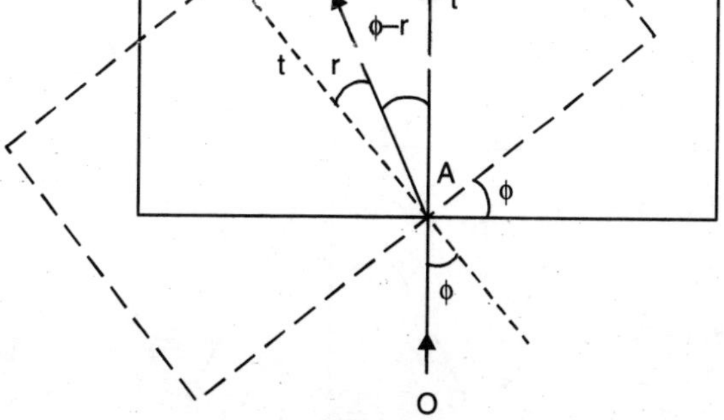

Fig. 12.5

$$\delta\Delta = \frac{nt}{\cos r} + t - \frac{t\cos(\phi - r)}{\cos r} - nt$$

$$= t\left[1 - n + \frac{n - \cos\phi\cos r - \sin\phi\sin r}{\cos r}\right]$$

By using the relation $n = \sin\phi/\sin r$, we have

$$\delta\Delta = t[1 - n - \cos\phi + \sqrt{n^2 - \sin^2\phi}]$$

When ϕ is small we may approximate $\sin\varphi \approx \phi$ and $\cos\varphi \approx 1 - \phi^2/2$. Thus, we have

$$\delta\Delta = t\left[1 - n - (1 - \phi^2/2) + n\left(1 - \frac{1}{2}\phi^2/n^2\right)\right]$$

$$= \frac{t\phi^2}{2n}(n - 1).$$

20. *(a) A ray of light travelling through the medium I is incident on the surface of separation of two media I and II of which the medium II is optically denser (Fig. 12.6). This light is partly reflected on to the rarer medium and partly transmitted into the denser medium. By reversing the reflected and the transmitted rays show that*

$$r' = -r$$

where r = amplitude reflection coefficient in the rarer medium I and r′ = amplitude reflection coefficient in the denser medium.

(b) Comment on the change of phase due to reflection of light.

Solution

(a) *Stoke's treatment*: Suppose PQ represents the separation surface of two media I and II of which the medium II is optically denser. A ray of light AO is incident on PQ. This light is partly reflected along OB and partly transmitted along OC. Suppose a is the amplitude of the incident light wave, r is the amplitude reflection coefficient in the rarer medium I and t is the amplitude transmission coefficient from the rarer medium I to the denser medium II. The amplitude of the reflected ray OB is ar and that of the transmitted ray OC is at. If there is no absorption in the medium, we have

$$r + t = 1$$

Now if the reflected and transmitted rays are reversed they should recombine along OA to give the original amplitude a.

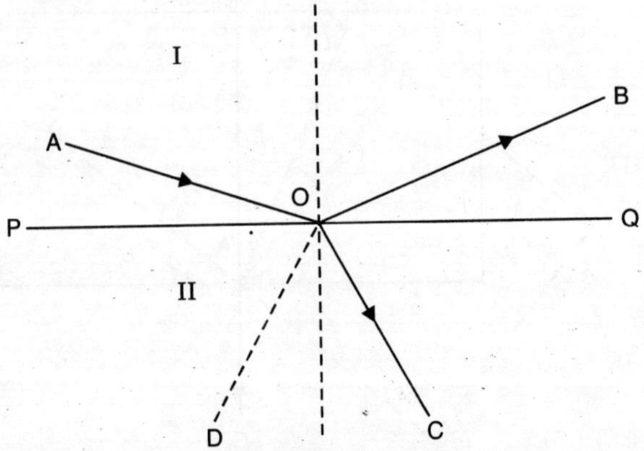

Fig. 12.6

If OB is reversed, the part of the amplitude reflected along OA is $arr = ar^2$ and that transmitted along OD is art. If OC is reversed, the part reflected along OD is atr' and transmitted along OA is att' where r' is the reflection coefficient in the denser medium II and t' is the transmission coefficient from denser to the rarer medium. Since the two amplitudes along OA combine together to give the original amplitude a, we have

$$ar^2 + att' = a, \text{ or, } r^2 + tt' = 1$$

The total amplitude along OD is zero, *i.e.*,

$$art + atr' = 0$$

or

$$r' = -r.$$

The negative sign indicates that one of the rays has a positive displacement and the other has a negative displacement.

(*b*) The coefficient r represents the reflection from a denser medium on to a rarer medium and the coefficient r' represents the reflection from a rarer medium on to a denser medium. The two rays differ in phase by π. Actually in some experiments (*e.g.*, Newton's rings and Lloyd's mirror experiments) a phase change of π is noticed in the former case (reflection from a denser on to a rarer medium). No change of phase is found to occur in the latter case. When reflection occurs from an interface beyond which the medium has a higher refractive index, the reflected wave undergoes a phase change of π; when the medium beyond the interface has a lower refractive index, there is no change of phase. The transmitted wave undergoes no change of phase in either case.

21. *Interference fringes are produced by two pencils of light one proceeding directly from the source S_1 and the other being reflected from a mirror MM′ (Fig. 12.7). Find the conditions for maximum and minimum brightness of fringes on the screen AB. (This arrangement for obtaining a double-slit interference pattern from a single slit is called Lloyd's mirror.)*

Solution

Here, the source S_1 and its virtual image S_2 act as two coherent sources.

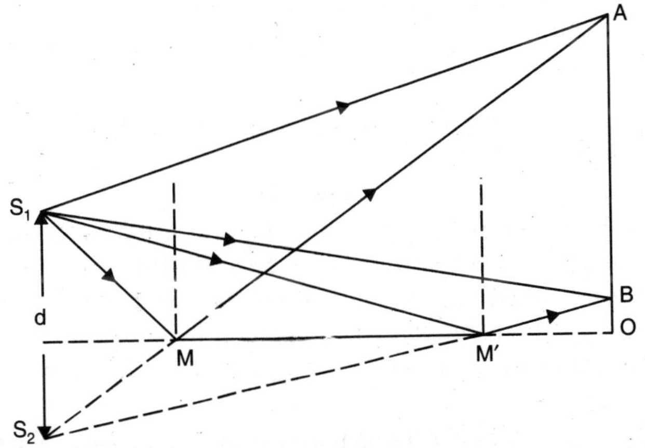

Fig. 12.7

Suppose d is the separation between the two sources. The ray reflected by the mirror has gone a phase change of π which results in a path difference of $\lambda/2$. Thus, the conditions for maximum and minimum brightness are

minima $d \sin\theta = m\lambda$, $m = 0, 1, 2, ...$

maxima $d \sin \theta = \left(m + \frac{1}{2}\right)\lambda,\ m = 0, 1, 2, ...$

The fringe width on the screen is given by

$$\beta = \frac{\lambda D}{d}.$$

The interference between the two wave trains occurs in the region of overlapping *AB*.

22. *Find the condition of formation of bright and dark fringes due to thin film interference.*

Solution

Light coming from the same source is reflected from the upper and lower surfaces of the plane-parallel film many times. These reflected rays originate from the same source. They are coherent and are in a position to interfere. The set of transmitted light rays obtained are also in a position to interfere.

(a) Interference due to reflected rays

Let us calculate the path difference between any two reflected rays. A ray of light from the source *S* is incident at *A* (Fig. 12.8) on the upper surface of a thin film of thickness *d* and of refractive index *n*.

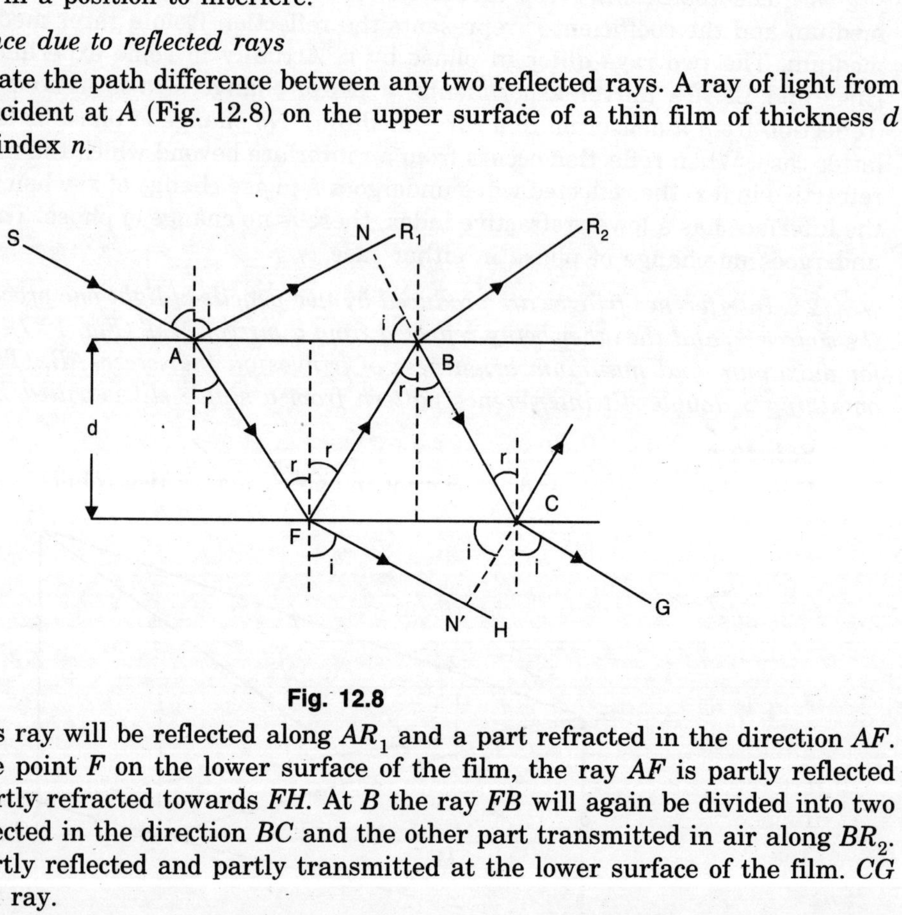

Fig. 12.8

A part of this ray will be reflected along AR_1 and a part refracted in the direction *AF*. After reaching the point *F* on the lower surface of the film, the ray *AF* is partly reflected towards *B* and partly refracted towards *FH*. At *B* the ray *FB* will again be divided into two parts—a part reflected in the direction *BC* and the other part transmitted in air along BR_2. The ray *BC* is partly reflected and partly transmitted at the lower surface of the film. *CG* is the transmitted ray.

We consider the two successive reflected rays AR_1 and BR_2. They are derived from the same source *S* and are in a position to interfere when they are brought together. Since the film is parallel sided, AR_1 and BR_2 will also be parallel to each other. A perpendicular *BN* is dropped from *B* on AR_1. Then the plane passing through *BN* and perpendicular to the plane of the paper is the new wavefront. The path difference between the rays AR_1 and BR_2 is

$$\Delta = n\,(AF + FB) - AN$$

$$= 2n\,AF - AN$$

$$= 2n\frac{d}{\cos r} - AB\sin i$$

$$= \frac{2nd}{\cos r} - 2d\tan r\sin i$$

$$= 2nd\left[\frac{1}{\cos r} - \frac{\sin^2 r}{\cos r}\right]$$

$$= 2nd\cos r$$

where we have used $n = \sin i/\sin r$. It should be noted that the two rays are reflected under different conditions. The first ray AR_1 is reflected from the optically denser medium, and the second ray *AFB* is reflected from the rarer medium. The first ray undergoes a phase change of π and no change of phase occurs in the second ray. Thus, an extra change of π, or, a path difference of $\lambda/2$ is introduced in Δ. Hence the correct result is

$$\Delta = 2nd\cos r \pm \lambda/2. \qquad ...(12.8)$$

The conditions for the formation of bright and dark fringes due to interference in the reflected rays are:

$$2nd\cos r \pm \frac{\lambda}{2} = m\lambda \qquad \text{(maxima)}$$

$$2nd\cos r \pm \frac{\lambda}{2} = \left(m + \frac{1}{2}\right)\lambda \qquad \text{(minima)}$$

These may be rewritten as

$$2nd\cos r = \left(m + \frac{1}{2}\right)\lambda \text{ (maxima)} \qquad ...(12.9a)$$

$$2nd\cos r = m\lambda \text{ (minima)} \qquad ...(12.9b)$$

with $m = 0, 1, 2, 3, ...$

For near normal incidence ($r \approx 0°$), we have

$$2nd = \left(m + \frac{1}{2}\right)\lambda \text{ (maxima)} \qquad ...(12.10a)$$

$$2nd = m\lambda \text{ (minima)} \qquad ...(12.10b)$$

with $m = 0, 1, 2, ...$

(b) Interference due to transmitted rays

Interference phenomenon occurs with the transmitted rays *FH* and *CG*. A perpendicular *CN′* is dropped from *C* on *FH* (Fig. 12.8). The path difference between the transmitted rays *FH* and *CG* is

$$\Delta = n(FB + BC) - FN'$$

$$= 2n\,FB - FC\sin i$$

$$= 2n\frac{d}{\cos r} - 2d\tan r\sin i$$

$$= 2nd\cos r$$

In the case of transmitted rays there is only internal reflection at B and this causes no additional change of phase. Thus, for the transmitted rays we have the following conditions for maxima and minima:

maxima $$2nd \cos r = m\lambda \tag{12.11a}$$

minima $$2nd \cos r = \left(m + \frac{1}{2}\right)\lambda \tag{12.11b}$$

with $m = 0, 1, 2, ...$

These conditions are just opposite to those obtained with the reflected rays. As a result a reflected and transmitted interference patterns are complementary to each other. This means that the position which corresponds to the maximum of the reflected system will be the position of minimum of the transmitted system and vice versa.

23. *A water film (n = 4/3) in air is 315 nm thick. If it is illuminated with white light at normal incidence, what colour will it appear to be in the reflected light?*

Solution

For maxima, we have the relation $\lambda = \dfrac{2nd}{\left(m + \frac{1}{2}\right)}$

When $m = 0$, $\lambda = 1680$ nm (infrared),
$m = 1$, $\lambda = 560$ nm (visible),
$m = 2$, $\lambda = 336$ nm (ultraviolet).

Only the maximum corresponding to $m = 1$ lies in the visible region. Light of wavelength 560 nm appears yellow-green. The water film will appear yellow-green in the reflected light.

24. *A lens is coated with a thin film of transparent substance magnesium fluoride (MgF_2) with n = 1.38 to reduce the reflection from the glass surface (n = 1.50). How thick a coating is needed to produce a minimum reflection at the centre of the visible spectrum (λ = 550 nm)?*

Solution

We assume that the light strikes the lens at near-normal incidence. We would like to find the thickness of the film that will bring about destructive interference between rays R_1 and R_2 (Fig. 12.9). Phase change of π is now associated with each ray and the condition of minimum intensity is now

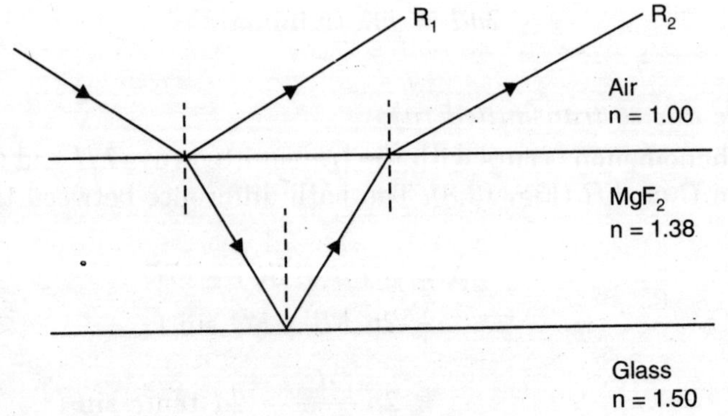

Fig. 12.9

$$2nd = \left(m+\frac{1}{2}\right)\lambda,\ m = 0,\ 1,\ 2,...$$

Thus, the thickness of the thin nest possible film is (for $m = 0$)

$$d = \frac{\lambda}{4n} = \frac{550}{4\times1.38}\text{ nm} = 99.64\text{ nm.}$$

25. *A soap film of thickness 5.5 × 10⁻⁵ cm is viewed at an angle of 45°. Its index of refraction is 1.33. Find the wavelengths of light in the visible spectrum which will be absent from the reflected light.*

Solution

Since $$n = \frac{\sin i}{\sin r},\ \cos r = 0.84696,$$

where i is the angle of incidence and r is the angle of refraction.

For minima, $$2nd\cos r = m\lambda$$

When $m = 1,\ \lambda = 12.39 \times 10^{-5}$ cm (infrared)

$m = 2,\ \lambda = 6.20 \times 10^{-5}$ cm (visible)

$m = 3,\ \lambda = 4.13 \times 10^{-5}$ cm (visible)

$m = 4,\ \lambda = 3.10 \times 10^{-5}$ cm (ultra violet)

Hence the absent wavelengths in the reflected light are 6.20×10^{-5} cm and 4.13×10^{-5} cm in the visible region of spectrum.

26. *A tanker leaks kerosene (n = 1.2) into the Persian Gulf, creating a large slick on top of the water (n = 1.3). (a) If you are looking straight down from an airplane onto a region of the slick where its thickness is 460 nm, for which wavelength(s) of visible light is the reflection the greatest? (b) If you are scuba-diving directly under this same region of the slick, for which wavelength(s) of visible light is the transmitted intensity the strongest.*

Solution

(*a*) For the reflected rays, the condition for maxima is [see problem 24]

$$2nd = m\lambda,$$

or $$\lambda = \frac{2nd}{m} = \frac{2\times1.2\times460}{m}\text{ nm}$$

The wavelength of visible light (for $m = 2$) is 552 nm.

(*b*) For the transmitted rays, the condition for maxima is

$$2nd = \left(m+\frac{1}{2}\right)\lambda$$

or $$\lambda = \frac{2nd}{m+\frac{1}{2}} = \frac{2\times1.2\times460}{m+\frac{1}{2}}\text{ nm.}$$

The wavelengths of visible light are 736 nm (when $m = 1$) and 441.6 nm (when $m = 2$). [For the transmitted rays an extra phase change of π is introduced in the ray *AFBCG* whereas no change of phase occurs in the ray *AFH* (Fig. 12.8) under the conditions stated in (*b*)].

27. *A plane wave of monochromatic light falls normally on a uniform thin film of oil that covers a glass plate. The wavelength of the source can be varied continuously. Complete destructive interference of the reflected light is observed for wavelengths of 500 and 700 nm and for no wavelengths between them. If the index of refraction of the oil is 1.3 and that of glass is 1.5, find the thickness of the oil film.*

Solution

The condition of destructive interference of the reflected light is (see problem 24)

$$2nd = (2m + 1)\lambda/2$$

or

$$4nd = (2m + 1)\lambda$$

Thus, $4 \times 1.3 \times d = 700\ (2m + 1) = 500[2(m + 1) + 1]$

which gives $m = 2$ and $d = 673.08$ nm.

28. *Find the conditions of formation of bright and dark fringes produced by a wedge-shaped thin film for near normal incidence.*

Solution

We consider two plane surfaces OA and OB inclined at an angle θ which is very small (Fig. 12.10). The thickness of the film increases from O to A. When the film is viewed with reflected monochromatic light, a system of equidistant interference fringes are observed which are parallel to the line of intersection of the two surfaces. A parallel beam of monochromatic light falls on the face OB almost normally. The components reflected from the upper and the lower surfaces will interfere to produce alternatively dark and bright bands.

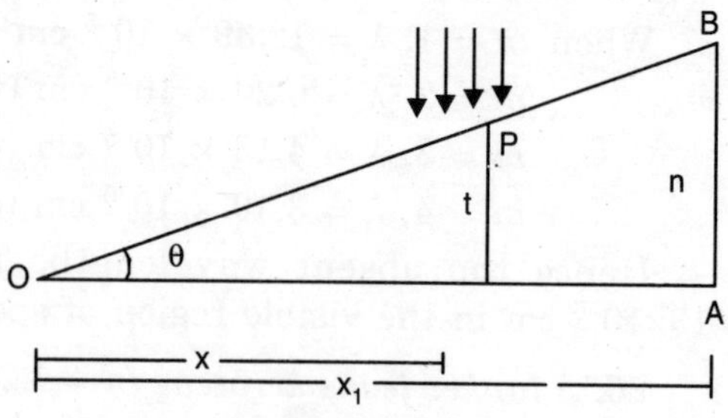

Fig. 12.10

Since the angle of wedge θ is very small, the beam incident normally to the face OB may very well be considered normal to the face OA too. The path difference between the two reflected beams is $(2nt + \lambda/2)$ at the point P where the thickness of the wedge is t. So at this position of the film there will be dark fringe when

$$2nt + \frac{\lambda}{2} = \left(m + \frac{1}{2}\right)\lambda, \qquad m = 0, 1, 2, \ldots$$

or

$$2nt = m\lambda \text{ (minima)}$$

where n is the refractive index of the film. For bright fringes, we have

$$2nt = \left(m + \frac{1}{2}\right)\lambda \text{ (maxima)}$$

We know $\theta = t/x$ or $t = x\theta$ (see Fig. 12.10). Thus mth dark fringe appears at P when

$$2nx\theta = m\lambda$$

Similarly, if $(m + p)$th dark fringe is formed at a distance x_1 from O, we have

$$2nx_1\theta = (m + p)\lambda$$

Thus,

$$2n(x_1 - x)\theta = p\lambda$$

We can count the number of fringes in a space $(x_1 - x)$ with a travelling microscope. Here $x_1 - x$ is the distance corresponding to p fringes. The fringe width β is given by

$$\beta = \frac{x_1 - x}{P} = \frac{\lambda}{2n\theta}.$$

If air is enclosed between OA and OB, $n = 1$, and we have

$$\beta = \frac{\lambda}{2\theta}.$$

29. *Two glass plates enclose a wedge-shaped air film, touching at one edge and are separated by a wire of 0.04 mm diameter at a distance of 10 cm from the edge. Calculate the fringe width. Monochromatic light of wavelength 589.3 nm from a broad source falls normally on the film.*

Solution

Referring to the figure 12.10, we have $OA = 10$ cm and $AB = 0.04$ mm so that

$$\theta = \frac{AB}{OA} = \frac{0.004}{10} = 0.0004 \text{ rad.}$$

Fringe width $$\beta = \frac{\lambda}{2\theta} = \frac{589.3\,\text{nm}}{2 \times 0.0004} = 0.074 \text{ cm.}$$

30. *A broad source of light (λ = 680 nm) illuminates normally two glass plates 120 mm long that touch at one end and are separated by a wire 0.048 mm in diameter at the other end. How many bright fringes appear over the 120 mm distance?*

Solution

Fringe width $$\beta = \frac{\lambda}{2\theta} = \frac{680 \times 120}{2 \times 0.048} \text{nm}$$

Number of bright fringes over the 120 mm distance

$$= \frac{120 \times 10^6}{\beta} = 141.$$

31. *Two wires with diameters 0.01 cm and 0.03 cm are laid parallel to each other and 2 cm apart on a piece of plane glass. Another piece of plane glass is laid on top. Monochromatic light of wavelength 546 nm falls normally on the glass plates. Calculate the distance between consecutive bright fringes.*

Solution

Angle of the wedge $$\theta = \frac{0.03 - 0.01}{2} = 0.01 \text{ rad.}$$

Fringe width $$\beta = \frac{\lambda}{2\theta} = \frac{546}{2 \times 0.01} \text{nm} = 27.3 \text{ μm.}$$

32. *In an air wedge formed by two plane glass plates touching each other along one edge there are 4001 dark lines observed when viewed by reflected monochromatic light. When the air between the plates is evacuated, only 4000 such lines are observed. Calculate the index of refraction of the air from this data.*

Solution

If mth dark fringe appears at a distance x from the edge then

$$x = m\lambda/(2n\theta)$$

In the present problem we have

$$x = \frac{4001\lambda}{2n\theta} = \frac{4000\lambda}{2\theta}$$

which gives $$n = \frac{4001}{4000} = 1.00025.$$

33. *Newton's rings: The curved surface of the plano-convex lens of large focal length is placed on an optically flat glass plate. A transparent liquid of refractive index n is placed between the lens and the glass plate so that a thin film is formed between the lens and the glass plate (n being less than the refractive index of the material of the lens or glass plate; the liquid may be air). Interference fringes are produced when monochromatic light is incident on the lens. Find the conditions of formation of bright and dark fringes.*

Solution

Let *LOL′* be the convex surface of a lens in contact with the plane surface *PP′* (Fig. 12.11). *O* is the point of contact. A liquid of refractive index *n* is placed between the glass plate and the lens. The point where the lens touches the glass plate, the thickness of the film is zero. The thickness of the film increases all around this point as we move away from *O*. The thin film of liquid thus enclosed is wedge shaped, and the loci of all points having the same thickness are circles. When monochromatic light is made to fall normally on such a film, a series of concentric bright and dark rings are visible all round the point of contact, both in the reflected and in the transmitted rays.

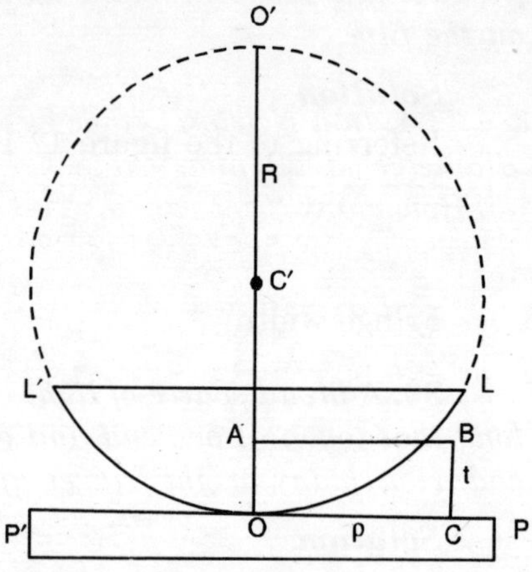

Fig. 12.11

Let *C* be a point on *PP′* at a distance ρ from the point of contact. The thickness of the film at *C* is *CB* = *t*. Thus, the path difference between the two reflected rays at this position is $(2nt + \lambda/2)$ for normal incidence (see problem 28). So at this position of the film there will be a dark fringe when

$$2nt + \lambda/2 = \left(m + \frac{1}{2}\right)\lambda, \quad m = 0, 1, 2, \ldots$$

or

$$2nt = m\lambda \quad \text{(minima)}$$

For bright fringes we have

$$2nt = \left(m + \frac{1}{2}\right)\lambda \quad \text{(maxima)}$$

Let *C′* be the centre of the spherical surface *LOL′*, having radius of curvature equal to *R* (Fig. 12.11). We have

$$AB^2 = OA.AO'$$

or

$$\rho^2 = t(2R - t) \approx 2Rt,$$

since *t* is very small.

Thus,

$$t = \frac{\rho^2}{2R},$$

The conditions for dark and bright fringes are

$$\rho_m^2 = \frac{m\lambda R}{n} \quad \text{for the } m\text{th dark ring, } m = 0, 1, 2, \ldots$$

and

$$\rho_m^2 = \frac{\left(m - \frac{1}{2}\right)\lambda R}{n} \quad \text{for the } m\text{th bright ring, } m = 1, 2, 3, \ldots$$

where ρ_m is the radius of the *m*th ring

Newton's rings may be obtained with the transmitted light also. In that case the ring system will be complimentary to the reflected one. The central spot will be bright in the transmitted system. For the transmitted light we have

$$\rho_m^2 = \frac{m\lambda R}{n} \text{ (bright rings), } m = 0, 1, 2,...$$

and

$$\rho_m^2 = \frac{\left(m - \frac{1}{2}\right)\lambda R}{n} \text{ (dark rings), } m = 1, 2, 3,...$$

34. *In a Newton's rings experiment the radius of curvature R of the lens is 5 m and its diameter is 20 mm. (a) How many bright rings are produced in the reflected rays? (b) How many rings would be seen if the arrangement were immersed in water (n = 1.33)? The wavelength of light used is 589 nm.*

Solution

(*a*) The radius of the *m*th bright ring is given by

$$\rho_m^2 = \left(m - \frac{1}{2}\right)\lambda R/n$$

In air, $$m - \frac{1}{2} = \frac{\rho_m^2}{\lambda R} = \frac{(10\times10^{-3})^2}{589\times10^{-9}\times5} = 33.956$$

34 rings are produced.

(*b*) In water, $$m - \frac{1}{2} = \frac{n\rho_m^2}{\lambda R} = 45.16$$

45 rings are produced.

35. *The convex surface of radius 200 cm of a plano-convex lens rests on a concave spherical surface of radius 400 cm and Newton's rings are viewed with reflected light of wavelength 600 nm. Calculate the diameter of the 8th bright ring.*

Solution

The thickness of the air-film at *B* (Fig. 12.12) is

$$t = BC = BD - CD = \frac{\rho^2}{2R_1} - \frac{\rho^2}{2R_2} = \frac{\rho}{2}\left(\frac{1}{R_1} - \frac{1}{R_2}\right)$$

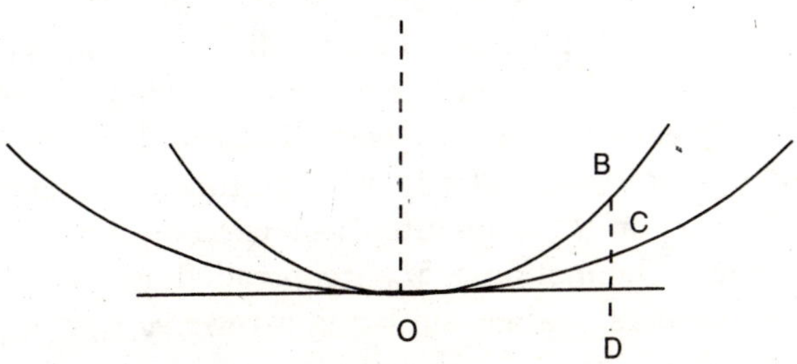

Fig. 12.12

where $\rho = OD$, R_1 = Radius of curvature of *OB* and R_2 = Radius of curvature of *OC*. Thus, the condition of the *m*th bright fringe is

$$\rho_m^2 = \frac{\left(m - \frac{1}{2}\right)\lambda/n}{\frac{1}{R_1} - \frac{1}{R_2}}$$

In the present problem, $m = 8$, $\lambda = 600$ nm, $n = 1$, $R_1 = 200$ cm, $R_2 = 400$ cm. We find $2\,\rho_8 = 0.85$ cm.

36. *Light from a source emitting two wavelengths λ_1 and λ_2 falls normally on a plano-convex lens of radius of curvature R resting on a glass plate. It is found that nth dark ring due to λ_1 coincides with (n + 1)th dark ring due to λ_2. Show that the radius of the nth dark ring for the wavelength λ_1 is given by*

$$\rho_n = [\lambda_1\lambda_2 R/(\lambda_1 - \lambda_2)]^{1/2}.$$

Solution

$$\rho_n = \text{the radius of the } n\text{th dark ring due to } \lambda_1 = \sqrt{n\lambda_1 R}$$

The radius of the $(n + 1)$th dark ring due to λ_2

$$= \sqrt{(n+1)\lambda_2 R}.$$

Thus, we have

$$\rho_n = \sqrt{n\lambda_1 R} = \sqrt{(n+1)\lambda_2 R}$$

which gives $$n\lambda_1 = (n + 1)\lambda_2$$

or $$n = \lambda_2/(\lambda_1 - \lambda_2)$$

and $$\rho_n = [\lambda_1\lambda_2 R/(\lambda_1 - \lambda_2)]^{1/2}.$$

37. *Describe the construction of Michelson's interferometer. If the movable mirror of Michelson's interferometer is moved through a small distance d, and the number of fringes that cross the field of view is m, then show that the wavelength of light is given by*

$$\lambda = 2d/m.$$

Solution

An instrument that is designed primarily to measure lengths or changes in length with degree of accuracy by means of interference fringes is known as an interferometer. One of the widely used interferometers is the one devised and built by A.A. Michelson.

The light from the source S falls on a beam splitter F (Fig. 12.13). F and G are two plane-parallel optical flats, which are usually cut from a single parallel plate so that they are identical. The surface of F nearest to G may be half silvered in order that the light falling upon it be 50% reflected and 50% transmitted, although this is not necessary for the operation of the interferometer. The mirror M_2 is mounted on a carriage, so that it can be moved along a precision-mechanical track. The motion of M_2 is controlled by a very fine micrometer screw. The mirror M_1 is held by springs against adjusting screws so that M_1 may be made exactly perpendicular to M_2,

Light from the source S is rendered parallel by the lens L. After entering F the light is divided into two parts: one part proceeds to the mirror M_1 which returns the light to the eye after reflection at F; the other part, reflecting inside F, proceeds to M_2, which reflects it back through F to E. From a single source, by division of amplitude, two beams are produced and they are recombined for the formation of interference fringes.

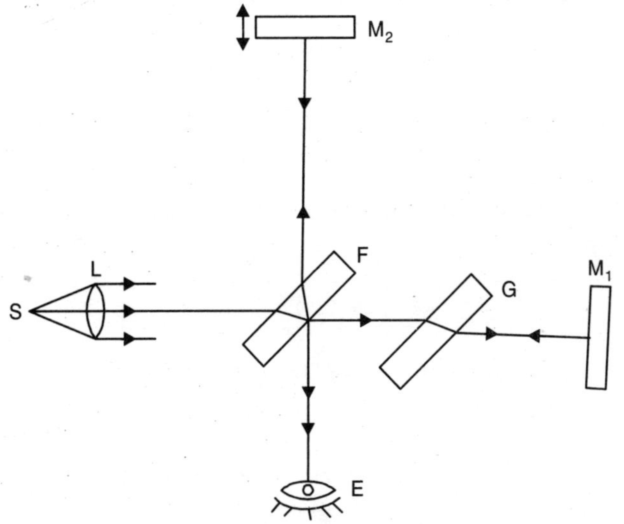

Fig. 12.13

The beam going towards the mirror M_2 and reflected back, has to pass twice through the glass plate F. Therefore to compensate for the path the plate G is used between the mirror M_1 and F. The light beam going towards the mirror M_1 and reflected back towards F also passes twice through the compensating plate G. Thus the paths of the two rays in glass are the same. Suppose the mirrors M_1 and M_2 are at distances d_1 and d_2 respectively from the glass plate F. In measuring d_1 and d_2 we do not consider the paths of the rays in glass. The path difference for the two waves when they recombine is $2d_2 - 2d_1$ and anything that changes this path difference will cause a change in the relative phase of the two waves as they enter the eye. If the mirror M_2 is moved by a distance $\frac{1}{2}\lambda$, the path difference changes by λ and the observer will see the fringe pattern shift by one fringe. If the mirrors M_1 and M_2 are exactly perpendicular to each other, the system is essentially equivalent to the interference from an air film whose thickness is equal to $d_2 - d_1$. Suppose when the mirror M_2 is moved through a distance d, the number of fringes that cross the field of view is m. Then we have

$$d = m\frac{\lambda}{2}$$

or

$$2d = m\lambda.$$

38. *Let m_1 and $m_1 + 1$ be the changes in the order at the centre of the field of view of Michelson's interferometer when the movable mirror is displaced through a distance d between two consecutive positions of maximum distinctness of the fringes of two neighbouring spectral lines with wavelengths λ_1 and λ_2 respectively. Show that*

$$\Delta\lambda = \lambda_1 - \lambda_2 \approx \frac{\lambda^2}{2d}$$

where λ is the mean of λ_1 and λ_2.

Solution

We have

$$2d = m_1\lambda_1 = (m_1 + 1)\lambda_2$$

which gives

$$m_1 = \frac{\lambda_2}{\lambda_1 - \lambda_2}$$

and

$$2d = \frac{\lambda_1\lambda_2}{\lambda_1 - \lambda_2}$$

where λ_1 and λ_2 are two neighbouring spectral lines, we may write $\lambda_1\lambda_2 \approx \lambda^2$, where λ is the mean value of λ_1 and λ_2.

Thus,

$$\Delta\lambda = \lambda_1 - \lambda_2 = \frac{\lambda^2}{2d}.$$

39. *How far must one of the mirrors of Michelson's interferometer be moved for 500 fringes of light of wavelength 500 nm to cross the centre of the field of view?*

Solution

$$d = m\frac{\lambda}{2} = \frac{500 \times 5000 \times 10^{-8}}{2} \text{ cm} = 0.0125 \text{ cm.}$$

40. *After obtaining the fringes in Michelson's interferometer with white light, the white light is replaced by sodium light. When one mirror of the interferometer is now moved through a distance of 0.15 mm the fringes are found to disappear. If the mean wavelength for the two components of the D lines of sodium light is 5893 Å, find the difference between their wavelengths.*

Solution

We have

$$2d = m_1\lambda_1 = \left(m_1 + \frac{1}{2}\right)\lambda_2,$$

which gives

$$m_1 = \frac{\lambda_2}{2(\lambda_1 - \lambda_2)}$$

and

$$\Delta\lambda = \lambda_1 - \lambda_2 = \frac{\lambda_1\lambda_2}{4d} \approx \frac{\lambda^2}{4d}$$

$$= \frac{5893 \times 5893 \times 10^{-8}}{4 \times 0.15 \times 10^{-1}} \text{ Å} = 5.79 \text{ Å.}$$

41. *An airtight chamber 5 cm long with glass windows is placed in one arm of a Michelson's interferometer. Light of wavelength λ = 480 nm is used. The air is slowly evacuated from the chamber using a vacuum pump. While the air is being removed 61 fringes are observed to pass through the field of view. Find the index of refraction of air at atmospheric pressure.*

Solution

In Michelson interferometer, light travels twice through the chamber. Thus, we have the formula,

$$2t = \frac{x\lambda}{(n-1)\beta} = \frac{61\beta\lambda}{(n-1)\beta}$$

or $$n - 1 = \frac{61\lambda}{2t} = \frac{61 \times 480 \times 10^{-9}}{2 \times 5 \times 10^{-2}} = 0.00029$$

giving $$n = 1.00029.$$

42. *A plane wavefront of light is incident on a plane mirror as shown in Fig. 12.14. Show that the intensity is maximum at P when*

$$\cos\theta = \frac{\lambda}{4d}, \frac{3\lambda}{4d}, \frac{5\lambda}{4d}, \ldots$$

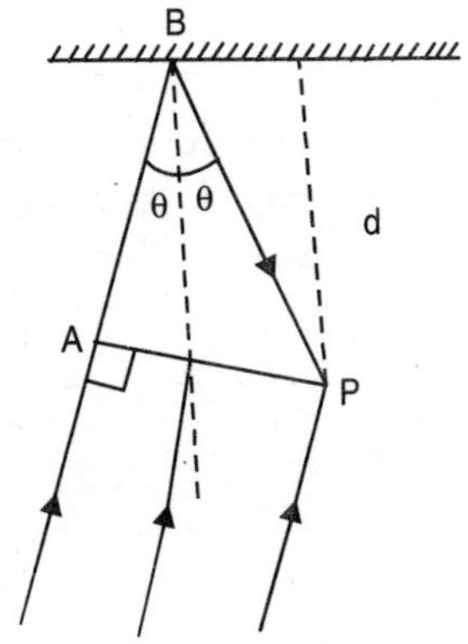

Fig. 12.14

Solution

The path difference between the disturbance reaching at point P directly and after reflection is

$$AB + BP + \frac{\lambda}{2} = BP\cos 2\theta + \frac{d}{\sin(90° - \theta)} + \frac{\lambda}{2}$$

$$= \frac{d}{\cos\theta}(\cos 2\theta + 1) + \frac{\lambda}{2}$$

$$= 2d\cos\theta + \frac{\lambda}{2}$$

Here $\lambda/2$ is due to reflection from the denser medium.

For maximum intensity at P, we have

$$2d\cos\theta + \frac{\lambda}{2} = n\lambda,\ n = 1, 2, 3, \ldots$$

or $$\cos\theta = \frac{(2n-1)\lambda}{4d},\ n = 1, 2, 3, \ldots$$

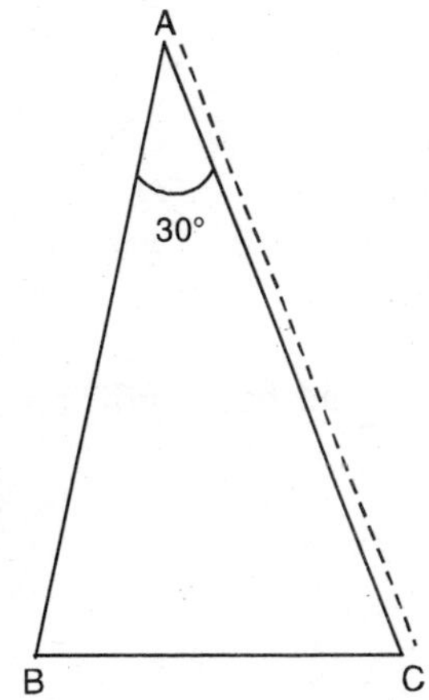

Fig. 12.15

43. *A prism of refracting angle 30° is coated with a thin film of transparent material of refractive index 2.2 on the face AC of the prism.*

A light of wavelength 5500 Å is incident on face AB such that the angle of incidence is 60°. Find

(a) the angle of emergence, and

(b) the minimum value of thickness of the coated film on the face AC for which the light emerging from the face has maximum intensity. [Given the refractive index of the material of the prism is $\sqrt{3}$ *]* (I.I.T. 2003)

Solution (*a*) From Fig. 12.16, we have

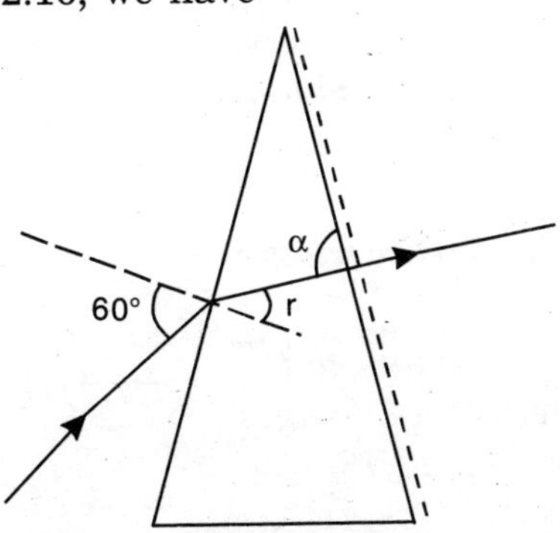

Fig. 12.16

$$1.\ \sin 60° = \sqrt{3} \sin r$$

or
$$\sin r = \frac{1}{2},$$
$$r = 30°,\ \alpha = 90°.$$

Angle of emergence is zero.

(*b*) The reflected ray 1 is reflected from the denser medium (Fig. 12.17). For the reflected ray 1 there is a phase change of π. There is no change of phase for the second reflected ray 2. Thus the condition of maxima for the reflected rays is [see problem 22].

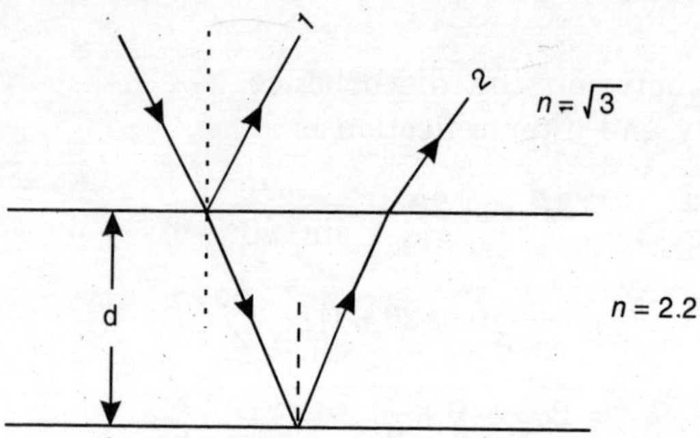

Fig. 12.17

$$2nd = \left(m + \frac{1}{2}\right)\lambda,\ m = 0, 1, 2,...$$

and the condition of maxima for the transmitted rays is

$$2nd = m\lambda.$$

$$d_{\min} = \frac{\lambda}{2n},\ \text{where we put } m = 1$$
$$= \frac{5500}{2 \times 2.2} \text{Å}$$
$$= 1250 \text{ Å}.$$

44. *A point source S emitting light of wavelength 600 nm is placed at a very small height h above a flat reflecting surface AB. The intensity of the reflecting light is 36% of the incident intensity. Interference fringes are observed on a screen placed parallel to the reflecting surface at a very large distance D from it (Fig. 12.18).*

(a) What is the shape of the interference fringes on the screen?

(b) Calculate the ratio of the minimum to the maximum intensities in the interference fringes found near the point P (shown in Fig. 12.18).

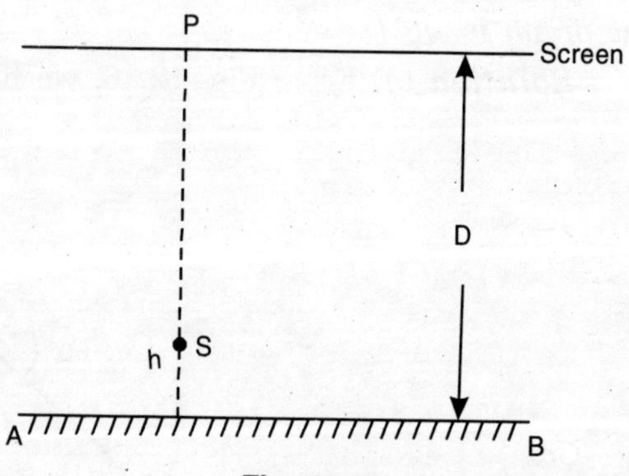

Fig. 12.18

(c) If the intensity at point P corresponds to a maximum, calculate the minimum distance through which the reflecting surface AB should be shifted so that the intensity at P again becomes maximum. (I.I.T. 2002)

Solution

(*a*) The screen is placed parallel to the mirror *AB*. The points of equal optical path for direct rays from *S* will lie on a circle with *P* as the centre. The points of equal optical path for the reflected rays will also lie on a circle with *P* as the centre. Therefore the fringes formed will be circular.

(*b*) Amplitude of the incident ray = $a_1 = \sqrt{I}$, where *I* is the intensity of the incident ray.

Amplitude of the reflected ray = $a_2 = \sqrt{0.36I} = 0.6\sqrt{I}$

$$\frac{I_{min}}{I_{max}} = \frac{(a_1 - a_2)^2}{(a_1 + a_2)^2} = \left(\frac{\sqrt{I} - 0.6\sqrt{I}}{\sqrt{I} + 0.6\sqrt{I}}\right)^2 = \left(\frac{0.4}{1.6}\right)^2 = \frac{1}{16}.$$

(*c*) When the reflecting surface is shifted in such a way that *P* becomes a maximum point again, the path difference between the direct and reflected rays should change by $n\lambda$. Here $n = 1$. When *AB* is moved by *x*, the path difference changes by 2*x*.

Thus,
$$2x = \lambda \text{ or } x = \frac{\lambda}{2} = 300 \text{ nm}.$$

45. Two *beams of light having intensities I and 4I interfere to produce a fringe pattern on a screen. The phase difference between the beams is $\frac{\pi}{2}$ at point A and π at point B. Then the difference between the resultant intensities at A and B is*

(a) 2I (b) 4I (c) 5I (d) 7I. (I.I.T. 2001)

Solution

Let
$$y_1 = a_1 \sin \omega t$$
$$y_2 = a_2 \sin(\omega t + \delta)$$
$$y_1 + y_2 = A \sin(\omega t + \phi)$$

where
$$A \cos\phi = a_1 + a_2 \cos\delta$$
$$A \sin\phi = a_2 \sin\delta$$

The resultant intensity is

$$I_R = A^2 = a_1^2 + a_2^2 + 2a_1a_2 \cos\delta$$

Here $a_1 = \sqrt{I}$, $a_2 = \sqrt{4I}$.

$$I_R = 5I + 4I \cos\delta$$

At *A*, $\delta = \frac{\pi}{2}$, $I_A = 5I$

At *B*, $\delta = \pi$, $I_B = I$

$$I_A - I_B = 4I$$

Correct Choice: *b*.

46. *In the Young's double slit experiment 12 fringes are observed to be formed in a certain segment of the screen when light of wavelength 600 nm is used. If the wavelength of* *light is changed to 400 nm, the number of fringes observed in the same segment of the screen is given by*

(a) 12 (b) 18 (c) 24 (d) 30. (I.I.T. 2001)

Solution

$$\text{Fringe width } \beta = \frac{\lambda D}{d}$$

$$\frac{\text{Length of the region}}{\lambda_1 D/d} = 12$$

$$\frac{\text{Length of the region}}{\lambda_2 D/d} = n$$

Thus,
$$\frac{n}{12} = \frac{\lambda_1}{\lambda_2} = \frac{600}{400}$$

$$n = 18$$

Correct Choice: *b.*

47. *A vessel ABCD of 10 cm width has two small slits S_1 and S_2 separated by a distance of 0.8 mm (Fig. 12.19). The source of light S is situated 2 m from the vessel. Calculate the position of the central bright fringe on the other wall CD with respect to the line OQ. Now a liquid is poured into the vessel and filled upto OQ. The central bright fringe is found to be at Q.*

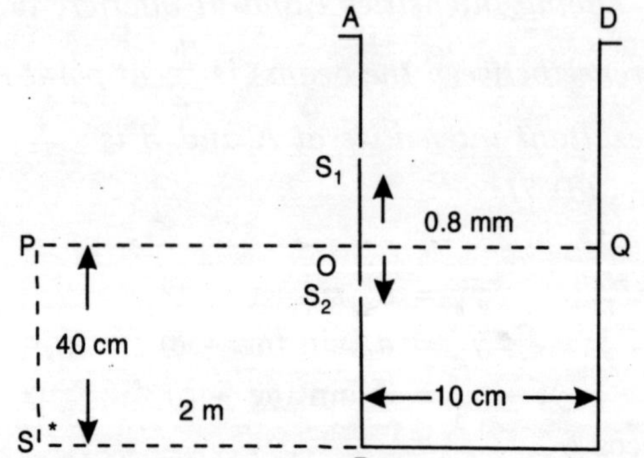

Fig. 12.19

Calculate the refractive index of the liquid. (I.I.T. 2001)

Solution

(*a*) For the Central fringe to be at *R* (Fig. 12.20), we have

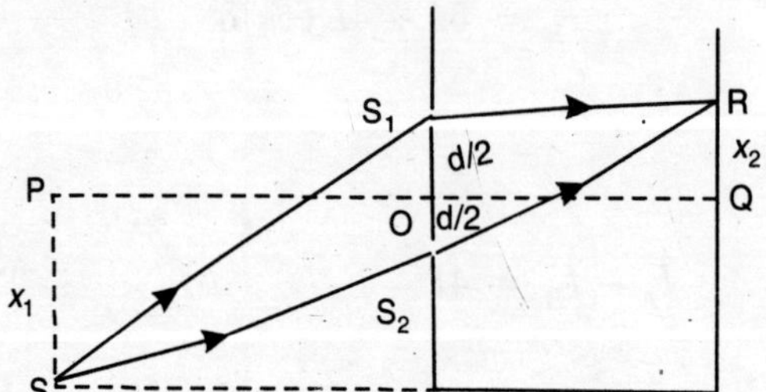

Fig. 12.20

$$SS_1 + S_1R = SS_2 + S_2R$$

or
$$SS_1 - SS_2 = S_2R - S_1R$$

$$SS_1^2 = D_1^2 + \left(x_1 + \frac{d}{2}\right)^2$$

$$SS_2^2 = D_1^2 + \left(x_1 - \frac{d}{2}\right)^2$$

$$S_2R^2 = D_2^2 + \left(x_2 + \frac{d}{2}\right)^2$$

$$S_1R^2 = D_2^2 + \left(x_2 - \frac{d}{2}\right)^2$$

We assume that $D_1 >> x_1$, d and $D_2 >> x_2$, d.

$$SS_1 = D_1\left[1 + \frac{x_1^2 + x_1 d + d^2/4}{D_1^2}\right]^{1/2}$$

$$\approx D_1 + \frac{1}{2}\frac{x_1^2}{D_1} + \frac{1}{2}\frac{x_1 d}{D_1} + \frac{1}{8}\frac{d^2}{D_1}$$

Similarly,

$$SS_2 \approx D_1 + \frac{1}{2}\frac{x_1^2}{D_1} - \frac{1}{2}\frac{x_1 d}{D_1} + \frac{1}{8}\frac{d^2}{D_1}$$

Thus,
$$SS_1 - SS_2 \approx \frac{x_1 d}{D_1}$$

Similarly,
$$S_2R - S_1R \approx \frac{x_2 d}{D_2}$$

Thus,
$$\frac{x_1 d}{D_1} = \frac{x_2 d}{D_2}$$

or
$$x_2 = \frac{D_2}{D_1}x_1 = \frac{10}{200} \times 40 = 2 \text{ cm.}$$

(*b*)
$$SS_1 + S_1Q = SS_2 + n\, S_2Q.$$

or
$$SS_1 - SS_2 = n\, S_2Q - S_1Q$$

$$\frac{x_1 d}{D_1} \sim n\left[D_2^2 + \frac{d^2}{4}\right]^{1/2} - \left[D_2^2 + \frac{d^2}{4}\right]^{1/2}$$

$$= (n - 1)\, D_2 \left[1 + \frac{d^2}{4D_2^2}\right]^{1/2}$$

$$\approx (n - 1)\, D_2$$

$$n - 1 = \frac{x_1 d}{D_1 D_2} = \frac{40 \times 0.08}{200 \times 10} = 0.0016$$

$$n = 1.0016.$$

48. *A glass plate of refractive index 1.5 is coated with a thin layer of thickness d and refractive index 1.8. Light of wavelength λ travelling in air is incident normally on the layer. It is partly reflected at the upper and lower surfaces of the layer and the two reflected rays interfere. Write the condition for their constructive interference. If λ = 648 nm obtain the least value of d for which the rays interfere constructively.* (I.I.T. 2000)

Solution

Glass-coating interference (Fig. 12.21) is just like air-glass interference. (see Fig. 12.8)

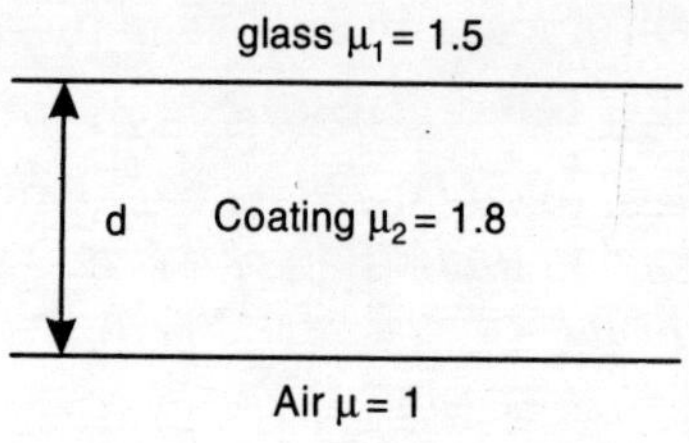

Fig. 12.21

The path difference between AR_1 and BR_2 is

$$\Delta = \mu_2 (AF + FB) - \mu_1 AN \pm \frac{\lambda}{2}.$$

The first ray AR_1 undergoes a phase change of π which corresponds to a path difference of λ/2.

$$\Delta = 2\mu_2 AF - \mu_1 AN \pm \frac{\lambda}{2}.$$

$$= 2\mu_2 \frac{d}{\cos r} - \mu_1 AB \sin i \pm \frac{\lambda}{2}$$

$$= 2\mu_2 \frac{d}{\cos r} - 2\mu_1 d \tan r \sin i \pm \frac{\lambda}{2}$$

$$\Delta = 2\mu_2 \frac{d}{\cos r} - 2\mu_1 d \frac{\sin r}{\cos r} \cdot \frac{\mu_2}{\mu_1} \sin r \pm \frac{\lambda}{2}$$

$$= 2\mu_2 d \cos r \pm \frac{\lambda}{2}$$

For constructive interference

$$2\mu_2 d \cos r \pm \frac{\lambda}{2} = m\lambda, \ m = 0, 1, 2, ...$$

or

$$2\mu_2 d \cos r = \left(m + \frac{1}{2}\right)\lambda$$

For normal incidence, cos r = 1 and

$$2\mu_2 d = \left(m + \frac{1}{2}\right)\lambda, \text{ maxima}$$

The least value of d is

$$d_{min} = \frac{\lambda}{4\mu_2} = \frac{648}{4 \times 1.8} = 90 \text{ nm}.$$

49. *A thin slice is cut out of a glass cylinder along a plane parallel to its axis. The slice is placed on a flat glass plate as shown in Fig. 12.22. The observed interference fringes from this combination shall be*

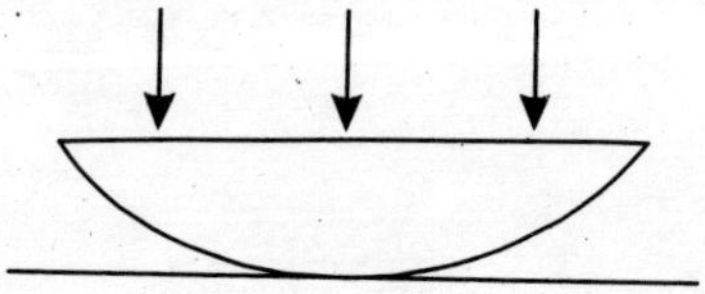

Fig. 12.22

(a) straight line

(b) circular

(c) equally spaced

(d) of type such that the fringe spacing increases as we go outwards. (I.I.T. 1999)

Solution

When a cylinder is placed on a glass plate with its curved surface touching the plane surface, a thin film is formed between the curved surface of the cylinder and the glass plate. The glass plate will touch the slice of the cylinder in a straight line parallel to the axis of the cylinder, and the thickness of the film increases as we move away from this straight line. The loci of all points having the same thickness are straight lines. Thus straight line fringes will appear in this combination. The fringe spacing will decrease as we go outwards.

Correct Choice: *a.*

50. *The Young's double slit experiment is done in a medium of refractive index 4/3. A light of 600 nm wavelength falls on the slits having 0.45 mm separation. The lower slit S_2 is covered by a thin glass sheet of thickness 10.4 μm and refractive index 1.5. The interference pattern is observed on a screen placed 1.5 m from the slits as shown in Fig. 12.23.*

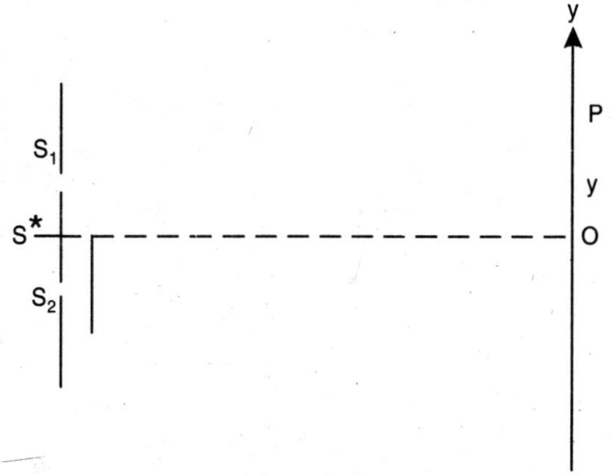

Fig. 12.23

(*a*) *Find the location of the central maximum (bright fringe with zero path difference) on the y-axis.*

(*b*) *Find the light intensity at point O relative to the maximum fringe intensity.*

(*c*) *Now, if 600 nm light is replaced by white light of range 400 to 700 nm, find the wavelengths of the light that form maxima exactly at point O.*

[*All wavelengths in this problem are for the given medium of refractive index 4/3. Ignore dispersion.*] (I.I.T. 1999)

Solution

(*a*) Optical path difference between the rays S_2P and S_1P in the absence of the thin film is

$$\mu S_2P - \mu S_1P = \mu \frac{yd}{D} \quad \text{(see problem 1)}$$

where μ = *r.i.* of the medium.

Suppose μ' = *r.i.* of the glass sheet. In the presence of the glass sheet the optical path difference between the rays is

$$\mu S_2P - (S_1P - t)\mu - \mu' t$$

where t is the thickness of the glass sheet.

If this optical path difference is zero, we have

$$\mu \frac{yd}{D} + \mu t - \mu' t = 0$$

or

$$y = \left(\frac{\mu'}{\mu} - 1\right) t \frac{D}{d}$$

$$= \left(\frac{1.5}{4/3} - 1\right) \times 10.4 \times 10^{-6} \times \frac{1.5}{0.45 \times 10^{-3}} \text{ m}$$

$$= 4.33 \text{ mm.}$$

(*b*) Optical path difference at O is

$$\mu S_1O - \mu (S_2O - t) - \mu' t = \mu t - \mu' t = (\mu - \mu')t$$

Here $\mu' > \mu$.

$$\text{Phase difference} = \delta = \frac{2\pi}{\lambda}(\mu' - \mu)t = 5.78\,\pi$$

$$I = I_m \cos^2 \frac{\delta}{2} = I_m \cos^2 (2.89\,\pi)$$

$$= 0.89\, I_m.$$

(*c*) For maxima at O we have

$$(\mu' - \mu)t = n\lambda,\ n = 0, 1, 2, \ldots$$

or

$$\lambda = \frac{1}{n}\,\frac{0.5}{3} \times 10.4 \times 10^{-6} \text{ m}$$

$$= \frac{1}{n} \times 1733.3 \text{ nm}$$

For $n = 3$, $\lambda = 577.3$ nm

For $n = 4$, $\lambda = 433.3$ nm.

51. *A coherent parallel beam of microwave of wavelength $\lambda = 0.5$ mm falls on a Young's double-slit apparatus. The separation between the slits is 1.0 mm. The intensity of microwaves is measured on a screen placed parallel to the plane of the slits at a distance of 1.0 m from it as shown in the Fig. 12.24.*

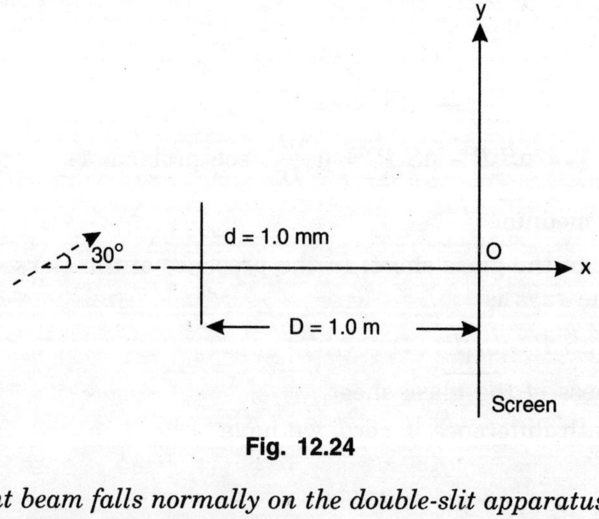

Fig. 12.24

(a) If the incident beam falls normally on the double-slit apparatus, find the coordinates of all the interference minima on the screen.

(b) If the incident beam makes an angle of 30° with the x-axis (as in the dotted arrow shown in the Fig. 12.24), find the y-coordinates of the first minima on either side of the central maximum. (I.I.T. 1998)

Solution

(*a*) For normal incidence the interference minima occur on the screen at a distance y from O where

$$y = \left(n+\frac{1}{2}\right)\frac{\lambda D}{d},\ n = 0,\ \pm 1,\ \pm 2,...$$

$$= \left(n+\frac{1}{2}\right) . \frac{0.5\times 10^{-3}\times 1}{1.0\times 10^{-3}}\ \text{m}$$

$$= 0.5\left(n+\frac{1}{2}\right)\ \text{m},\ n = 0,\ \pm 1,\ \pm 2,...$$

(*b*) The incident beam makes an angle of 30° with the x-axis (Fig. 12.25). The angle of incidence = i = 30°. From Q we draw a perpendicular QN on the incident ray EP. Now Q and N are in the same phase. $NP = d \sin i$

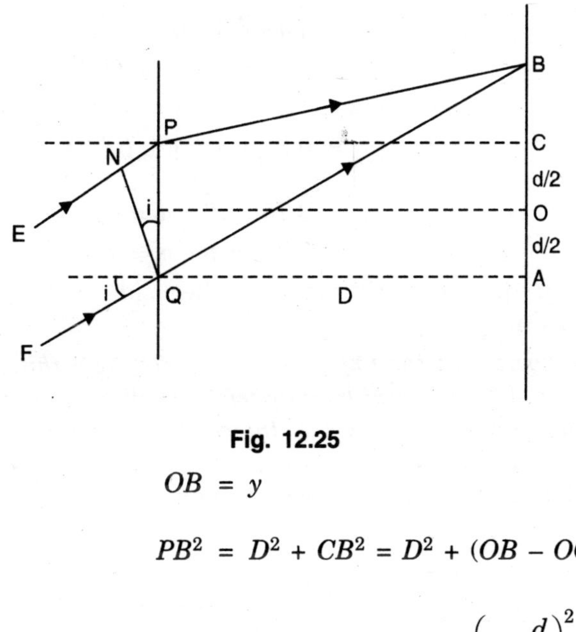

Fig. 12.25

$$OB = y$$

$$PB^2 = D^2 + CB^2 = D^2 + (OB - OC)^2 = D^2 + \left(y-\frac{d}{2}\right)^2$$

$$QB^2 = D^2 + AB^2 = D^2 + \left(y+\frac{d}{2}\right)^2$$

$$QB \approx D\left[1+\frac{1}{2}\frac{\left(y+\frac{d}{2}\right)^2}{D^2}\right]$$

$$PB \approx D\left[1+\frac{1}{2}\frac{\left(y-\frac{d}{2}\right)^2}{D^2}\right]$$

$$QB - PB_o = \frac{yd}{D}$$

$$\text{Path difference} = QB - (PB + NP) = \frac{yd}{D} - d \sin i$$

For maxima, $\frac{yd}{D} - d \sin i = n\lambda$

or $$y = \frac{n\lambda D}{d} + D \sin i = 0.5\ n + 0.5\ D.$$

The position of the central maximum, $y = 0.5$ m, when $n = 0$. Due to inclined incident beam the central maximum moves upwards. It is given by $y = D \sin i$

The position of the minima

$$\frac{yd}{D} - d \sin i = \left(n + \frac{1}{2}\right)\lambda,$$

or $$y = \frac{\left(n + \frac{1}{2}\right)\lambda D}{d} + D \sin i, \; n = 0, \pm 1, \pm 2, \ldots$$

$$= 0.5\left(n + \frac{1}{2}\right) + 0.5\ D$$

$$y = 0.75 \text{ m for } n = 0$$

and $$y = 0.25 \text{ m for } n = -1.$$

The coordinates of the first minima on either side of the central maximum are $y = 0.25\ m$ and $y = 0.75$ m.

52. *In a Young's experiment the upper slit is covered by a thin glass plate of refractive index 1.4 while the lower slit is covered by another glass plate, having the same thickness as the first one but having refractive index 1.7. Interference pattern is observed using light of wavelength 5400 Å. It is found that the point P on the screen where the central maximum (n = 0) fell before the glass plates were inserted now has 3/4 the original intensity (Fig. 12.26).*

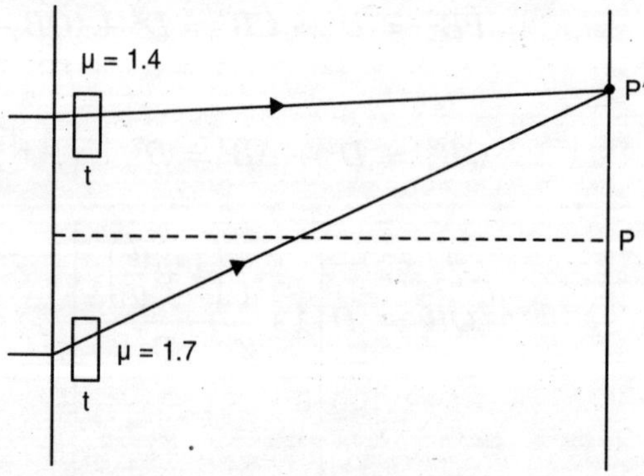

Fig. 12.26

It is further observed that what used to be the fifth maximum earlier, lies below the point P while the sixth minimum lies above P. Calculate the thickness of the glass plate. (Absorption of light by glass plate may be neglected) (I.I.T. 1997)

Solution

The optical path difference developed due to insertion of two glass plates is

$$x = (\mu_2 - \mu_1)t = (1.7 - 1.4)t = 0.3t$$

where t is the thickness of each glass plate.

Phase difference $\delta = \dfrac{2\pi}{\lambda}x.$

The intensity distribution equation is

$$I = I_0 \cos^2\left(\frac{\delta}{2}\right)$$

At P $$\frac{3}{4}I_0 = I_0 \cos^2 \frac{\delta}{2}$$

or $$\cos\frac{\delta}{2} = \pm\frac{\sqrt{3}}{2}.$$

or $$\frac{\delta}{2} = n\pi \pm \frac{\pi}{6}$$

or $$\delta = 2n\pi \pm \frac{\pi}{3}$$

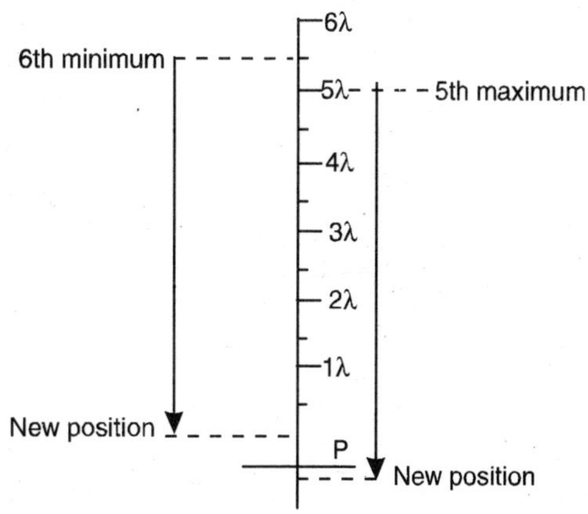

Fig. 12.27

The corresponding path difference is $n\lambda \pm \dfrac{1}{6}\lambda.$

After insertion of glass plates the 5th maximum goes below the point P and the 6th minimum lies above the point P. (Fig. 12.27). Thus due to insertion of glass plates the change of path difference is $(5 + \varepsilon)\,\lambda$ where $\varepsilon > 0$. Again the 6th minimum lies above P. The change in path difference is $(5 + \frac{1}{2} - \eta)\,\lambda,\ \eta > 0$

Thus, $$(5 + \varepsilon)\,\lambda = \left(5 + \frac{1}{2} - \eta\right)\lambda$$

or $$\varepsilon + \eta = \frac{1}{2} \text{ or } \varepsilon < \frac{1}{2}.$$

The change of optical path must be $\left(5+\dfrac{1}{6}\right)\lambda.$

Thus, $$0.3t = \left(5+\frac{1}{6}\right)\lambda = \frac{31\,\lambda}{6}$$

or $$t = \frac{31\times 5400\times 10^{-10}}{6\times 0.3}\text{m}$$

$$= 9.3 \times 10^{-6} \text{ m.}$$

53. *In Young's experiment the source is red light of wavelength 7×10^{-7} m. When a thin glass plate of refractive index 1.5 at this wavelength is put in the path of one of the interfering beams, the central bright fringe shifts by 10^{-3} m to the position previously occupied by the 5th bright fringe. Find the thickness of the plate. When the source is now changed to green light of wavelength 5×10^{-7}m, the central fringe shifts to the position initially occupied by the 6th* *bright fringe due to red light. Find the refractive index of glass for the green light. Also estimate the change in fringe width due to the change in wavelength.* (I.I.T. 1997)

Solution

For red light, fringe width $\beta_r = \dfrac{\lambda_r D}{d}$.

Due to insertion of the glass plate the central fringe shifts by x:

$$x = (n_r - 1)t\, D/d$$ [see problem 16]

The position of the 5th bright fringe is given by

$$x = 5\beta_r = \frac{5\lambda_r D}{d} = (n_r - 1)t\, D/d$$

or

$$t = \frac{5\lambda_r}{n_r - 1} = \frac{5\times7\times10^{-7}}{1.5-1} = \mathbf{7 \times 10^{-6}\ m}$$

For green light

$$x = 6\beta_r = \frac{6\lambda_r D}{d} = (n_g - 1)t\, D/d$$

or

$$6\lambda_r = (n_g - 1)t$$

or

$$n_g - 1 = \frac{6\lambda_r}{t} = \frac{6\times7\times10^{-7}}{7\times10^{-6}} = 0.6$$

Thus,

$$n_g = \mathbf{1.6}$$

Change in fringe width $= \beta_r - \beta_g = (\lambda_r - \lambda_g)\, D/d$

Also, $(n_r - 1)\, t\, D/d = x = 10^{-3}$ m

or

$$\frac{D}{d} = \frac{10^{-3}}{(n_r - 1)t} = \frac{10^3}{3.5}$$

$$\beta_r - \beta_g = (7 \times 10^{-7} - 5 \times 10^{-7}) \times \frac{10^3}{3.5}\ \text{m}$$

$$= \mathbf{5.7 \times 10^{-5}\ m.}$$

54. *In Young's double slit experiment the angular position of a point above the central maximum whose intensity is one-fourth of maximum intensity is*

(a) $\sin^{-1}\left(\dfrac{\lambda}{d}\right)$ *(b)* $\sin^{-1}\left(\dfrac{\lambda}{2d}\right)$ *(c)* $\sin^{-1}\left(\dfrac{\lambda}{3d}\right)$ *(d)* $\sin^{-1}\left(\dfrac{\lambda}{4d}\right)$. (I.I.T. 2005)

Solution

$$I = I_{\max}\cos^2\left(\frac{\delta}{2}\right)$$

where

$$\delta = \frac{2\pi}{\lambda}\, d \sin\theta$$

$$\frac{1}{4} = \cos^2\frac{\delta}{2} \text{ or } \frac{\delta}{2} = \frac{\pi}{3} \text{ or } \delta = \frac{2\pi}{3}.$$

Thus,

$$\frac{2\pi}{3} = \frac{2\pi}{\lambda}\, d\sin\theta$$

or $$\sin\theta = \frac{\lambda}{3d}$$

or $$\theta = \sin^{-1}\left(\frac{\lambda}{3d}\right).$$

Correct Choice: *c.*

55. *In YDSE an electron beam is used to obtain interference pattern. If the speed of the electron is increased then*

(a) no interference pattern will be observed

(b) distance between two consecutive fringes will increase

(c) distance between two consecutive fringes will decrease

(d) distance between two consecutive fringes remain same. (I.I.T. 2005)

Solution

$$\lambda = \frac{h}{mv}$$

when v is increased λ decreases.

Fringe width $$\beta = \frac{\lambda D}{d}.$$

Thus when v increases, β decreases.

Correct Choice: *c.*

56. *The figure shows a surface XY separating two transparent media, medium 1 and medium 2 (Fig. 12.28).*

The lines ab and cd represent wavefronts of a light wave travelling in medium 1 and incident on XY. The lines ef and gh represent wave fronts of the light wave in medium-2 after refraction.

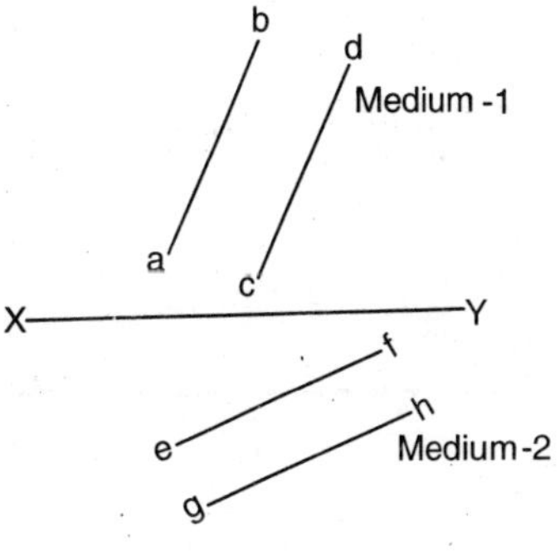

Fig. 12.28

(a) Light travels as a

(A) parallel beam in each medium

(B) convergent beam in each medium

(C) divergent beam in each medium

(D) divergent beam in one medium and convergent beam in the other medium.

(b) The phases of the light wave at c, d, e, f are ϕ_c, ϕ_d, ϕ_e *and* ϕ_f *respectively. It is given that* $\phi_c \neq \phi_f$. *Then*

(A) ϕ_c *cannot be equal to* ϕ_d

(B) ϕ_d *cannot be equal to* ϕ_e

(C) $(\phi_d - \phi_f)$ *is equal to* $(\phi_c - \phi_e)$

(D) $(\phi_d - \phi_c)$ *is not equal to* $(\phi_f - \phi_e)$.

(c) Speed of light is

(A) the same in medium 1 and medium 2

(B) larger in medium 1 than in medium 2

(C) larger in medium 2 than in medium 1

(D) different at b and d.

Solution

(*a*) The ray of light is perpendicular to the wavefront.

Correct choice: A

(*b*) The particles are in same phase on a wavefront:

Thus $\phi_c = \phi_d$ and $\phi_e = \phi_f$

$$\phi_d - \phi_f = \phi_c - \phi_e$$

Correct Choice: *c*.

(*c*) The ray of light goes towards normal after refraction (Fig. 12.29). Thus the medium 2 is denser than medium 1. The velocity of light decreases in the denser medium.

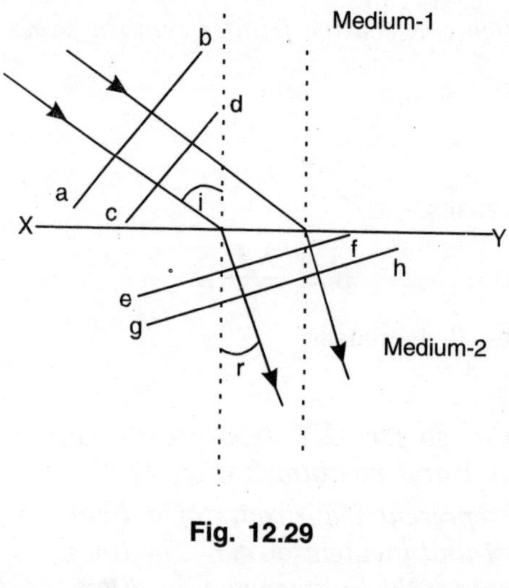

Fig. 12.29

Correct Choice: *b*.

SUPPLEMENTARY PROBLEMS

1. Find the sum of the following disturbances

$$y_1 = 9 \sin \omega t$$
$$y_2 = 8 \sin (\omega t + 30°)$$

2. Find the resultant of the following disturbances

$$E_1 = 10 \sin \omega t$$
$$E_2 = 15 \sin (\omega t + 30°)$$
$$E_3 = 5 \sin (\omega t - 45°)$$

3. In Young's double slit experiment the slit separation is 0.12 mm, the slit-screen separation is 50 cm and the wavelength of light is 540 nm. What is the linear distance on the screen between adjacent maxima?

4. In Young's double slit experiment the separation *d* of the two narrow slits is doubled. In order to maintain the same spacing of the fringes, how must the distance *D* of the screen from the slits be altered? (The wavelength of the light remains unchanged.)

5. What is the phase difference between the waves from the two slits at the 10th bright fringe in Young's double-slit experiment?

6. In a double-slit arrangement the slits are separated by a distance equal to 100 times of the wavelength of the light passing through the slits. What is the angular separation in radians between the central maximum and the adjacent maximum?
7. Light passes through two narrow slits 0.07 cm apart. If on a screen 65 cm away, the distance between two second order maxima is 0.2 cm, what is the wavelength of light?
8. A double slit is illuminated by light containing two wavelengths 480 nm and 600 nm. What is the lowest order at which a maximum of one wavelength falls on a minimum of the other?
9. In a double-slit experiment the distance between the slits is 5 mm and the slits are 1 m from the screen. Two interference patterns can be seen on the screen, one due to light with wavelength 500 nm and the other due to light with wavelength 600 nm. What is the separation on the screen between the fifth order interference fringes of the two different patterns?
10. If the distance between the first and tenth minima of a double-slit pattern is 18 mm and the slits are separated by 0.2 mm with the screen 60 cm from the slits, what is the wavelength of the light used?
11. One of the slits of a double-slit system is wider than the other, so that the amplitude of the light reaching the central part of the screen from one slit, acting alone, is twice that from the other slit, acting alone. Derive an expression for the intensity I in terms of θ.
12. In Fig. 12.30 S_1 and S_2 are two point sources of radiation, excited by the same oscillator. They are coherent and in phase with each other. Placed 3.5 m apart, they emit equal amount of power in the form of 1 m wavelength electromagnetic waves. Find the positions of the first (that is, the nearest), the second, and the third maxima of the received signal, as the detector is moved out along OX.
13. In Young's interference experiment in a large ripple tank the coherent sources are placed 10 cm apart. The distance between maxima 2 m away is 20 cm. If the speed of ripples is 25 cm/s, find the frequency of the vibrators.

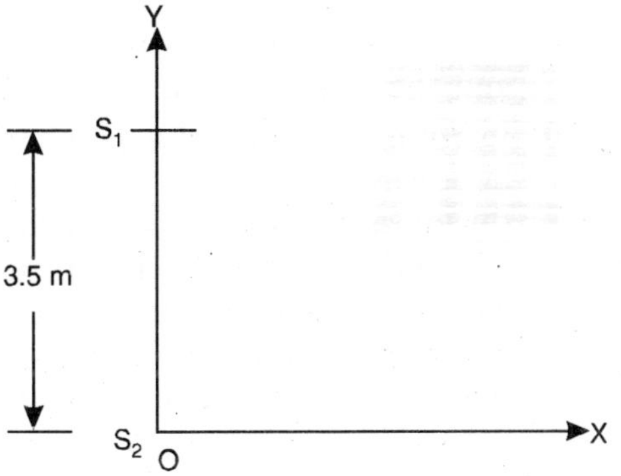

Fig. 12.30

14. Two loud speakers are emitting sound of same frequency and also same intensity in all directions. The intensity of sound at a point which is 3 m from one loud speaker and 3.5 m from the other, is zero. If the two loud speakers are in phase, calculate the minimum frequency of sound emitted by the loud speakers. Velocity of sound in air is 330 m/s.
15. In the Young's double slit experiment, the interference pattern is found to have an intensity ratio between bright and dark fringes, as 9. This implies that
    (*a*) intensities at the screen due to the two slits are 5 units and 4 units respectively.
    (*b*) the intensities at the screen due to the two slits are 4 units and 1 unit respectively.
    (*c*) the amplitude ratio is 3.
    (*d*) the amplitude ratio is 2. *(I.I.T. 1982)*

    [**Hints:** $a_1 + a_2 = 3$ and $a_1 - a_2 = 1$]

16. In Young's double slit experiment the separation between the slits is halved and the distance between the slits and screen is doubled. The fringe width is (*a*) unchanged (*b*) halved (*c*) doubled (*d*) quadrupled. *(I.I.T. 1981)*

17. White light is used to illuminate the two slits in a Young's double slit experiment. The separation between slits is b and the screen is at a distance d ($\gg b$) from the slits. At a point on the screen directly in front of one of the slits, certain wavelengths are missing. Some of these missing wavelengths are

(*a*) $\lambda = b^2/d$ (*b*) $\lambda = 2b^2/d$ (*c*) $\lambda = b^2/3d$ (*d*) $\lambda = 2b^2/3d$. *(I.I.T. 1984)*

$\left[\textbf{Hints: } \frac{b}{2} = \left(m + \frac{1}{2}\right)\frac{\lambda d}{b},\ m = 0, 1, 2, \ldots\right]$

18. Two coherent monochromatic light beams of intensities I and $4I$ are superimposed. The maximum and minimum possible intensities in the resulting beam are

(*a*) $5I$ and I (*b*) $5I$ and $3I$ (*c*) $9I$ and I (*d*) $9I$ and $3I$. *(I.I.T. 1988)*

[**Hints:** $a_1 = 2\sqrt{I}$ and $a_2 = \sqrt{I}$]

19. A beam of light consisting of two wavelengths 6500 Å and 5200 Å is used to obtain interference fringes in a Young's double slit experiment.

(*i*) Find the distance of the third bright fringe on the screen from the central maximum for the wavelength 6500 Å.

(*ii*) What is the least distance from the central maximum where the bright fringe due to both the wavelengths coincide?

The distance between the slits is 2 mm and the distance between the plane of the slits and the screen is 120 cm. *(I.I.T. 1985)*

20. In Young's double slit experiment using monochromatic light, the fringe pattern shifts by a certain distance on the screen when a mica sheet of refractive index 1.6 and thickness 1.964 micron is introduced in the path of one of the interfering waves. The mica sheet is then removed and the distance between the slits and the screen is doubled. It is found that the distance between successive maxima (or minima) now is the same as the observed fringe shift upon the introduction of the mica sheet. Calculate the wavelength of monochromatic light used in the experiment. *(I.I.T. 1983)*

[**Hints:** $(n-1)\,t\,D/d = \lambda\,(2D)/d$ and 1 micron = 10^{-6} m]

21. In a modified Young's double slit experiment, a monochromatic uniform and parallel beam of light of wavelength 6000 Å and intensity $(10/\pi)$ Wm^{-2} is incident normally on two circular apertures A and B of radii 0.001 m and 0.002 m respectively. A perfect transparent film of thickness 2000 Å and refractive index 1.5 for the wavelength of 6000 Å is placed in front of aperture A, see Fig. 12.31. Calculate the power (in Watts) received at the focal spot F of the lens. The lens is symmetrically placed with respect to the aperture. Assume that 10% of the power received by each aperture goes in the original direction and is brought to the focus spot. *(I.I.T. 1989)*

[**Hints:** Intensity of light in the original direction going through A and brought to the focus spot F is $I_A = \frac{1}{10} \times \left(\frac{10}{\pi}\right) \pi(0.001)^2 = 10^{-6}$ W. Similarly, $I_B = 4 \times 10^{-6}$ W. Path difference = $(n-1)t$ = 1000 Å and phase difference $\delta = \frac{2\pi}{6000} \times 1000 = \frac{\pi}{3}$. Intensity at the point F is $I_F = I_A + I_B + 2\sqrt{I_A I_B}\cos\delta$]

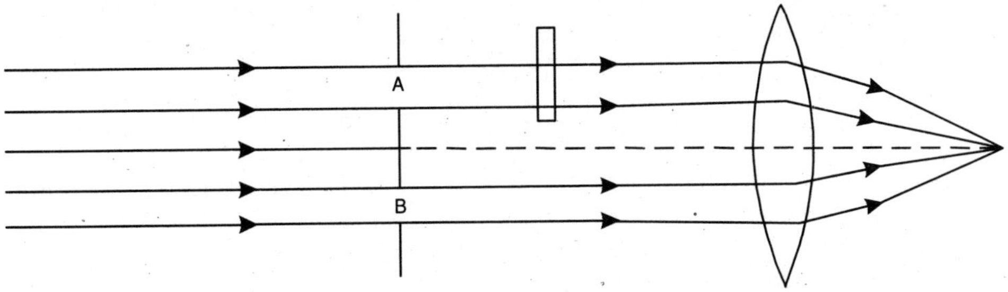

Fig. 12.31

22. A double-slit apparatus is immersed in a liquid of refractive index 1.33. It has slit separation of 1 mm, and distance between the plane of slits and screen is 1.33 m. The slits are illuminated by a parallel beam of light whose wavelength is 6300 Å.

(*i*) Calculate the fringe-width.

(*ii*) One of the slits of the apparatus is covered by a thin glass sheet of refractive index 1.53. Find the smallest thickness of the sheet to bring the adjacent minimum on the axis. (*I.I.T. 1996*)

[**Hints:** (*i*) Fringe-width $\beta = \frac{\lambda D}{nd}$. (*ii*) From Fig. 12.4, we have $n[S_2P_0' - (S_1P_0' - t)] - n_g t = 0$ and $S_2P_0' - S_1P_0' = \frac{xd}{D}$. Thus, $t = \frac{x\lambda}{\beta(n_g - n)}$].

23. Interference fringes are produced using a biprism having refracting angles of 4° each and refractive index 1.5. The slit is kept at a distance of 10 cm from the biprism and is illuminated with light of wavelength 589 nm. Calculate the fringe width at 85 cm from the biprism.

24. In an experiment with a biprism the distance between the focal plane of the eyepiece and the plane of the interfering sources is 99 cm and the width of 10 fringes is 9.73 mm. The distances between the two images for the two positions of the lens in the displacement method are 0.4 mm and 0.9 mm. Find the wavelength of the light used.

25. Interference fringes are produced using white light with a double-slit arrangement. A piece of parallel-sided mica of refractive index 1.6 is placed in front of one of the slits, as a result of which the centre of the fringe system moves to a distance subsequently shown to accommodate 30 dark bands when light of wavelength 540 nm is used. What is the thickness of the mica?

26. On placing a thin film of mica of thickness 7.8×10^{-5} cm in the path of one of the interference beams in the biprism arrangement, it is found that the central fringe shifts a distance equal to the spacing between two successive bright fringes. If $\lambda = 5893$ Å, find the refractive index of mica.

27. A Lloyd's mirror is illuminated with light of wavelength 550 nm from a narrow slit whose height with respect to mirror-plane is 0.10 cm. Find the separation of the interference fringes at a distance 100 cm from the slit.

28. A soap film of refractive index $\frac{4}{3}$ and of thickness 2.1×10^{-4} cm is illuminated by white light incident at an angle of 45°. The light reflected by it is examined spectroscopically in which it is found that a dark band corresponds to a wavelength of 474.3 nm. Find the order of the dark band.

29. A thin film in air is 0.43 μm thick and is illuminated by white light normal to its surface. Its index of refraction is 1.5. What wavelenghts within the visible spectrum will be intensified in the reflected beam?

30. A thin coating of refractive index 1.28 is applied to a glass camera lens to minimize the intensity of the light reflected from the lens. In terms of λ, the wavelength in air of the incident light, what is the smallest thickness of the coating that is needed? Assume normal incidence of light.

31. A sheet of glass having an index of refraction of 1.40 is to be coated with a film of material having a refractive index of 1.56 such that green light (λ = 520 nm) is preferentially transmitted. (*a*) What is the minimum thickness of the film that will achieve the result? (*b*) Will the transmission of any wavelength in the visible spectrum be sharply reduced? Assume normal incidence of light.

32. A thin film of acetone (index of refraction = 1.24) is coated on a thick glass plate (n = 1.50). Plane light waves of variable wavelengths are incident normal to the film. When one views the reflected wave, it is noted that complete destructive interference occurs at 600 nm and constructive interference at 700 nm. Calculate the thickness of the acetone film.

33. Two similar rectangular plates are placed in contact along one edge and separated by a strip of paper along the opposite edge, thus forming air wedge of very small angle between them. When the wedge is illuminated normally by light from a sodium lamp, it appears to be crossed by bright bands with a spacing of 1 mm. Calculate the angle of the wedge.

34. Two rectangular grass-plates are laid one upon another with a thin wire between them at one edge so as to enclose a thin wedge-shaped air film. The plates are illuminated by sodium light. Bright and dark fringes are formed, there being 9 of each per cm length of the wedge, measured normal to the edge in contact. Find the angle of the wedge.

35. A glass wedge of angle 0.001 radian is illuminated by monochromatic light of wavelength 600 nm falling normally on it. The index of refraction of glass = 1.5. At what distance from the edge of the wedge will the 10th dark fringe be observed by reflected light?

[**Hints:** $2nx\theta = m\lambda$]

36. Fringes of equal thickness are observed in a thin glass wedge of refractive index 1.5. The fringe spacing is 2 mm and wavelength of light is 589.3 nm. Calculate the angle of the wedge in seconds of arc.

37. A perfectly flat piece of glass (n = 1.5) is placed over a perfectly flat piece of black plastic (n = 1.2). They touch at one edge. Light of wavelength 600 nm is incident normally from above. Six dark fringes are observed in the reflected light. (*a*) How thick is the space between the glass and the plastic at the other end? (*b*) Water (n = 1.33) seeps into the region between the glass and plastic. How many dark fringes are seen when all the air has been displaced by water?

38. The diameters of the mth and $(m + p)$th dark or bright rings of the Newton's rings experiment with a liquid of r.i. n in between the lens and the glass plate are d_m and

d_{m+p} respectively. Show that the wavelength of the monochromatic light is given by

$$\lambda = \frac{n}{4R}\left[\frac{d_{m+p}^2 - d_m^2}{p}\right].$$

39. In the Newton's rings experiment the diameters of the mth and $(m + p)$th dark/bright fringes with air film in between the lens and the plate are d_m and d_{m+p} respectively. A small quantity of the transparent liquid is placed between the glass plate and the lens and the diameters of the mth and $(m+p)$th dark/bright fringes are measured again. They are found to be D_m and D_{m+p} respectively. Show that r.i. of the liquid is

$$n = \frac{d_{m+p}^2 - d_m^2}{D_{m+p}^2 - D_m^2}.$$

40. In Newton's rings experiment with a liquid of r.i. n in between the glass plate and the lens show that the difference in radii between adjacent rings (maxima) is given by

$$\Delta\rho = \rho_{m+1} - \rho_m \approx \frac{1}{2}\sqrt{\frac{\lambda R}{nm}}$$

assuming $m \gg 1$.

41. In Newton's rings experiment show that the area between adjacent rings (maxima) is given by

$$A = \pi(\rho_{m+1}^2 - \rho_m^2) = \frac{\pi\lambda R}{n}.$$

42. A thin equiconvex lens of focal length 3 m and refractive index 1.5 rests on and in contact with an optical flat, and using light of wavelength 600 nm, Newton's rings are viewed normally by reflection. What is the diameter of the 10th bright ring?

[**Hints:** For equiconvex lens use $\frac{1}{f} = (\mu - 1)2/R$]

43. Newton's rings are formed by reflected light of wavelength 5893 Å with a liquid between the plane and the curved surfaces. If the diameter of the 8th bright ring is 3.6 mm and the radius of curvature of the curved surface is 100 cm, calculate the refractive index of the liquid.

44. The convex surface of radius R_1 of a plano-convex lens rests on a convex spherical surface of radius of curvature R_2. Newton's rings are observed with reflected light of wavelength λ. Show that the radius ρ_m of the mth bright ring is given by

$$\rho_m = \left[\left(m - \frac{1}{2}\right)\lambda \Big/ \left(\frac{1}{R_1} + \frac{1}{R_2}\right)\right]^{1/2}.$$

45. Newton's rings by reflection are formed between two biconvex lenses with radii of curvature of 100 cm each. Calculate the distance between the 9th and 16th dark rings using monochromatic light of wavelength 500 nm.

46. If the movable mirror in Michelson's interferometer is moved through 0.23 mm, 790 fringes are counted with a light meter. What is the wavelength of light?

47. In an experiment of Michelson's interferometer with sodium light the distance through which the mirror is moved between two consecutive positions of maximum distinctness is 0.294 mm. If the mean wavelength of the D_1 and D_2 lines of sodium is 5893 Å, find the difference between their wavelengths.

48. When one leg of a Michelson interferometer is lengthened slightly, 150 dark fringes sweep through the field of view. If the light used has λ = 480 nm, how far was the mirror in that leg moved?

49. A thin film with n = 1.40 for light of wavelength 600 nm is placed in one arm of a Michelson interferometer. If a shift of 8 fringes occurs, what is the thickness of the film?

50. In the ideal double-slit experiment when a glass-plate (refractive index 1.5) of thickness t is introduced in the path of one of the interfering beams (wavelength λ), the intensity at the position where the central maximum occurred previously remains unchanged. The minimum thickness of the glass plate is

(*a*) 2λ (*b*) $2\lambda/3$ (*c*) $\lambda/3$ (*d*) λ. *(I.I.T. 2002)*

[**Hints:** $t = x\lambda/(n-1)\beta$ where $x = \beta$]

51. In a double slit experiment instead of taking slits of equal widths, one slit is made twice as wide as the other. Then in the interference pattern.

(*a*) the intensities of both the maxima and minima increase.

(*b*) the intensity of maxima increases and the minima have zero intensity.

(*c*) the intensity of maxima decreases and that of minima increases.

(*d*) the intensity of maxima decreases and the minima have zero intensity.

(I.I.T. 2000)

[**Hints:** Previously $a_1 = a$, $a_2 = a$, $I_{max} = 4a^2$, $I_{min} = 0$, Now, $a_1 = 2a$, $a_2 = a$, $I_{max} = 9a^2$, $I_{min} = a^2$]

52. In a Young's double slit experiment two wavelengths of 500 nm and 700 nm are used. What is the minimum distance from the central maximum where their maxima coincide again?

Take $D/d = 10^3$. Symbols have their usual meanings.

(I.I.T. 2004)

13 Diffraction

13.1 DIFFRACTION

When a light wave encounters an obstacle or hole whose size is comparable to its wavelength the light wave spreads out to some extent into the region of the geometrical shadow. This bending of light round an obstacle is an example of diffraction of light. Diffraction is a convincing evidence of the wave theory of light. According to Huygens' principle, the wave is divided at the obstruction into infinitesimal wavelets which interfere with each other as they proceed. We consider diffraction effects that occur at a large distance from the obstruction (Fraunhofer diffraction). Sometimes this effect is studied experimentally by interposing a lens so that patterns that would otherwise occur at infinite distance are focussed onto a screen at the focal plane of the lens.

13.2 SINGLE-SLIT DIFFRACTION

When a beam of parallel rays is incident normally on a long narrow slit of width a, the emergent rays produce single-slit diffraction pattern with a central maximum together with minima corresponding to the diffraction angles θ that satisfy

$$a \sin \theta = m\lambda, \; m = \pm 1, \pm 2, \pm 3, \ldots \text{ (minima)}$$

or

$$\alpha = m\pi \text{ with } \alpha = \frac{\pi a}{\lambda} \sin \theta.$$

The diffracted intensity for a given diffraction angle θ is

$$I = I_m \sin^2 \alpha / \alpha^2$$

where I_m is the intensity of the central maximum ($\theta = 0$) or the principal maximum.

13.3 DIFFRACTION BY A CIRCULAR APERTURE

Diffraction by a circular aperture or lens with diameter d produces a central maximum with a first minimum at a diffraction angle θ_1 given by

$$\sin \theta_1 = 1.22 \frac{\lambda}{d} \qquad \text{(first minimum)}$$

When θ_1 is small we may write

$$\theta_1 \approx 1.22\lambda/d.$$

13.4 RAYLEIGH CRITERION

According to Lord Rayleigh two adjacent spectral lines are just resolved when the principal maximum of one falls on the first minimum of the other. When two objects are viewed through a telescope or microscope they are on the verge of resolvability if the central diffraction maximum of one is at the first minimum of the other. Due to diffraction through a circular aperture their angular separation must be at least

$$\theta_R = 1.22\lambda/d$$

where d is the diameter of the objective lens.

13.5 DOUBLE-SLIT DIFFRACTION

A beam of parallel rays passing normally through two slits, each of width a, whose centres are d distance apart, produces diffraction pattern whose intensity I at various diffraction angles θ is given by

$$I = I_m \cos^2 \beta \sin^2 \alpha/\alpha^2$$

with $\beta = \dfrac{\pi d}{\lambda}\sin\theta$ and $\alpha = \dfrac{\pi a}{\lambda}\sin\theta$.

13.6 MULTIPLE-SLIT DIFFRACTION

Diffraction by N identical slits results in principle maxima whenever

$$d \sin \theta = m\lambda,\ m = 0, \pm 1, \pm 2,... \qquad \text{(maxima)}$$

The half angular width ($\Delta\theta$) of the mth principal maximum is given by

$$\Delta\theta = \frac{\lambda}{Nd\cos\theta}.$$

13.7 DIFFRACTION GRATING

A diffraction grating consists of a large number (N) of parallel equidistant slits. The principal maxima are given by

$$d \sin \theta = m\lambda,\ m = 0, \pm 1, \pm 2,...$$

A grating is characterised by two parameters, the dispersion D and the resolving power R.

When a composite pencil of light is incident on a grating the different colours are separated from each other forming the diffraction spectrum. This separation of colours or wavelengths, known as dispersion, is given by

$$D = \frac{\Delta\theta}{\Delta\lambda} = \frac{m}{d\cos\theta}.$$

The resolving power of a grating is its power of separating two very close spectral lines. According to Rayleigh's criterion two lines are just resolved in the mth order when

$$\frac{\lambda}{\Delta\lambda} = Nm.$$

Here $\Delta\lambda$ is the smallest difference of wavelengths that can be resolved at the wavelength λ. The quantity $\lambda/\Delta\lambda$ is called the resolving power R of the grating.

13.8 X-RAY DIFFRACTION

The wavelength of X-rays or thermal neutrons is of the same order as the distances between atoms in a crystal. The regular array of atoms in a crystal is considered to be a three-dimensional diffraction grating for waves of short wavelengths. The atoms can be visualised as being arranged in planes with characteristic interplanar spacing d. Diffraction maxima occur if the incident direction of the wave, measured from the surface of a plane of atoms, satisfies Bragg's law:

$$2d \sin\theta = m\lambda, \quad m = 1, 2, 3,...$$

where λ is the wavelength of the radiation, and θ is the angle which the incident beam makes with the lattice plane.

SOLVED PROBLEMS

1. *Single-slit diffraction:* *When a monochromatic beam of parallel rays of wavelength λ falls normally on a long narrow slit of width a, the emergent rays produce single-slit diffraction pattern. Show that the diffracted intensity corresponding to the diffraction angle θ is given by*

$$I = I_m \sin^2 \alpha/\alpha^2$$

where

$$\alpha = \frac{\pi a}{\lambda} \sin \theta,$$

and I_m is the intensity of the central maximum ($\theta = 0$).

Solution

A beam of parallel rays is incident normally on the single slit of width a (Fig. 13.1). Let ds be an element of the width of the wavefront in the plane of the slit, at a distance s from the centre O of the slit. The secondary waves which travel normal to the slit will be focussed at P_0 by the lens L and those which travel at the diffraction angle θ will reach P. The amplitude at P due to the spherical wavelet emitted by the element ds situated at O is directly proportional to the lengths ds and inversely proportional to the distance x. Thus the infinitesimal displacement at P due to this element may be written as

$$dy_0 = B\frac{ds}{x} \sin(\omega t - kx)$$

where B is the proportionality constant.

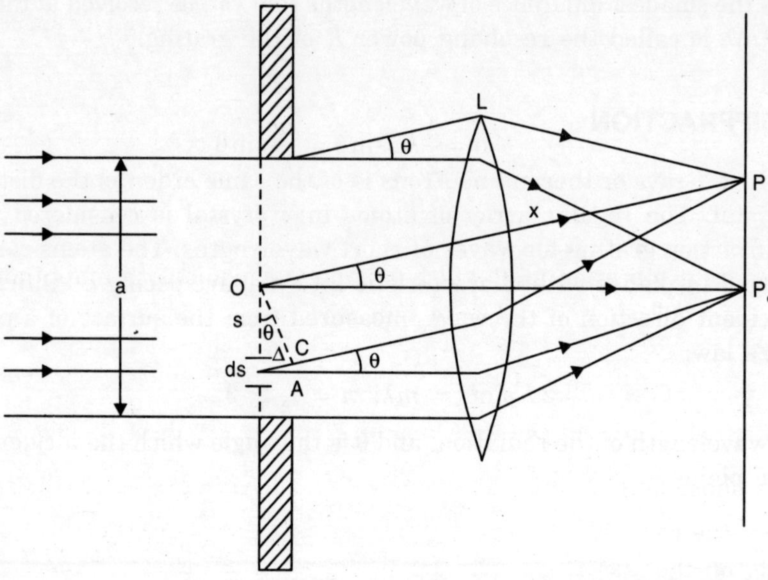

Fig. 13.1

As the position of *ds* is changed the displacement it produces at *P* will have different phase. When it is at *A* which is at a distance *s* below the origin, the contribution will be

$$dy_s = \frac{Bds}{x}\sin[\omega t - k(x + \Delta)].$$

where Δ is the path difference between the corresponding rays coming from *O* and *A*. From *O* a perpendicular *OC* is dropped on *AC*.

Now, $\theta + \angle P_0OC = \angle P_0OC + \angle COA = 90°.$

Thus, $\angle COA = \theta$ and $\Delta = AC = OA \sin\theta = s \sin\theta,$

and
$$dy_s = \frac{Bds}{x}\sin[\omega t - kx - ks\sin\theta].$$

We now wish to sum the effects of all elements from $s = -\frac{a}{2}$ to $\frac{a}{2}$. The contribution from the pair of elements symmetrically placed at *s* and *–s* is

$$\begin{aligned} dy &= dy_s + dy_{-s} \\ &= \frac{Bds}{x}[\sin(\omega t - kx - ks\sin\theta) + \sin(\omega t - kx + ks\sin\theta)] \\ &= \frac{2Bds}{x}\sin(\omega t - kx)\cos(ks\sin\theta) \end{aligned}$$

which must be integrated from $s = 0$ to $\frac{a}{2}$.

In doing so we regard *x* as a constant since the change in amplitude due to small change of *x* is negligible. Thus the resultant vibration at *P* is given by

$$y = \frac{2B}{x}\sin(\omega t - kx)\int_0^{a/2}\cos(ks\sin\theta)ds$$

$$= \frac{2B}{x}\sin(\omega t - kx)\sin\left(\frac{ka}{2}\sin\theta\right)\Big/(k\sin\theta).$$

If we define

$$\alpha = \frac{1}{2}ka\sin\theta = \frac{\pi a}{\lambda}\sin\theta \qquad ...(13.1)$$

then
$$y = \frac{aB}{x}\sin(\omega t - kx)\sin\alpha/\alpha.$$

Thus the resultant vibration at P will be simple harmonic with amplitude

$$A = A_0 \sin \alpha/\alpha \qquad ...(13.2)$$

where
$$A_0 = \frac{aB}{x}.$$

The path difference between the rays coming from two edges of the slit is $a \sin \theta$ and the corresponding phase difference is $\frac{2\pi}{\lambda} a \sin \theta = 2\alpha$. Thus α signifies one-half of the phase difference between the contributions coming from the two edges of the slit.

The intensity on the screen is given by

$$I \propto A^2 = A_0^2 \sin^2\alpha/\alpha^2.$$

At the point P_0, $\theta = 0$, $\alpha = 0$, $\sin \alpha/\alpha = 1$ and all the secondary wavelets arrive in phase at this point. We get the position of the central maximum or the principal maximum. If I_m is the intensity at the central maximum, then we have

$$I = I_m \sin^2 \alpha/\alpha^2 \qquad ...(13.3)$$

2. *Find the positions and intensities of the maxima and the positions of the minima of the single-slit diffraction pattern (problem 1).*

Solution

From Eqn. 13.3 the position of the central maximum or the principal maximum is given by $\alpha = 0$ or $\theta = 0$.

Minima: The intensity falls to zero at $\alpha = \pm\pi, \pm 2\pi, \pm 3\pi,....$ or, in general $\alpha = m\pi$ with $m = \pm 1, \pm 2,.....$So on one side of the diffraction pattern the minima are equispaced.

Maxima: Between two minima we have secondary maxima. The secondary maxima do not lie half way between the minima points, but are displaced towards the centre of the pattern by an amount which decreases with increasing m. The exact values of α for these maxima may be found by differentiating Eqn. (13.2) with respect to α and equating to zero:

$$\frac{dA}{d\alpha} = A_0\left[\frac{\cos\alpha}{\alpha} - \frac{\sin\alpha}{\alpha^2}\right] = 0$$

or
$$\alpha = \tan \alpha,\ \alpha \neq 0 \qquad ...(13.4)$$

The solutions of Eqn. (13.4) gives us the secondary maxima points. These points are

$$\alpha = \pm 1.43\pi, \pm 2.46\pi, \pm 3.47\pi,....$$

The positions of maxima and minima are given below:

Maxima $\alpha = m\pi$, $m = 0, \pm 1.43, \pm 2.46, \pm 3.47,....$

Minima $\alpha = m\pi$, $m = \pm 1, \pm 2, \pm 3,....$

Since $\alpha = \frac{\pi a}{\lambda} \sin\theta = m\pi$, we have

$$\sin\theta = \frac{m\lambda}{a} \text{ or, } \theta = \frac{m\lambda}{a}$$

with

$$m = 0, \pm 1.43, \pm 2.46, \pm 3.47, ... \qquad \text{(maxima)}$$

$$m = \pm 1, \pm 2, \pm 3, ... \qquad \text{(minima)}$$

The angular width of the pattern for a given wavelength is inversely proportional to the slit width a. If the width a is made larger, the pattern shrinks to a small scale. When the width is very large compared to the wavelength of light, the diffraction pattern disappears.

From Eqn. (13.2) we find that the amplitude A decreases with increasing α as α occurs in the denominator. Therefore, the intensity decreases with increasing α (Fig. 13.2). The width of the central maximum (α ranges form $-\pi$ to π) is $\Delta\alpha = 2\pi$ while the width of the secondary maximum is $\Delta\alpha = \pi$. Thus the central maximum is twice as great as that of the fainter secondary maximum.

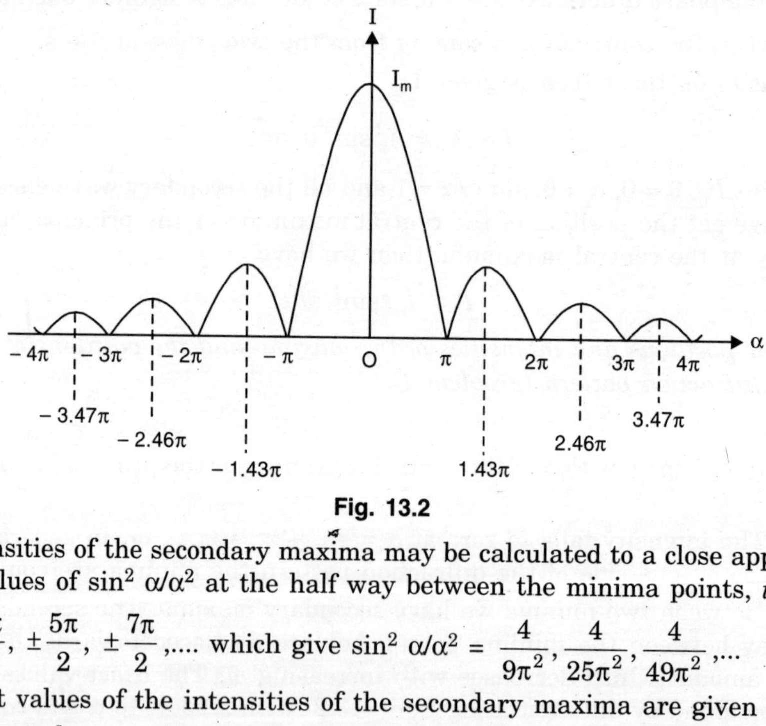

Fig. 13.2

The intensities of the secondary maxima may be calculated to a close approximation by finding the values of $\sin^2 \alpha/\alpha^2$ at the half way between the minima points, *i.e.*,

$$\alpha = \pm\frac{3\pi}{2}, \pm\frac{5\pi}{2}, \pm\frac{7\pi}{2}, ... \text{ which give } \sin^2 \alpha/\alpha^2 = \frac{4}{9\pi^2}, \frac{4}{25\pi^2}, \frac{4}{49\pi^2}, ...$$

The exact values of the intensities of the secondary maxima are given below:

(*i*) 1st secondary maximum: $\alpha = \pm 1.43\pi$ and $\sin^2 \alpha/\alpha^2 = 0.0496$. Thus the intensity of 1st secondary maximum is 4.96% of the intensity of the central maximum.

(*ii*) 2nd secondary maximum: $\alpha = \pm 2.46\pi$ and $\sin^2 \alpha/\alpha^2 = 0.0168$. The intensity is 1.68% of that of central maximum.

(*iii*) 3rd secondary maximum: $\alpha = \pm 3.47\pi$ and $\sin^2 \alpha/\alpha^2 = 0.0083$. The intensity is only 0.83% of the intensity of the central maximum.

Linear distance on the screen between successive minima: Suppose for the 1st and 2nd minima, the values of the angle θ are θ_1 and θ_2 respectively. Since the angle θ is very small we may write

$$\theta_1 = \frac{\lambda}{a} \text{ and } \theta_2 = \frac{2\lambda}{a}.$$

Let the distance between the slit and the screen be f which is also equal to the focal length of the lens if the lens is placed close to the slit. Let the linear distance on the screen between successive minima corresponding to the angular separation $\theta_2 - \theta_1$ $(= \lambda/a)$ be d (Fig. 13.3)

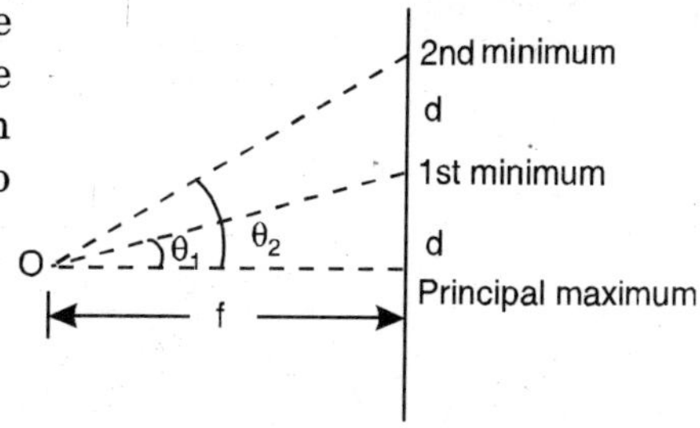

Fig. 13.3

Thus, we have
$$\frac{d}{f} = \frac{\lambda}{a}$$

or
$$d = \frac{\lambda f}{a},$$

which gives the linear distance on the screen between successive minima of the diffraction pattern.

3. *Find the half angular width of the central bright maximum in the Fraunhofer diffraction pattern of a slit of width 0.5 mm when the slit is illuminated by monochromatic light of wavelength 589.3 nm.*

Solution

We have $\sin\theta = \lambda/a$ where θ is the half angular width of the central maximum.

$$\sin\theta = \frac{589.3\times10^{-9}}{0.5\times10^{-3}} = 1.1786 \times 10^{-3}$$

or
$$\theta = 1.18 \times 10^{-3} \text{ rad.}$$

4. *When monochromatic light is incident on a slit 0.02 mm wide, the first diffraction minimum is observed at an angle of 1.5° from the direction of the direct beam. What is the wavelength of the incident light?*

Solution

$$\lambda = a \sin\theta = 0.02 \times 10^{-3} \sin(1.5°) \text{ m} = 523.5 \text{ nm.}$$

5. *Light of wavelength 630 nm is incident on a narrow slit. The angle between the first minimum on one side of the central maximum and the first minimum on the other side is 1°. What is the width of the slit?*

Solution

$$a = \frac{\lambda}{\sin\theta} = \frac{630 \text{ nm}}{\sin(0.5°)} = 72.19\ \mu\text{m.}$$

6. *The distance between the first and fifth minima of a single-slit pattern is 0.35 mm with the screen 40 cm away from the slit, using light having a wavelength of 560 nm. (a) Calculate the diffraction angle θ of the first minimum. (b) Find the width of the slit.*

Solution

(*a*) Let the angle of diffraction be θ_5 for the fifth minimum, and the corresponding distance on the screen from the central maximum be d_5. Thus,

$$\theta_5 = 5\theta = \frac{d_5}{D} \text{ and } \theta = \frac{d}{D}$$

where θ = diffraction angle of the first minimum

d = distance of the first minimum on the screen from the central maximum

D = distance of the screen from the slit.

So, we have

$$4\theta = \frac{d_s - d}{D} = \frac{0.35\text{ mm}}{400\text{ mm}}$$

or $$\theta = 2.19 \times 10^{-4}\text{ rad.}$$

(*b*) $$a = \frac{\lambda}{\sin\theta} = \frac{560\text{ nm}}{2.19\times10^{-4}} = 2.56\text{ mm.}$$

7. *A parallel beam of light is incident normally on a narrow slit of width 0.22 mm. The Fraunhofer diffraction pattern is observed on a screen which is placed at the focal plane of a convex lens whose focal length is 70 cm. Calculate the distance between the first two secondary maxima on the screen. The wavelength of light = 600 nm and the lens is placed very close to the slit.*

Solution

The first and second maxima occur at $\alpha = 1.43\pi$ and 2.46π. Thus,

$$a \sin\theta = 1.43\lambda \text{ and } 2.46\lambda$$

or $$\sin\theta = \frac{1.43\lambda}{a} = 3.9 \times 10^{-3} \text{ for 1st maximum}$$

and $$\sin\theta = \frac{2.46\lambda}{a} = 6.71 \times 10^{-3} \text{ for 2nd maximum.}$$

Consequently the maxima will be separated on the screen by the distance given by

$$(6.71 - 3.9) \times 10^{-3} \times 70 = 0.2\text{ cm.}$$

8. *Sound waves with frequency 3000 Hz diffract out of a speaker cabinet with a 0.3 m diameter opening into a large auditorium. Velocity of sound in air is 343 m/s. Where does a listener standing against a wall 100 m from the speaker have the most difficulty in hearing? Assume single slit diffraction from a slit of width 0.3 m.*

Solution

The first minima occur at *B* and *B′* on two sides of the central maximum *A* (Fig. 13.4).

Now, $$\lambda = \frac{343}{3000}\text{ m}$$

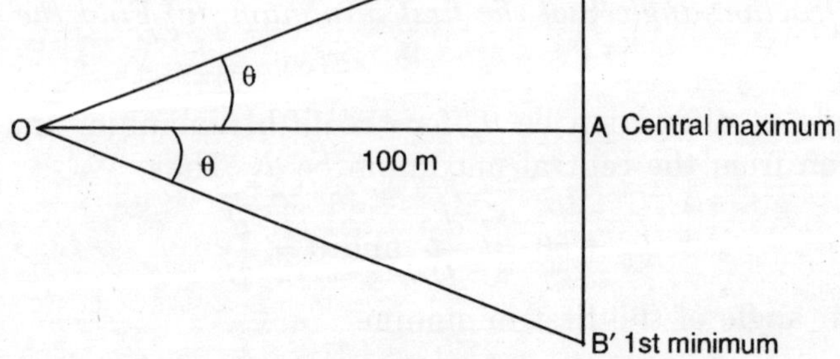

Fig. 13.4

and $$\sin\theta = \frac{\lambda}{a} = \frac{343}{3000 \times 0.3}$$

or $$\theta = 22.4°$$

$$\tan\theta = \frac{AB}{100} = 0.412$$

or $$AB = 41.2 \text{ m and } AB' = 41.2 \text{ m}.$$

9. *The full width at half maximum (FWHM) of the central diffraction maximum is defined as the angle between the two points in the pattern where the intensity is one-half of that at the centre of the pattern. (a) Show that FWHM is*

$$\Delta\theta = 2 \sin^{-1} (0.442\lambda/a)$$

where a is the width of the single slit. (b) Calculate the FWHM of the central maximum for slits whose widths are 1.5λ and 10λ.

Solution

(*a*) From Eqn. (13.3), we find that $I = \frac{1}{2} I_m$ when $\sin^2 \alpha = \frac{1}{2}\alpha^2$. This transcendental equation has a solution $\alpha = 1.39$ radians.

Thus, we have $\frac{\pi a}{\lambda} \sin\theta = 1.39$

or $$\sin\theta = \frac{1.39\lambda}{\pi a} = 0.442\lambda/a,$$

and *FWHM* is $\Delta\theta = 2 \sin^{-1} (0.442\, \lambda/a)$.

(*b*) When $a = 1.5\lambda$, $\Delta\theta = 34.28°$

When $a = 10\lambda$, $\Delta\theta = 5.07°$.

10. *Find the smallest angular separation of two narrow slit sources which could be resolved theoretically (according to Rayleigh's criterion of just resolution) by a rectangular slit of width a.*

Solution

Two narrow sources form real images on the screen after passing through the rectangular aperture. Each image consists of a single-slit diffraction pattern. The angular separation of the sources ψ is equal to the angular separation of the central maxima. According to Rayleigh's criterion for just resolution two images appear to be just resolved when the principal maximum of one falls on the 1st minimum of the other *i.e.,*

$$\psi = \theta_1$$

where θ_1 is the diffraction angle of the 1st minimum

$$\sin\theta_1 \approx \theta_1 = \lambda/a.$$

Thus at just resolution of two images the angular separation of two narrow slits is

$$\psi = \lambda/a.$$

11. *Find the smallest angular separation of two narrow slit sources which could be resolved by a rectangular slit of width 1″. [Take the mean value of λ = 6000 Å]*

Solution

Minimum angle of resolution is

$$\frac{\lambda}{a} = \frac{6 \times 10^{-5}}{2.54} \text{ rad} = 4.87''.$$

12. *Circular aperture:* *When a monochromatic beam of parallel rays of wavelength λ falls normally on a circular aperture of diameter a, the emergent rays produce circular diffraction pattern.*

(a) Show that the diffracted intensity corresponding to the diffraction angle θ is given by

$$I = I_m J_1^2(2\alpha)/\alpha^2$$

where
$$\alpha = \frac{\pi a \sin\theta}{2\lambda},$$

$J_1(2\alpha)$ is the Bessel function of order 1, and I_m is the intensity of the central maximum (θ = 0).

(b) Find the positions of the maxima and the minima of the circular diffraction pattern.

Solution

OAB is one quadrant of a circular aperture of diameter *a*, with its plane in *yz*-plane (Fig. 13.5). The plane waves are incident on the aperture in the *x*-direction. The circular aperture is divided into elementary diffracting areas *s ds dϕ*, each acting as a secondary source of disturbance. On account of the presence of circular symmetry we choose a point P_θ having coordinates $(x_0, y_0, 0)$ in the *xy*-plane. The distance of P_θ from the centre of the aperture is *r* and θ is the angle of diffraction. We wish to find the resultant disturbance at P_θ. We have

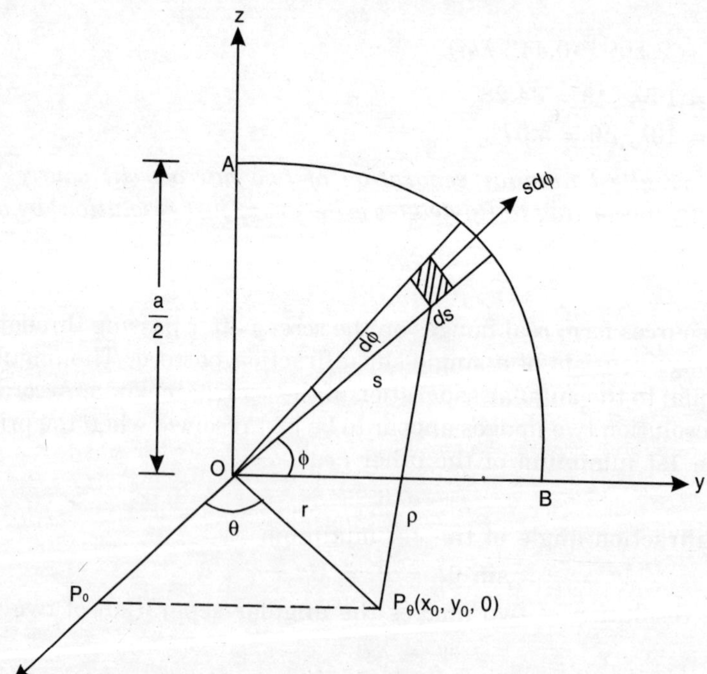

Fig. 13.5

$$x_0 = r\cos\theta \text{ and } y_0 = r\sin\theta.$$

Let the distance of P_θ from the elementary area *s ds dϕ* be ρ. A disturbance originating at the elementary area produces a displacement *du* at P_θ proportional to the area and is given by

$$du = B\,(s\,ds\,d\phi)\,\sin(\omega t - k\rho), \quad ...(13.5)$$

where B is the proportionality constant, $\omega = \dfrac{2\pi}{T}$ and $k = \dfrac{2\pi}{\lambda}$. Eqn. (13.5) can be written as

$$du = Bs\,ds\,d\phi \sin 2\pi\left(\frac{t}{T} - \frac{\rho}{\lambda}\right) \quad ...(13.6)$$

The resultant disturbance at P_θ is obtained by integrating the above expression

$$u = B\int_0^{2\pi}\int_0^{a/2} \sin 2\pi\left(\frac{t}{T} - \frac{\rho}{\lambda}\right) s\,ds\,d\phi \quad ...(13.7)$$

The coordinates of the elementary area are $(0, s\cos\phi, s\sin\phi)$ and those of P_θ are $(r\cos\theta, r\sin\theta, 0)$. Thus,

$$\rho^2 = r^2\cos^2\theta + (r\sin\theta - s\cos\phi)^2 + s^2\sin^2\phi$$
$$= r^2 - 2rs\sin\theta\cos\phi + s^2$$

Since r is very large for Fraunhofer diffraction and s is small compared to r, we obtain, expanding by binomial theorem

$$\rho \approx r\left[1 - \frac{2s\sin\theta\cos\phi}{r}\right]^{1/2}$$
$$\approx r - s\sin\theta\cos\phi.$$

From Eqn. (13.7), we have

$$u = B\int_0^{2\pi}\int_0^{a/2} \sin 2\pi\left(\frac{t}{T} - \frac{r}{\lambda} + \frac{s.\sin\theta\cos\phi}{\lambda}\right) s\,ds\,d\phi$$

$$= B\int_0^{2\pi}\int_0^{a/2} s\,ds\,d\phi\left[\sin 2\pi\left(\frac{t}{T} - \frac{r}{\lambda}\right)\cos\left(\frac{2\pi s.\sin\theta\cos\phi}{\lambda}\right)\right.$$
$$\left. + \cos 2\pi\left(\frac{t}{T} - \frac{r}{\lambda}\right)\sin\left(\frac{2\pi s.\sin\theta\cos\phi}{\lambda}\right)\right]$$

The ϕ-integration of the second term gives:

$$\int_0^{2\pi}\sin(ps\cos\phi)d\phi = \int_0^{\pi}\sin(ps\cos\phi)d\phi + \int_{\pi}^{2\pi}\sin(ps\cos\phi)d\phi$$

where $$p = \frac{2\pi\sin\theta}{\lambda}.$$

In the second integral we put $\beta = \phi - \pi$ so that it becomes

$$\int_0^{\pi}\sin(-ps\cos\beta)dB = -\int_0^{\pi}\sin(ps\cos\beta)d\beta$$

and the value of the integral $\int_0^{2\pi}\sin(ps\cos\phi)d\phi$ is seen to be zero.

Thus, we have

$$u = B \sin 2\pi \left(\frac{t}{T} - \frac{r}{\lambda}\right) \int_0^{2\pi} d\phi \int_0^{a/2} s\, ds \cos(ps\cos\phi) \qquad \text{...(13.8)}$$

For the s-integration, integrating by parts, we obtain

$$\int_0^{a/2} s\cos(ps\cos\phi)ds = \left[\frac{s\sin(ps\cos\phi)}{p\cos\phi}\right]_0^{a/2} - \int_0^{a/2} \frac{\sin(ps\cos\phi)}{p\cos\phi}\, ds$$

or

$$\int_0^{a/2} s\cos(ps\cos\phi)ds = \frac{\frac{a}{2}\sin\left(\frac{ap}{2}\cos\phi\right)}{p\cos\phi} + \left[\frac{\cos(ps\cos\phi)}{p^2\cos^2\phi}\right]_0^{a/2}$$

$$= \frac{\frac{a}{2}\sin\left(\frac{ap}{2}\cos\phi\right)}{p\cos\phi} + \frac{\cos\left(\frac{ap}{2}\cos\phi\right)}{p^2\cos^2\phi} - \frac{1}{p^2\cos^2\phi}$$

$$= \frac{a}{2p\cos\phi}\left[\frac{ap}{2}\cos\phi - \frac{a^3p^3}{3!8}\cos^2\phi + \ldots\right]$$

$$+ \frac{1}{p^2\cos^2\phi}\left[1 - \frac{a^2p^2}{2!4}\cos^2\phi + \frac{a^4p^4}{4!16}\cos^4\phi - \ldots\right] - \frac{1}{p^2\cos^2\phi}$$

$$= \frac{a^2}{4} - \frac{a^4p^2}{96}\cos^2\phi + \ldots - \frac{a^2}{8} + \frac{a^4p^2}{16\times 24}\cos^2\phi - \ldots$$

$$= \frac{a^2}{8} - \frac{a^4p^2}{128}\cos^2\phi + \ldots$$

Thus, Eqn. (13.8) becomes

$$u = B \sin 2\pi\left(\frac{t}{T} - \frac{r}{\lambda}\right)\int_0^{2\pi} d\phi \left[\frac{a^2}{8} - \frac{a^4p^2}{128}\cos^2\phi + \ldots\right]$$

Integrating term by term with respect to ϕ, we obtain

$$u = B \sin 2\pi \left(\frac{t}{T} - \frac{r}{\lambda}\right)\left[\frac{2\pi a^2}{8} - \frac{a^4p^2}{128}\pi + \ldots\right]$$

$$= B \sin 2\pi \left(\frac{t}{T} - \frac{r}{\lambda}\right)\left[\frac{\pi a^2}{4} - \frac{\pi^3 a^4}{32\lambda^2}\sin^2\theta + \ldots\right]$$

$$= B \sin 2\pi \left(\frac{t}{T} - \frac{r}{\lambda}\right)\left[\frac{a\lambda}{2\sin\theta}\left\{\alpha - \frac{\alpha^3}{1!2!} + \ldots\right\}\right] \qquad \text{...(13.9)}$$

Here, $\alpha = \dfrac{\pi a\sin\theta}{2\lambda}$.

The terms within the parentheses in Eqn. (13.9) can be summed up:

$$\alpha - \frac{\alpha^3}{1!2!} + \frac{\alpha^5}{1!2!3!} - \ldots = J_1(2\alpha)$$

where J_1 is the Bessel function of order 1. Thus, we obtain finally

$$u = \frac{Ba\lambda}{2\sin\theta} J_1(2\alpha)\sin 2\pi\left(\frac{t}{T} - \frac{r}{\lambda}\right). \quad ...(13.10)$$

The amplitude of the resultant simple harmonic motion is

$$A = \frac{Ba\lambda}{2\sin\theta} J_1(2\alpha) = \frac{B\pi a^2}{4} \frac{J_1(2\alpha)}{\alpha}.$$

when $\theta = 0$, $\alpha = 0$ and $\frac{J_1(2\alpha)}{\alpha} = 1$. The amplitude at the central maximum (at P_0 in Fig. 13.5) is

$$A_m = \frac{B\pi a^2}{4}.$$

Therefore, $A = A_m J_1(2\alpha)/\alpha$.

The intensity is maximum (I_m) at the central bright spot ($\theta = 0$). The intensity I at P_θ is seen to be

$$I = I_m J_1^2(2\alpha)/\alpha^2 \quad ...(13.11)$$

(*b*) The function J_1 (2α) has an infinite number of zeros, going through positive and negative values alternately as α increases with diminishing ordinates at the maxima. The principal maximum occurs at $\alpha = 0$ or $\theta = 0$. The secondary maxima are at

$$2\alpha = 1.64\pi,\ 2.67\pi,\ 3.69\pi,\ 4.72\pi \text{ etc.}$$

or

$$\sin\theta = \frac{1.64\lambda}{a}, \frac{2.67\lambda}{a}, \frac{3.69\lambda}{a}, \frac{4.72\lambda}{a} \text{ etc.}$$

The minima occur at

$$2\alpha = 1.22\pi,\ 2.23\pi,\ 3.24\pi,\ 4.26\pi \text{ etc.}$$

or

$$\sin\theta = \frac{1.22\lambda}{a}, \frac{2.23\lambda}{a}, \frac{3.24\lambda}{a}, \frac{4.26\lambda}{a} \text{ etc.}$$

The first minimum of the diffraction pattern corresponds to the diffraction angle θ_1 at which

$$\sin\theta_1 = \frac{1.22\lambda}{a}$$

or

$$\theta_1 = \frac{1.22\lambda}{a} \quad ...(13.12)$$

since θ_1 is a very small angle.

We get a circular diffraction pattern with a very bright central disc which is called Airy's disc, followed by a series of alternate dark and bright rings of decreasing intensity. The ratio

$$\frac{I}{I_m} = \frac{\text{Intensity at the secondary maximum}}{\text{Intensity at the central maximum}}$$

for the 1st, 2nd and 3rd secondary bright rings are 0.0175, 0.00416, 0.00160 respectively. The intensity of the rings fades away rapidly as the rings recede from the centre. So, the pattern will be limited only to a small region near the centre.

If a convex lens of focal length f is placed very close to the aperture and the pattern is seen on a screen, the linear radius d of the first dark ring is given by

$$d = \theta_1 f = \frac{1.22\lambda f}{a}.$$

13. *Find the smallest angular separation of two stars which could be resolved by a telescope of diameter1". [Mean λ = 600 nm]*

Solution

The minimum angle of resolution is

$$\frac{1.22\lambda}{a} = \frac{1.22 \times 6 \times 10^{-5}}{2.54} \text{ rad} = 2.88 \times 10^{-5} \text{ rad}$$

$$\approx 6 \text{ seconds.}$$

14. *For the human eye the diameter of the pupil is about 3 mm. Find the smallest angular separation of two objects which could be resolved by the human eye.*

[Mean λ = 600 nm]

Solution

The minimum angle of resolution (θ_1)

$$= \frac{1.22 \times 6 \times 10^{-5}}{0.3} \text{ rad} \approx 50 \text{ seconds.}$$

[*Note:* Actually an average person cannot resolve objects less than about 1 minute apart. This is because the separation of two images is decreased by refraction of the rays as they enter the eye (n = 1.33 and θ_1 = 50 × 1.33 seconds)]

15. *A converging lens 3 cm in diameter has a focal length of 20 cm. (a) What angular separation must two distant point objects have to satisfy Rayleigh's criterion? (b) How far apart are the centres of the diffraction patterns in the focal plane of the lens? [Assume λ = 550 nm]*

Solution

(*a*) $\theta_R = \frac{1.22\lambda}{a} = \frac{1.22 \times 5.5 \times 10^{-5}}{3} = 2.24 \times 10^{-5}$ rad

(*b*) The linear separation is

$$\Delta x = f\theta_R = 20 \times 2.24 \times 10^{-5} \text{ cm} = 4.48 \times 10^{-4} \text{ cm.}$$

16. *A laser beam was fired from the Air Force Optical Station on Maui, Hawaii, and reflected back from the shuttle Discovery as it sped by, 220 miles overhead. The diameter of the central maximum of the beam at the shuttle position was said to be 28 ft and the beam wavelength was 540 nm. What is the effective diameter of the laser aperture at the Maui ground station? Assume circular exit aperture of the laser beam.*

Solution

The first minimum of the diffraction pattern due to circular aperture corresponds to the diffraction angle θ_1 at which we have

$$\theta_1 = \frac{1.22\lambda}{a}$$

where a is the diameter of the laser aperture.

Again, $$\theta_1 = \frac{14 \text{ ft}}{220 \times 1760 \times 3 \text{ ft}}.$$

Thus, $$a = \frac{1.22 \times 220 \times 1760 \times 3 \times 5.4 \times 10^{-5}}{14} \text{ cm} = 5.47 \text{ cm}$$

17. *Double-slit diffraction: A beam of parallel rays passing normally through two slits each of width a, whose centres are distance d apart, produces double-slit diffraction pattern. Show that the intensity I of the double-slit diffraction pattern at the diffraction angle θ is given by*

$$I = I_m \cos^2 \beta \sin^2 \alpha/\alpha^2$$

where $\beta = \frac{\pi d}{\lambda} \sin\theta$, $\alpha = \frac{\pi a}{\lambda} \sin\theta$ *and* I_m *is the intensity in the forward direction* ($\theta = 0$).

Solution

A double slit consists of two narrow slits AB and CD arranged parallel to each other (Fig. 13.6). Each slit is of width a and separated by an opaque space BC of width b. The origin O is chosen at the mid-point of BC.

Now, $$d = \frac{a}{2} + b + \frac{a}{2} = a + b$$

and $$OA = OD = a + \frac{b}{2} = \frac{d}{2} + \frac{a}{2},$$

$$OB = OC = \frac{b}{2} = \frac{d}{2} - \frac{a}{2}.$$

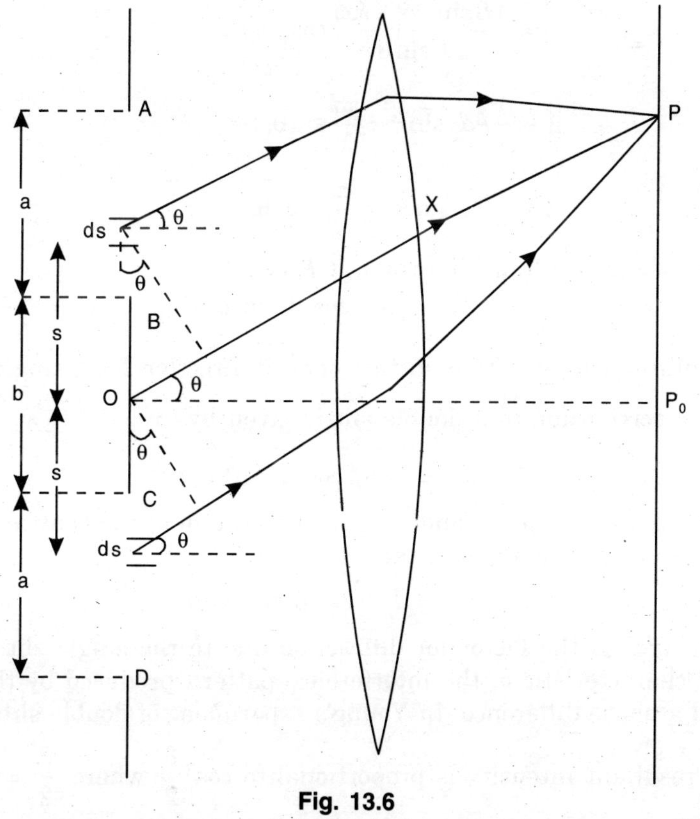

Fig. 13.6

The amplitude at a point P on the screen due to the elementary length ds of the slit at a distance s from O is

$$dy_s = \frac{Bds}{x} \sin[\omega t - k(x \pm s \sin \theta)]$$

where B is a constant and x is the distance of P and O. We sum the contributions from the corresponding elementary lengths ds on both sides of O at a distance s from O:

$$dy = \frac{Bds}{x} [\sin \{\omega t - k(x + s \sin \theta)\} + \sin \{\omega t - k(x - s \sin \theta)\}]$$

$$= \frac{2Bds}{x} \sin (\omega t - kx) \cos (ks \sin \theta)$$

This must be integrated from $s = \frac{d}{2} - \frac{a}{2}$ to $s = \frac{d}{2} + \frac{a}{2}$ in order to find the total contribution:

$$y = \frac{2B}{x} \sin (\omega t - kx) \int_{\frac{d}{2}-\frac{a}{2}}^{\frac{d}{2}+\frac{a}{2}} \cos (ks \sin \theta)\, ds$$

$$y = \frac{2B \sin(\omega t - kx)}{xk \sin \theta}\left[\sin\left\{\frac{1}{2}k\,(d+a) \sin \theta\right\} - \sin\left\{\frac{1}{2}k\,(d-a) \sin \theta\right\}\right]$$

$$= \frac{4B \sin (\omega t - kx)}{xk \sin \theta} \cos \beta \sin \alpha.$$

Here, $\beta = \frac{1}{2}kd \sin \theta = \frac{\pi d}{\lambda} \sin \theta,$

and $\alpha = \frac{1}{2} ka \sin \theta = \frac{\pi a}{\lambda} \sin \theta.$

The amplitude of the resultant vibration at P is

$$A = 2A_0 \cos \beta \sin \alpha/\alpha \qquad \text{...(13.13)}$$

where $A_0 = \frac{aB}{x}$ = diffraction amplitude in the forward direction for a single slit of width a. The intensity on the screen due to a double slit is given by

$$I \propto A^2 = 4A_0^2 \cos^2 \beta \sin^2 \alpha/\alpha^2 \qquad \text{...(13.14)}$$

At the point P_0, $\theta = 0$, $\alpha = \beta = 0$ and we get the position of the central maximum. If I_m is the intensity at the central maximum then we have

$$I = I_m \cos^2 \beta \sin^2 \alpha/\alpha^2 \qquad \text{...(13.15)}$$

The factor $\sin^2 \alpha/\alpha^2$ is the factor for diffraction due to the single slit of width a. The factor $\cos^2 \beta$ is the characteristic of the interference pattern produced by the two beams of equal intensity and a phase difference. In Young's experiment of double slits [see problem 1 of chapter 12] the resultant intensity is proportional to $\cos^2 \frac{\delta}{2}$ where $\frac{\delta}{2} = \frac{\pi d \sin \theta}{\lambda} = \beta$.

18. *Find the positions of the maxima and the minima of the double-slit diffraction pattern of problem 17.*

Solution

Positions of minima: The resultant intensity I [Eqn. (13.15)] is zero when either of the factors $\cos^2 \beta$ or $\sin^2 \alpha/\alpha^2$ is zero.

(*i*) Minima of the interference pattern: $\cos^2 \beta = 0$

or $$\beta = \pm\left(m+\frac{1}{2}\right)\pi, \quad m = 0, 1, 2,...$$

or $$d \sin \theta = \pm\left(m+\frac{1}{2}\right)\lambda, \quad m = 0, 1, 2,... \qquad ...(13.16)$$

(*ii*) Minima of the diffraction pattern: $\sin^2 \alpha/\alpha^2 = 0$,

or $$\alpha = \pm p\pi, \; p = 1, 2, 3,...$$

or $$a \sin\theta = \pm p\lambda, \; p = 1, 2, 3,... \qquad ...(13.17)$$

Positions of maxima: The exact positions of the maxima are not given by any simple relation, but their approximate positions may be found by neglecting the variation of the factor $\sin^2\alpha/\alpha^2$, a justifiable assumption only when d is large compared to a and the maxima only near the centre of the pattern are considered. For the same small value of θ near the centre of the pattern, α is small compared to β. In that case, $\sin^2\alpha/\alpha^2 \approx 1$ and the positions of the maxima will be determined solely by the $\cos^2\beta$ factor, which has maxima for $\beta = \pm m\pi$, $m = 0, 1, 2,...$

or $$d \sin \theta = \pm m\lambda, \; m = 0, 1, 2,... \qquad ...(13.18)$$

Since θ is small we may write

$$\theta = \pm m\lambda/d, \; m = 0, 1, 2,... \qquad ...(13.19)$$

Thus, the maxima near the centre are equispaced.

When the slit width a is not very small, the variation of the factor $\sin^2 \alpha/\alpha^2$ with θ must be taken into account. The complete double slit pattern is the product of two factors $\cos^2\beta$ and $\sin^2 \alpha/\alpha^2$. In this case the positions of the maxima are slightly different from those given by Eqn. (13.18) except for the central maximum ($m = 0$).

If the slit width a is much small compared to d, we get many interference maxima within the central maximum of the diffraction pattern. In the central diffraction maximum, $\sin\theta$ varies from $-\lambda/a$ to λ/a or $\Delta\theta \approx 2\lambda/a$. The width of a bright interference fringe $= \lambda/d$. In general, the number of interference maxima within the central diffraction maximum = $\frac{2\lambda}{a}\Big/\frac{\lambda}{d} = \frac{2d}{a}$. But sometimes the interference maximum falls on the diffraction minimum and it cannot be seen. In that case the number of interference maxima occurring under the central diffraction maximum is $\left(\frac{2d}{a}-1\right)$.

Missing orders: It is sometimes seen that certain interference maxima are missing. These so-called missing orders occur when the condition for a maximum of the interference and the condition for a minimum of the diffraction are satisfied for the same value of θ, *i.e.*,

$$d \sin \theta = m\lambda$$

and $$a \sin \theta = p\lambda$$

or $$\frac{d}{a} = \frac{m}{p},$$

i.e., $\frac{d}{a}$ is in the ratio of the two integers. When $\frac{d}{a} = 2$, orders $m = 2, 4, 6,...$ are missing and the number of interference maxima under the central diffraction maximum = 1 + 2 × 1 = 3. When $\frac{d}{a} = 3$, order $m = 3, 6, 9,...$ are missing and the number of interference maxima under the central diffraction maximum = 1 + 2 × 2 = 5.

19. *In a double-slit experiment the distance D of the screen from the slits is 60 cm, the wavelength λ is 500 nm, the slit separation d is 0.12 mm, and the slit width a is 0.025 mm. (a) What is the spacing between adjacent fringes? (b) What is the distance from the central maximum to the first minimum of the fringe envelope (diffraction factor)? (c) How many fringes are there in the central peak of the diffraction envelope?*

Solution

(*a*) The spacing between the adjacent fringes

$$= \Delta y = \frac{\lambda}{d} D = \frac{(500 \times 10^{-9})(60 \times 10^{-2})}{0.12 \times 10^{-3}} \text{ m}$$

$$= 2.5 \text{ mm}$$

(*b*) The minimum of the diffraction factor is given by

$$\sin\theta = \frac{\lambda}{a} = \frac{500 \times 10^{-9}}{0.025 \times 10^{-3}} = 0.02$$

Since sinθ is very small we can put

$$\sin\theta \approx \tan\theta \approx \theta = 0.02.$$

Thus the distance of the first minimum of the diffraction factor from the central maximum is

$$y = D \tan\theta = 60 \times 0.02 \text{ cm} = 1.2 \text{ cm}.$$

(*c*) $\frac{2 \times y}{\Delta y} = \frac{2 \times 1.2}{0.25} = 9.6$

No. of fringes = 9

$$\left[\text{Note that } \frac{2d}{a} = 9.6\right]$$

20. *What requirements must be met for the central maximum of the envelope of the double-slit interference pattern to contain exactly 9 fringes?*

Solution

The required condition will be met if the fifth minimum of the interference factor coincides with the first minimum of the diffraction factor. The fifth minimum of the interference factor occurs when $\beta = \frac{9}{2}\pi$. The first minimum in the diffraction term occurs for $\alpha = \pi$. Thus, we get

$$\frac{\beta}{\alpha} = \frac{d}{a} = \frac{9}{2} \text{ as the required condition.}$$

21. *In problem 20 the envelope of the central peak contains 9 fringes. How many fringes lie between the first and the second minima of the envelop?*

Solution

We have $$\frac{\beta}{\alpha} = \frac{d}{a} = \frac{9}{2}.$$

The first minimum in the diffraction term occurs at $\alpha = \pi$ or, $\sin\theta = \lambda/a$. The second minimum occurs at $\alpha = 2\pi$ or, $\sin\theta = 2\lambda/a$.

The interference maxima occur at $\sin\theta = \dfrac{m\lambda}{d} = \dfrac{2m\lambda}{9a}$, $m = 0, 1, 2,...$

The interference maxima lying between the $\sin\theta = \lambda/a$ and $\sin\theta = 2\lambda/a$ are $\dfrac{10\lambda}{9a}, \dfrac{12\lambda}{9a}, \dfrac{14\lambda}{9a}, \dfrac{16\lambda}{9a}$. So, four fringes lie between the first and the second minima of the envelope.

22. *A parallel beam of monochromatic light is incident normally on N number of parallel and equidistant slits (diffraction grating). Show that the intensity of the rays diffracted at an angle θ in the Fraunhofer pattern is given by*

$$I = A_0^2 \frac{\sin^2\alpha}{\alpha^2}\frac{\sin^2 N\beta}{\sin^2\beta};$$

where $\alpha = \dfrac{\pi a}{\lambda}\sin\theta$, $\beta = \dfrac{\pi d}{\lambda}\sin\theta$,

a = width of each slit, d = distance between the centres of two consecutive slits, and $A_0^2 \sin^2\alpha/\alpha^2$ *represents the intensity due to diffraction by a single slit.*

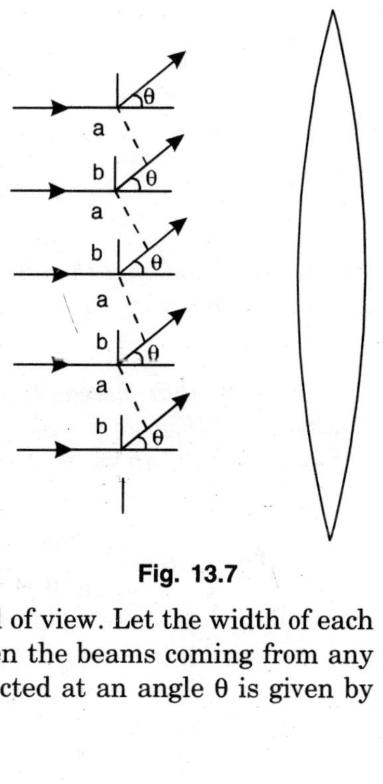

Fig. 13.7

Solution

A parallel beam of light rays is allowed to be incident normally on N number of parallel and equidistant slits (Fig. 13.7). When the diffracted beams are collected by the telescope we find a number of intensely bright lines in the field of view. Let the width of each opaque space be b, then $d = a + b$. The path difference between the beams coming from any two consecutive corresponding points in the N-slits and diffracted at an angle θ is given by

$$\Delta = d \sin\theta,$$

and the corresponding phase difference δ is

$$\delta = \frac{2\pi}{\lambda}\Delta = \frac{2\pi}{\lambda} d \sin\theta.$$

$$\text{Let } \beta = \frac{\delta}{2} = \frac{\pi d}{\lambda}\sin\theta = \frac{1}{2}kd\sin\theta.$$

The amplitudes contributed by the individual slits are all of equal magnitude. The phase will change by equal amounts δ from one slit to the next. Suppose that the magnitude of the amplitude of the contribution of each slit is B. Then, the resultant complex amplitude is the sum of the series

$$B[1 + e^{i\delta} + e^{2i\delta} + ... + e^{i(N-1)\delta}]$$

$$= B\frac{1-e^{iN\delta}}{1-e^{i\delta}}$$

$$= A\,e^{i\phi}, \text{ say.}$$

The intensity is found by multiplying this expression by its complex conjugate:

$$A^2 = B^2\frac{(1-e^{iN\delta})(1-e^{-iN\delta})}{(1-e^{i\delta})(1-e^{-i\delta})}$$

$$= B^2\,\frac{1-\cos N\delta}{1-\cos\delta} = B^2\frac{\sin^2 N\beta}{\sin^2\beta}.$$

The factor B^2 represents the intensity due to diffraction by a single slit [see Eqn. 13.2].

$$B^2 = A_0^2 \sin^2\alpha/\alpha^2;$$

$$\alpha = \frac{1}{2}\,ka\,\sin\theta = \frac{\pi a}{\lambda}\,\sin\theta.$$

Thus, the intensity on the Fraunhofer diffraction pattern of an array of N slits is given by

$$I \sim A^2 = A_0^2\frac{\sin^2\alpha}{\alpha^2}\frac{\sin^2 N\beta}{\sin^2\beta} \qquad ...(13.20)$$

23. *(a) Show that the mth principal maximum of the N-slit diffraction pattern of problem 22 occurs when*

$$d\ \sin\theta_m = \pm m\lambda,\ m = 0,\ 1,\ 2,...$$

(b) Show that there will be (N – 1) points of zero intensity between two adjacent principal maxima. (c) When do we get absent spectra? (d) What is the effect of the diffraction envelope $\sin^2\alpha/\alpha^2$ *on the intensity of the principal maxima?*

Solution

(*a*) The factor $\frac{\sin^2 N\beta}{\sin^2\beta}$ in Eqn. (13.20) may be said to represent the interference term for N slits. It possesses maximum value N^2 for $\beta = \pm\, m\pi$, $m = 0, 1, 2,...,$ because

$$\lim_{\beta\to\pm m\pi}\frac{\sin N\beta}{\sin\beta} = \lim_{\beta\to\pm m\pi}\frac{N\cos N\beta}{\cos\beta} = \pm N.$$

Thus, for the mth principal maximum we have

$$\beta = \frac{\pi d}{\lambda}\sin\theta_m = \pm m\pi$$

or

$$d\ \sin\theta_m = \pm m\lambda,\ m = 0,\ 1,\ 2,...$$

(*b*) The function $\frac{\sin^2 N\beta}{\sin^2\beta}$ vanishes when $\beta = \frac{p\pi}{N}$ excluding $p = mN$, $m = 0, 1, 2,...$ and $p = 1, 2, 3,....$

when $p = mN$, $\beta = m\pi$ and we have principal maxima. Hence, the conditions for interference minima are

$$d \sin \theta = \frac{\lambda}{N}, \frac{2\lambda}{N}, \ldots, \frac{(N-1)\lambda}{N}, \frac{(N+1)\lambda}{N}, \ldots$$

and the conditions for principal maxima are

$$d \sin \theta = 0, \lambda, 2\lambda, \ldots$$

Thus between two adjacent principal maxima there will be $(N - 1)$ points of zero intensity.

(*c*) It is sometimes seen that certain orders of N-slit diffraction spectra are absent even if they satisfy the condition of interference maxima. We know that the intensity of the diffraction pattern is governed by two factors, one due to diffraction by a single slit ($\sin^2 \alpha/\alpha^2$), and the other due to interference ($\sin^2 N\beta/\sin^2 \beta$). It may happen that for a particular N-slit spectra, the direction in which the interference factor gives maximum, the diffraction factor vanishes:

$$d \sin \theta = m\lambda,\ m = 1, 2, 3, \ldots \text{interference maxima}$$

$$a \sin \theta = p\lambda,\ p = 1, 2, 3, \ldots \text{diffraction minima.}$$

If these two conditions are simultaneously satisfied, mth order maxima will be missing. Thus, the condition for absent spectra is

$$\frac{d}{a} = \frac{m}{p} \quad \text{or} \quad \frac{a+b}{a} = \frac{m}{p}.$$

If $\frac{d}{a} = 2$, $m = 2p$ and $m = 2, 4, 6, \ldots$ order spectra are missing.

If $\frac{d}{a} = 3$, $m = 3p$ and $m = 3, 6, 9, \ldots$ order spectra are missing.

(*d*) The relative intensities of the different orders m are governed by the single-slit diffraction envelope $\sin^2\alpha/\alpha^2$. Due to this term the central maximum has the maximum intensity and the intensity decreases with the increase in the order m. For large m, the intensity will be very low.

24. *Show that half angular width ($\Delta\theta_m$) of the mth principal maximum of problem 23 is given by*

$$\Delta\theta_m = \frac{\lambda}{Nd \cos \theta_m}.$$

Solution

The direction of the mth principal maximum is given by

$$d \sin \theta_m = m\lambda.$$

Let $\theta_m + \Delta\theta_m$ and $\theta_m - \Delta\theta_m$ be the directions of the first minima on the two sides of the mth principal maximum (Fig. 13.8). Then,

$$d \sin (\theta_m \pm \Delta\theta_m) = m\lambda \pm \frac{\lambda}{N}.$$

Thus, we have

$$\frac{d \sin (\theta_m \pm \Delta\theta_m)}{d \sin \theta_m} = \frac{m\lambda \pm \frac{\lambda}{N}}{m\lambda}$$

Since $\Delta\theta_m$ is a very small angle, we may write, [$\cos\Delta\theta_m \cong 1$, $\sin\Delta\theta_m \cong \Delta\theta_m$],

$$\frac{\sin\theta_m \pm \cos\theta_m \Delta\theta_m}{\sin\theta_m} = 1 \pm \frac{1}{mN}$$

or
$$\Delta\theta_m = \frac{1}{mN\cot\theta_m} = \frac{\sin\theta_m}{mN\cos\theta_m} = \frac{m\lambda/d}{mN\cos\theta_m}$$

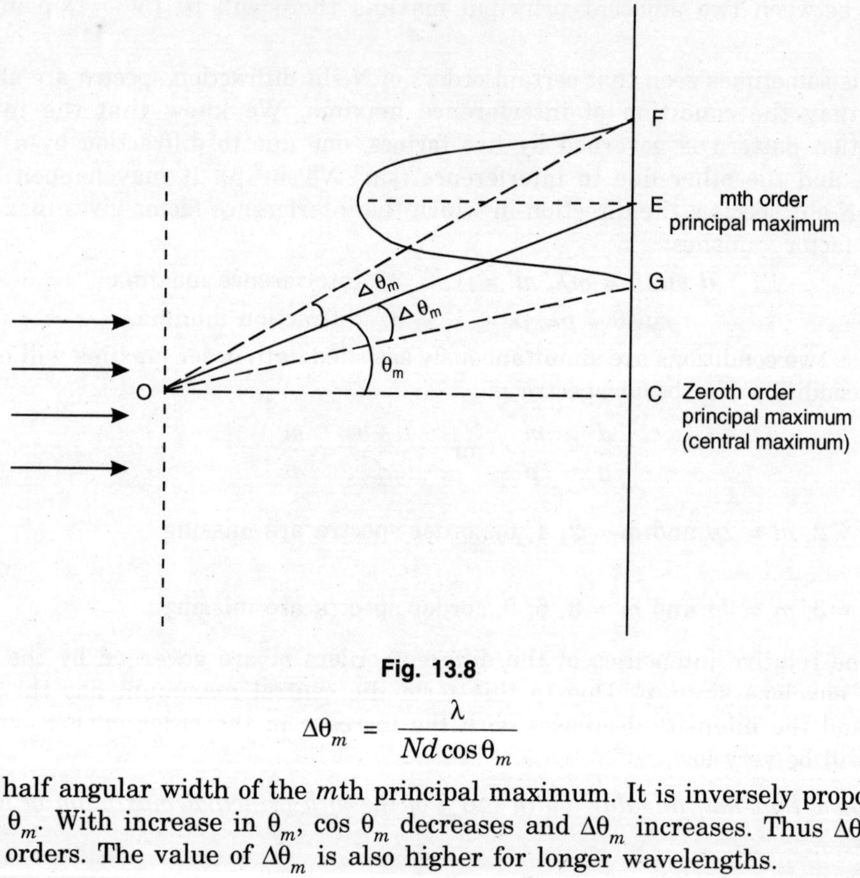

Fig. 13.8

$$\Delta\theta_m = \frac{\lambda}{Nd\cos\theta_m} \quad ...(13.21)$$

$\Delta\theta_m$ is the half angular width of the mth principal maximum. It is inversely proportional to N and $\cos\theta_m$. With increase in θ_m, $\cos\theta_m$ decreases and $\Delta\theta_m$ increases. Thus $\Delta\theta_m$ is more for higher orders. The value of $\Delta\theta_m$ is also higher for longer wavelengths.

25. *A parallel beam of monochromatic light is incident normally on a plane transmission grating having 2990 lines per cm and a third order spectral line is observed to be deviated through 30°. Calculate the wavelength of the spectral line.*

Solution

We have $d \sin\theta = m\lambda$.

Here, d = the grating element = $\frac{1}{2990}$ cm,

$\sin 30° = \frac{1}{2}$ and $m = 3$.

Thus, $$\lambda = \frac{d\sin\theta}{m} = \frac{1}{2990} \times \frac{1}{2} \times \frac{1}{3}\text{ cm} = 5.574 \times 10^{-5}\text{ cm}$$

$$= 557.4 \text{ nm.}$$

26. *A plane transmission grating having 5000 lines/cm is used to obtain a spectrum of light from a sodium lamp in the second order. Calculate the angular separation between the D_1 and D_2 lines of sodium whose wavelengths are 5890 Å and 5896 Å.*

Solution

We have $$d \sin \theta_1 = 2\lambda_1$$

and $$d \sin \theta_2 = 2\lambda_2$$

with $$d = \frac{1}{5000} \text{ cm}.$$

Thus,
$$\sin \theta_1 = 2 \times 5.890 \times 10^{-5} \times 5000 = 0.5890$$
$$\sin \theta_2 = 2 \times 5.896 \times 10^{-5} \times 5000 = 0.5896$$
$$\theta_1 = 36.086° \text{ and } \theta_2 = 36.129°.$$

Angular separation $= \theta_2 - \theta_1 = 0.043° = 2.58$ minutes of an arc.

27. *A grating has 315 rulings/mm. For what wavelengths in the visible spectrum (400–700 nm) can fifth-order diffraction be observed?*

Solution

The grating element $= d = \dfrac{1}{315}\text{ mm} = \dfrac{10^{-3}}{315}\text{ m}$

We have $$d \sin \theta = 5\lambda.$$

Thus, $$\lambda = \frac{10^{-3}}{5 \times 315} \times \sin \theta = 634.92 \sin \theta \text{ nm}$$

Maximum value of θ is 90°. Thus the required wavelengths in the visible spectrum are all wavelength between 400 nm and 634.92 nm.

28. *Two spectral lines have wavelengths λ and $\lambda + \Delta\lambda$, respectively, where $\Delta\lambda << \lambda$. Show that their angular separation $\Delta\theta$ in a grating spectrometer is given approximately by*

$$\Delta\theta = \frac{\Delta\lambda}{\left[\left(\frac{d}{m}\right)^2 - \lambda^2\right]^{1/2}}$$

where d is the separation of adjacent slit centres and m is the order at which the lines are observed.

Solution

We have $$d \sin \theta = m\lambda$$

and $$d \sin (\theta + \Delta\theta) = m(\lambda + \Delta\lambda)$$

or $$d (\sin \theta + \cos \theta\, \Delta\theta) = m\lambda + m\Delta\lambda.$$

Thus, we get $$d \cos \theta\, \Delta\theta = m\Delta\lambda$$

or $$\Delta\theta = \frac{m\Delta\lambda}{d[1 - m^2\lambda^2/d^2]^{1/2}} = \frac{\Delta\lambda}{\left[\left(\frac{d}{m}\right)^2 - \lambda^2\right]^{1/2}}.$$

29. *A parallel beam of monochromatic light of wavelength 500 nm is incident normally on a plane transmission grating having 6000 lines per cm, Find the highest order of the grating spectrum.*

Solution
$$d = \frac{1}{6000} \text{ cm and } d \sin\theta = m\lambda$$

For the highest order $\theta = 90°$ and

$$m = \frac{d}{\lambda} = \frac{1}{6000 \times 5 \times 10^{-5}} = 3.33$$

Thus the highest order is $m = 3$.

30. *Light of wavelength 5890 Å and 5896 Å falls on a grating having 5000 lines/cm. If a lens of focal length 100 cm is used to form spectra on a screen, find the distance between the lines in the third order.*

Solution

$$d = \frac{1}{5000} \text{ cm.}$$

$$\sin\theta = \frac{3\lambda}{d} \text{ for the third order.}$$

When $\lambda = 5890$ Å, $\sin\theta = 3 \times 5.890 \times 10^{-5} \times 5000 = 0.8835$

and $\theta = 62.0675°$

When $\lambda = 5896$ Å, $\sin\theta' = 3 \times 5.896 \times 10^{-5} \times 5000 = 0.8844$

and $\theta' = 62.1778°$

Therefore the linear distance on the screen is

$$\Delta y = 100 \times (\tan\theta' - \tan\theta) = 0.88 \text{ cm.}$$

31. *What is the maximum width, a, of any clear space in a grating having 10^4 lines per inch, if complete second order visible spectrum is to be formed?*

Solution

$$d = \frac{2.54}{10^4} = 2.54 \times 10^{-4} \text{ cm.}$$

The maximum angle θ in the second order visible spectrum ($\lambda = 700$ nm) is given by

$$\sin\theta = \frac{2 \times 7.0 \times 10^{-5}}{2.54 \times 10^{-4}} = 0.5512.$$

The half angular breadth of the central diffraction band must accommodate the complete second order spectrum, *i.e.*,

$$\sin\theta = \frac{\lambda}{a} = \frac{7.0 \times 10^{-5}}{a} = 0.5512$$

or
$$a = \frac{7 \times 10^{-5}}{0.5512} \text{ cm} = 1.27 \times 10^{-4} \text{ cm.}$$

Note that this is just half the grating element d.

32. *Show that the dispersion D and the resolving power R of a grating are given respectively*

$$D = \frac{d\theta}{d\lambda} = \frac{m}{d\cos\theta},$$

and $$R = \frac{\lambda}{\Delta\lambda} = Nm,$$

where d is grating element, m is the order number and N is the total number of rulings on the grating.

Solution

If the incident rays of wavelengths λ and $\lambda + d\lambda$ are diffracted in the direction θ and $\theta + d\theta$ respectively for the mth order of the bright fringe, the dispersive power of the grating is given by $\frac{d\theta}{d\lambda}$ *i.e.*, by the rate of variation of the angle of diffraction with the wavelength.

Differentiating the grating equation

$$d \sin \theta = m\lambda$$

with respect to λ, we get

$$d \cos \theta \frac{d\theta}{d\lambda} = m.$$

Thus, the dispersive power D of the grating is

$$D = \frac{d\theta}{d\lambda} = \frac{m}{d\cos\theta}. \quad ...(13.22)$$

A spectrum for which $\frac{d\theta}{d\lambda}$ is constant is known as the rational or normal spectrum.

Let us consider two spectral lines of wavelengths λ and $\lambda + \Delta\lambda$. Owing to dispersion, the angular separation between the lines in the mth order is given by

$$\Delta\theta = \frac{m}{d\cos\theta}\Delta\lambda.$$

Again, half angular width of the mth principal maximum is (see problem 24)

$$\Delta\theta_m = \frac{\lambda}{Nd\cos\theta}.$$

According to Rayleigh's criterion, two lines are just resolved when the principal maximum of one falls on the first minimum of the other *i.e.*, when the angle between the centres of two images is half the angular width of the principal maximum, *i.e.*, $\Delta\theta = \Delta\theta_m$.

Thus, we have $$\frac{m}{d\cos\theta}\Delta\lambda = \frac{\lambda}{Nd\cos\theta}$$

or $$R_m = \frac{\lambda}{\Delta\lambda} = mN. \quad ...(13.23)$$

The quantity R_m is called the resolving power in the mth order, $\Delta\lambda$ is the smallest difference of wavelength for which the spectral lines $\lambda - \frac{1}{2}\Delta\lambda$ and $\lambda + \frac{1}{2}\Delta\lambda$ are resolveable. Note that the resolving power is the same for all wavelengths.

33. *A plane transmission grating has 5000 lines per cm. (a) Find the dispersive power of the grating for the second order at the wavelength 5893 Å. (b) Calculate the angular separation between the D_1 and D_2 lines of sodium whose wavelengths are 5890 Å and 5896 Å. [see problem 26].*

Solution

(*a*) $$d = \frac{1}{5000}\ \text{cm} = \frac{1}{5000} \times 10^{-2}\ \text{m}.$$

$$\sin\theta = \frac{2\lambda}{d} = 0.5893 \text{ and } \cos\theta = 0.8079$$

Dispersive power, $D = \dfrac{m}{d\cos\theta} = 1.238 \times 10^6$ rad m^{-1}.

(*b*) $$\Delta\theta = \Delta\lambda D = 6 \times 10^{-10} \times 1.238 \times 10^6 = 7.428 \times 10^{-4}\ \text{rad}$$
$$= 0.043°.$$

34. *What should be the minimum number of lines in a grating that will just resolve in the second order the D_1 and D_2 lines of sodium?*

Solution

Resolving power, $$R = \frac{\lambda}{\Delta\lambda} = mN.$$

Thus, $$N = \frac{5893}{6} \times \frac{1}{2} = 491.08.$$

The grating must have at least 492 lines.

35. *A section through the NaCl cell lattice is shown in Fig. 13.9. The unit cells, each of dimension a_0, in this section are represented by small dots. The dashed sloping lines represent an arbitrary family of planes with characteristic interplanar spacing d. (a) Show that $d = a_0/\sqrt{5}$ (b) Derive Bragg's law: $2d \sin\theta = m\lambda$, $m = 1, 2, 3,...$*

Solution

(*a*) From Fig. 13.9, we have

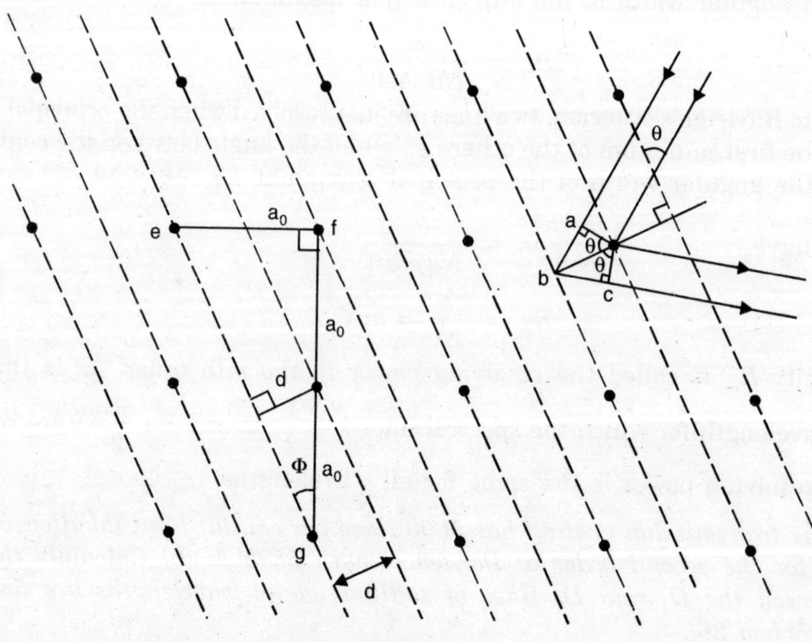

Fig. 13.9

$$\sin \phi = \frac{ef}{ge} = \frac{a_0}{\sqrt{5}a_0} = \frac{d}{a_0}.$$

Thus $$d = \frac{a_0}{\sqrt{5}} \quad ...(13.24)$$

(*b*) To have a constructive interference in the beam diffracted in the direction θ (Fig. 13.9) from the entire family of planes, the rays from the separate planes must reinforce each other. Thus the path difference for rays from adjacent planes (*abc* in Fig. 13.9) must be an integral number of wavelengths,

or $$2d \sin \theta = m\lambda, \; m = 1, 2, 3,... \quad ...(13.25)$$

This is known as Bragg's law. It should be noted that the angle θ is measured from the surface of the crystal, not from the normal to that surface.

36. *At what angles must an x-ray beam with λ = 1 Å fall on the family of planes represented in Fig. 13.9 if a diffracted beam is to exist? Assume the material to be sodium chloride (a_0 = 5.63 Å).*

Solution

The interplanar spacing d is given by

$$d = \frac{a_0}{\sqrt{5}} = \frac{5.63 \text{ Å}}{\sqrt{5}} = 2.5178 \text{ Å}.$$

Bragg's law gives

$$\sin \theta = \frac{m\lambda}{2d} = \frac{m \times 1}{2 \times 2.5178} = 0.1986 \times m.$$

When m = 1, θ = 11.46°; m = 2, θ = 23.40°; m = 3, θ = 36.57°; m = 4, θ = 52.60°; m = 5, θ = 83.22°. Higher order beams cannot exist because they require that sinθ > 1. Actually the unit cell in cubic crystals such as NaCl has diffraction properties such that the intensity of diffracted x-ray beams corresponding to odd values of m is zero. Thus the angles are θ = 23.40° (m = 2) and θ = 52.60° (m = 4).

37. *Monochromatic x-rays are incident on a set of crystal planes whose interplanar spacing is 40 pm. When the beam is rotated 60° from the normal, first-order Bragg reflection is observed. What is the wavelength of the x-rays?*

Solution

Here θ = 30°, hence

$$\lambda = 2d \sin 30° = d = 40 \text{ pm}.$$

38. *In comparing the wavelengths of two monochromatic x-ray lines, it is noted that line A gives a first-order reflection maximum at a glancing angle of 24° to the smooth face of a crystal. Line B, known to have a wavelength of 96 pm, gives a third-order reflection maximum at an angle of 60° from the same face of the same crystal. (a) Calculate the interplanar spacing. (b) Find the wavelength of line A.*

Solution

We have $$2d \sin 24° = \lambda_A$$

and $$2d \sin 60° = 3 \times 96 \text{ pm},$$

which give (*a*) d = 166.28 pm (*b*) λ_A = 135.26 pm.

39. *Yellow light is used in a single slit diffraction experiment with slit width of 0.6 mm. If yellow light is replaced by x-rays, the observed pattern will reveal*

(a) that the central maximum is narrower

(b) more number of fringes

(c) less number of fringes

(d) no diffraction pattern. (I.I.T. 1999)

Solution

Slit width = a = 0.6 mm = 0.6×10^{-3} m

Wavelength of x-rays = $\lambda \approx 1$ Å = 10^{-10} m

$$\frac{a}{\lambda} = 0.6 \times 10^{7} >> 1$$

a is very large compared to the wavelength λ.

In this case, the diffraction pattern disappears.

Correct Choice: *d.*

40. *A parallel monochromatic beam of light is incident normally on a narrow slit. A diffraction pattern is formed on a screen placed perpendicular to the direction of the incident beam. At the first minimum of the diffraction pattern the phase difference between the rays coming from the two edges of the slit is*

(a) 0 (b) $\pi/2$ (c) π (d) 2π. (I.I.T. 1998)

Solution

Path difference for the rays coming from the two edges of the slit is

$$\Delta = a \sin\theta, \; a = \text{slit width.}$$

For the first minimum, $\alpha = \pi$

where

$$\alpha = \frac{\pi a}{\lambda} \sin\theta = \pi$$

or

$$a \sin\theta = \lambda$$

$$\text{Phase difference} = \frac{2\pi}{\lambda}\Delta = 2\pi$$

Correct Choice: *d.*

SUPPLEMENTARY PROBLEMS

1. A slits of width a is illuminated by white light. For what value of a will the first minimum for red light (λ = 650 nm) fall at $\theta = 10°$?
2. In problem 1 what is the wavelength of the light whose first secondary diffraction maximum falls at 10°?
3. Assuming that the secondary maxima lie approximately half way between the minima, calculate the intensities of the first three secondary maxima in the single-slit diffraction pattern, measured relative to the intensity of the central maximum.

4. A plane wave, wavelength 600 nm falls on a slit with a = 0.4 mm. A thin converging lens, focal length = 75 cm, is placed behind the slit and focuses the light on a screen. (*a*) How far is the screen behind the lens? (*b*) What is the linear distance on the screen from the centre of the pattern to the first minimum?
5. A convex lens of focal length 40 cm is placed after a slit of width 0.4 mm. If a plane wave of wavelength 5000 Å falls normally on the slit, calculate the separation between the second minima on either side of the central maximum.
6. Find the smallest angular separation of two stars which could be theoretically resolved by a telescope of diameter 200". [Mean λ = 600 nm]
7. Calculate the least value of the angular separation of two stars which can be resolved by a telescope of 200 cm aperture. [Mean λ = 550 nm]
8. Two stars at a distance of 9 light years are viewed through a telescope having a lens of 20 cm diameter. What is the minimum separation of these stars for which they would still be distinguishable as separate objects? [Mean λ = 600 nm]
9. Calculate the aperture of the objective of a telescope which may be used to resolve stars separated by 6×10^{-6} radian for light of wavelength 500 nm.
10. An astronaut in a satellite claims he can just barely resolve two point sources on the earth, 160 km below him. Calculate their (*a*) angular and (*b*) linear separation, assuming ideal conditions. Take λ = 500 nm, and the pupil diameter of the astronaut's eye to be 5 mm.
11. The wall of a large room is covered with acoustic tile in which small holes are drilled 5 mm from cèntre to centre. How far can a person be from such a tile and still distinguish the individual holes, assuming ideal conditions? Assume the diameter of the pupil of the observer's eye to be 4 mm and the wavelength to be 540 nm.
12. Under ideal conditions, estimate the linear separation of two objects on the planet Mars that can be resolved by an observer on earth using the 200 inch Mount Palomar telescope. Use the following data: distance from the earth to Mars = 8.0×10^7 km and wavelength of light = 550 nm.
13. A navy cruiser employs radar with a wavelength of 1.6 cm. The circular antenna has a diameter of 2.3 m. At a range of 6.2 km, what is the smallest distance that two speed boats can be from each other and still be resolved as two separate objects by the radar system?
14. A plane wave is incident on a convex lens of focal length 100 cm. If the diameter of the lens is 5 cm, calculate the radius of the first Airy disc. Assume λ = 500 nm.
15. By putting b = 0 (no opaque space) or $d = a$ in the expression of the double slit diffraction equation [Eqn. (13.13)], derive the single-slit diffraction equation with slit width = $2a$.
16. (*a*) Design a double-slit system in which the fourth fringe, not counting the central maximum, is missing.

    (*b*) What other fringes, if any, are also missing?
17. (*a*) What requirements must be met for the central maximum of the envelope of the double-slit interference pattern to contain exactly 11 fringes?

    (*b*) How many fringes lie between the first and second minima of the envelope?
18. (*a*) How many complete fringes appear between the first minima of the fringe envelope to either side of the central maximum for a double-slit pattern if λ = 550 nm, d = 0.15 mm, and a = 0.30 mm?

(b) What is the ratio of the intensity of the third fringe to the side of the centre to that of the central fringe?

19. Show that for a three-slit grating the diffracted intensity at an angle θ is given by

$$I = I_m(1 + 4 \cos i + 4 \cos^2 i)/9,$$

where
$$i = \frac{2\pi}{\lambda} d \sin \theta$$

[Assume $\sin^2 \alpha/\alpha^2 \approx 1$].

20. By putting $N = 1$ and $N = 2$ in Eqn. (13.20) obtain single slit and double slit diffraction patterns.

21. Show that between two adjacent principal maxima of the N-slit diffraction pattern there are $(N - 1)$ secondary minima and $(N - 2)$ secondary maxima.

22. A diffraction grating has 1.25×10^4 rulings uniformly spaced over 25 mm. It is illuminated at normal incidence by yellow light from a sodium vapour lamp. This light contains two closely spaced lines of wavelengths 589 nm and 589.59 nm. (a) At what angles will the first order maxima occur for these wavelengths? (b) What is the angular separation between these two lines in first order?

23. A diffraction grating 20 mm wide has 6000 rulings. At what angles will maximum intensity beams occur if the incident radiation has a wavelength of 590 nm?

24. A diffraction grating has 200 rulings/mm, and a strong diffracted beam is noted at $\theta = 30°$. What are the possible wavelengths of the incident light in the visible region of the spectrum?

25. A diffraction grating 3 cm wide produces a deviation of 30° in the second order with light of wavelength 600 nm. What is the total number of lines on the grating?

26. Light of wavelength 500 nm is incident normally on a diffraction grating. Two adjacent principal maxima occur at $\sin\theta = 0.2$ and $\sin\theta = 0.3$ respectively. The fourth order is missing. (a) What is the separation between adjacent slits? (b) What is the smallest possible individual slit width? (c) Name all orders actually appearing on the screen with the values derived in (a) and (b).

27. Assume that the limits of the visible spectrum are chosen as 400 and 700 nm. Calculate the number of rulings per mm of a grating that will spread the first-order spectrum through an angular range of 20°.

28. White light (400 nm $< \lambda <$ 700 nm) is incident on a grating. Show that, no matter what the value of the grating spacing d, the second and third-order spectra overlap.

29. A wire grating is made of 200 wires per cm placed at equal distances apart. The diameter of each wire is 0.025 mm. Calculate the angle of diffraction for the third order spectrum for light of wavelength 600 nm and also find the order of absent spectra, if any.

30. Assume that light is incident on a grating at an angle of incidence i. Show that the condition for a diffraction maximum is

$$d (\sin i + \sin \theta) = m\lambda, \; m = 0, 1, 2,...$$

where i and θ are on the same side of the normal.

31. A diffraction grating used at normal incidence gives a green line (λ = 560 nm) in a certain order superimposed on the violet line (λ = 420 nm) of the next higher order. If the angle of diffraction is 30°, how many lines are there to the centimetre in the grating.

32. Show that the dispersive power of a grating is given by

$$(a)\ \frac{\Delta\theta}{\Delta\lambda} = \frac{\tan\theta}{\lambda} \qquad (b)\ \frac{\Delta\theta}{\Delta\lambda} = mn \sec\theta$$

where n = number of rulings per cm.

33. What should be the minimum number of lines in a grating which will just resolve in the first order the lines whose wavelengths are 5890 Å and 5896 Å?

34. Calculate the least width that a grating must have to resolve two components of sodium D lines in the first order, the grating having 600 lines per cm.

35. Light containing a mixture of two wavelength 540 nm and 600 nm, is incident normally on a plane transmission grating. It is desired (*i*) that the second principal maximum for each wavelength appear at $\theta \leq 30°$, (*ii*) that the dispersion be as high as possible, and (*iii*) that the third order for 600 nm be a missing order. (*a*) What should be the separation between adjacent slits? (*b*) What is the smallest possible individual slit width? (*c*) Name all orders for 600 nm that actually appear on the screen.

36. In the second order spectrum of a grating a spectral line appears at 10°; another of wavelength 4×10^{-9} cm greater appears at 3" farther. Find the wavelengths of the lines and the minimum grating width required to resolve them in the second order.

37. A diffraction grating has a resolving power $R = \lambda/\Delta\lambda = mN$. (*a*) Show that the corresponding frequency range $\Delta\nu$ that can just be resolved is given by $\Delta\nu = c/(mN\lambda)$. (*b*) From Fig. 13.7 show that the "times of flight" of the two extreme rays differ by an amount $\Delta t = (Nd/c)\sin\theta$. (*c*) Show that $\Delta\nu\ \Delta t = 1$ [Assume $N >> 1$].

38. The x-ray wavelength 0.11 nm is found to reflect in the second order from the face of lithium fluoride crystal at a Bragg angle of 27.8°. Find the distance between adjacent crystal planes.

39. A beam of x-rays of wavelength 30 pm is incident on a calcite crystal of lattice spacing 0.32 nm. What is the smallest angle between the crystal planes and x-ray beam that will result in constructive reflection of the x-rays?

40. Monochromatic high energy x-rays are incident on a crystal. If first-order reflection is observed at Bragg angle 3.5°, at what angle would second-order reflection be expected?

41. Prove that it is not possible to determine both wavelength of radiation and spacing of Bragg reflecting planes in a crystal by measuring the angles for Bragg reflection in several orders.

42. Consider an infinite two-dimensional square lattice as in Fig. 13.9. One interplanar spacing is obviously a_0 itself. (*a*) Calculate the next five smaller interplanar spacings by sketching figures similar to Fig. 13.9. (*b*) Show that your answer obeys the general formula

$$d = \frac{a_0}{\sqrt{h^2 + k^2}}$$

where h and k are both relatively prime integers that have no common factor other than unity.

43. Angular width of central maximum in the Fraunhofer diffraction pattern of a slit is measured. The slit is illuminated by light of wavelength 6000 Å. When the slit is illuminated by light of another wavelength, the angular-width decreases by 30%. Calculate the wavelength of this light. The same decrease in the angular-width of the central maximum is obtained when the original apparatus is immersed in a liquid. Find refractive index of the liquid. *(I.I.T. 1996)*

$$\left[\textbf{Hints: } \theta = \frac{2\lambda}{a} \text{ and } 0.7\theta = \frac{2\lambda'}{a} = \frac{2\lambda}{na}\right]$$

44. A slit of width d is placed in front of a lens of focal length 0.5 m and is illuminated normally with light of wavelength 5.89×10^{-7} m. The first diffraction minima on either side of the central diffraction maximum are separated by 2×10^{-3} m. The width d of the slit is ....m.
(I.I.T. 1997)

[**Hints:** $y = \lambda f/d$, $2y = 2 \times 10^{-3}$ m]

45. A beam of light of wavelength 600 nm from a distant source falls on a single slit 1.00 mm wide and the resulting diffraction pattern is observed on a screen 2 m away. The distance between the first dark fringes on either side of the central bright fringe is

(*a*) 1.2 cm (*b*) 1.2 mm (*c*) 2.4 cm (*d*) 2.4 mm *(I.I.T. 1994)*

[**Hints:** $\frac{d}{D} = \frac{\lambda}{a}$; find the value of $2d$]

Answers to Supplementary Problems

CHAPTER 1

1. 5 cm, π s, 20 cm/s^2

3. 2.4 π m/s, 0.144 J

5. (*a*) $\dfrac{2\sqrt{2}}{3\pi}A\omega$ (*b*) $\dfrac{2\left(4-\sqrt{2}\right)}{3\pi}A\omega$

6. $\dfrac{1}{3}$

8. (*a*) $\dfrac{\pi}{2}$ s (*b*) 1 m

9. $\nu > 503.29$ Hz

10. 3.0 s

11. (*a*) 24.8 cm (*b*) 2.49 Hz

12. 7.54 m/s

13. (*a*) 3.5 N/m (*b*) 0.67 s

14. (*d*)

15. (*a*) 0.66 s (*b*) 0.315 J (*c*) 0.07875 J, 0.23625 J

16. 0.8

17. 707.9 N/m

18. (*a*) 0.02 m (*b*) 0.02 m

19. 5 N

20. (*a*) 56 cm/s (*b*) 0.9 s

21. 0.02 J

22. 15.3 m

23. (*a*) $mv/(m + M)$ (*b*) $mv\big/\sqrt{k(M+m)}$

24. (*a*) 0.79 N/m (*b*) 5 cm (*c*) 5.03 cm/s, 4.74 cm/s^2

26. (*a*) : (1) $\dfrac{10}{3}$ kN/m (2) 1.088 s (3) 0.058 m/s (4) 0.33 m/s^2

(*b*) : (1) 15 kN/m (2) 0.513 s (3) 0.122 m/s (4) 1.50 m/s^2

27. 5.63×10^{-3} Hz

28. $3k$

30. $\pi\sqrt{2}/20\,\text{s}$

31. $k_2L/(k_1 + k_2)$, $k_1L/(k_1 + k_2)$, $T = 2\pi\sqrt{m/(k_1+k_2)}$

32. $T = 2\pi\sqrt{2m/(9k)} = 0.094\,\text{s}$

33. $k_1 = (n + 1)k/n$; $k_2 = (n + 1)\,k$

35. $A/\sqrt{3}$

36. 0.127 J, 1.59 m/s

37. 0.99 m

38. $T_e/T_m = \sqrt{g_m/g_e} = 0.408$

40. (*b*) and (*c*)

41. $2\pi\left[\dfrac{l}{\sqrt{v^4/R^2+g^2}}\right]^{1/2}$

42. $2\pi[l\rho/[g(\rho - \sigma)]]^{1/2}$

44. $2\sqrt{2}$ s

47. $\frac{2\pi}{g}\sqrt{W/(A\rho)}$

48. $2\pi\sqrt{M/(2\,mg)}$; 2 mg

50. 0.63 s

51. $2\pi\left[\frac{l}{g - Eq/m}\right]^{1/2}$

52. $8\pi^2 \times 10^{-3}$ N/m; 0.09 cm

53. $\sqrt{51}/10$

54. $y = 3 \sin (30\,\pi t)$ cm

55. (*a*) $x = x_0 \cos \omega t + \frac{v_0}{\omega} \sin \omega t$; $A = \left(x_0^2 + \frac{v_0^2}{\omega^2}\right)^{1/2}$; $v_{max} = \left(v_0^2 + x_0^2\omega^2\right)^{1/2}$

(*b*) $x = x_0 \cos \omega t - \frac{v_0}{\omega} \sin \omega t$;

The amplitude and maximum speed are the same as in part (*a*).

56. (*a*) $x = 2 (\cos 2t - \sin 2t)$ m; (*b*) $2\sqrt{2}$ m; π s (*c*) 0

57. $x = 0.1 \cos (4t + \pi/4)$ m

59. One equilibrium point at $x = 2/a$; stable equilibrium; $T = 2\pi e/\left(a\sqrt{m}\right)$.

60. 0.0792 kg. m^2

61. 7.77 s

62. $1/\sqrt{3\pi}$

63. $T = 2\pi\sqrt{d/g} = 0.2$ s

64. 4.9 pF to 42 pF.

65. 6×10^{-2} m.

66. (*b*)

CHAPTER 2

1. $\sqrt{a^2 + b^2}$

2. (*a*) $x^2/A^2 + y^2/B^2 = 1$, clockwise, (*b*) $\vec{a} = -\omega^2 \vec{r}$

3. (*a*) $y^2 = 4x^2 (1 - x^2/a^2)$, (*b*) $y = a(1 - 2x^2/a^2)$

4. $\frac{4y^2}{b^2}\left(\frac{y^2}{b^2} + \frac{x}{a} \sin \delta - 1\right) + \left(\frac{x}{a} - \sin \delta\right)^2 = 0$

6. $x = a \cos \omega t$, $y = \frac{v_0}{2\omega} \sin 2\omega t$, where $\omega = \sqrt{k/m}$.

The path is a Lissajous figure having the shape of "figure eight" as shown in Fig. 2.10.

7. 255.9 Hz.

8. 512.2 Hz; 512.1 Hz or 511.9 Hz

9. (256.1 Hz, 255.8 Hz) or, (255.9 Hz, 255.8 Hz)

12. $x_1 = -\frac{m_2 a(1 - \cos \omega t)}{4(m_1 + m_2)}$, $x_2 = \frac{a(4m_1 + 3m_2 - m_1 \cos \omega t)}{4(m_1 + m_2)}$ with $\omega^2 = k\left(\frac{1}{m_1} + \frac{1}{m_2}\right)$

13. $x_1 = \frac{3}{5}\cos t - \frac{8}{5}\cos\sqrt{6}\,t,\ x_2 = \frac{6}{5}\cos t + \frac{4}{5}\cos\sqrt{6}\,t$

14. $x_1 = b + \frac{a}{2}(\cos\omega_2 t - \cos\omega_1 t),\ x_2 = b - \frac{a}{2}(\cos\omega_2 t - \cos\omega_1 t)$

with $b = l - mg/2k,\ \omega_1^2 = 2\ k/m,\ \omega_2^2 = 6\ k/m$

15. $\omega_1 = \sqrt{k_1/m}$

$\omega_2 = [(k_1 + 2k_2)/m]^{1/2}$

For mode 1, $x_1 = x_2$ and for mode 2, $x_1 = -x_2$,

The general solution is

$x_1 = A\cos\omega_1 t + A'\sin\omega_1 t + B\cos\omega_2 t + B'\sin\omega_2 t$

$x_2 = A\cos\omega_1 t + A'\sin\omega_1 t - B\cos\omega_2 t - B'\sin\omega_2 t$

16. (*i*) $m\ddot{x}_1 = k(x_2 - x_1)$

$M\ddot{x}_2 = k(x_2 - x_1) + k(x_3 - x_2)$

$m\ddot{x}_3 = -k(x_3 - x_2)$

(*iii*) $\omega = 0$ corresponds to pure translation of the system:

$x = x_2 = x_3$.

$\omega = \omega_2$ gives $x_2 = 0$ and $x_1 = -x_3$.

$\omega = \omega_3$ gives $x_1 = x_3$ and $x_2 = -2x_1\frac{m}{M} = -2x_3\frac{m}{M}$.

(*iv*) 1.915

17. $\omega = \pi, 2\pi$

$x = iA\exp(i\pi t) - iB\exp(i\,2\pi t)$

$y = A\exp(i\pi t) + B\exp(i\,2\pi t)$

18. 385 Hz

19. 501, 503, 508 Hz or, 505, 507, 508 Hz.

20. (*a*) 10 (*b*) 4

21. 0.0201.

CHAPTER 3

1. (*a*) $\beta > 4\sqrt{2}$, (*b*) $\beta < 4\sqrt{2}$, (*c*) $\beta = 4\sqrt{2}$.

2. (*a*) $x = e^{-t}\left(2\cos 2t + \frac{1}{2}\sin 2t\right)$ (*b*) damped oscillatory motion.

3. $2m/\beta$

4. 0.5 Hz, 2 s, 0.693

5. $\ddot{x} + 0.693\dot{x} + 158.03x = 0$; 0.173

6. (*a*) $\ddot{x} + 10\dot{x} + 25x = 0$; $x(0) = 1$ m and $\dot{x}(0) = 0$

(*b*) critically damped (*c*) $x = e^{-5t}(1 + 5t)$

7. $3\pi/2$

8. (*a*) a_0, $a_0\omega$; (*b*) $t_n = \frac{1}{\omega}\left[\tan^{-1}\left(\frac{\omega}{b}\right) + n\pi\right]$, $n = 0, 1, 2, \ldots$

10. 11

12. 0.435 s.

13. 0.000019

14. 0.18 s; – 0.62 m

15. (*a*) 4.95 s, (*b*) 2.48 s, (*c*) 0.385

16. 0.067 Ω

17. 5.516 μC, 1.928 μC

18. (*a*) $\sqrt{\dfrac{C}{I}} > \dfrac{\beta}{2I}$, (*b*) 0.0248

19. 4532.4.

CHAPTER 4

6. (*a*) 3.03×10^{-4} cm (*b*) zero. $Q = 628.3$

7. 500, π rad/s

9. (*a*) $x = e^{-2t}(2\cos 2t + \sin 2t) + \sin 2t - 2\cos 2t$

(*b*) Amplitude = $\sqrt{5}$, period = π, frequency = 1/π.

10. Acceleration amplitude = $\dfrac{f}{\left[\left(\dfrac{\omega^2 - p^2}{p^2}\right)^2 + \dfrac{4b^2}{p^2}\right]^{1/2}}$; $p = \dfrac{\omega^2}{(\omega^2 - 2b^2)^{1/2}}$

11. 318.3 Hz.

12. 200 mA, – 29.4°, 85.69 Hz

13. $\omega_0 = \dfrac{1}{\sqrt{LC}}; \dfrac{E_0}{R}$; $\omega_1 = \dfrac{-\sqrt{3}\,RC + \sqrt{3R^2C^2 + 4LC}}{2LC}$; $\omega_2 = \dfrac{\sqrt{3}\,RC + \sqrt{3R^2C^2 + 4LC}}{2LC}$;

$$\frac{\Delta\omega}{\omega_0} = R\sqrt{\frac{3C}{L}}$$

CHAPTER 5

1. π m; $\dfrac{2}{\pi}$ Hz; 2 m s^{-1}; negative *x*-direction

3. (*a*) 1 m (*b*) 0.4 πm^{-1} (*c*) 5 m (*d*) 0.2 s (*e*) 25 m s^{-1} (*f*) 5 Hz

4. (*i*) $\vec{k} = 0.2\,\hat{i} - 0.3\,\hat{j} + 0.4\,\hat{k}$ (*ii*) 0.928 units

5. $\omega/\sqrt{k_1^2 + k_2^2 + k_3^2}$

6. 20 Hz, 1100 m/s

7. (*a*), (*b*), (*c*), (*d*)

8. (*b*)

9. 42.6 m s^{-1}

10. 70.08 m s^{-1}

11. 9.9 m s^{-1}

12. 187.5 N

13. (*a*) 15 m s^{-1} (*b*) 3.6 N

14. $y = 1.2 \times 10^{-4} \sin 2\pi\left(\dfrac{50}{3}x + 100\,t\right) m$

15. (*a*) $\dfrac{2}{\pi}$ Hz (*b*) 0.2π m (*c*) 0.4 m s^{-1} (*d*) 0.064 N (*e*) $y = 0.05 \sin(10x - 4t)$ m

16. 353.5 m s^{-1}

19. (*a*) 346.96 m/s (*b*) 11.57 m, 23.13×10^{-3} m

20. 2.4×10^9 N/m^2

21. 293 m s^{-1}

22. 407.4 m/s

23. (*a*) 2×10^9 N/m^2 (*b*) 1.41 km/s

24. 9.8×10^{-6} m

25. 2.13 m W/m^2

26. (*b*)

27. (*c*)

28. (*a, c, d*)

29. (*b*)

30. (*d*)

31. π, 2.5×10^{-5} m

32. (*d*)

33. (*a*)

34. (*d*)

CHAPTER 6

3. 1.41 A

4. $\lambda = 2\sqrt{d^2 + 4(H+h)^2} - 2\sqrt{d^2 + 4H^2}$

5. 19736.8 Hz

6. (*b*) 254.4 Hz

7. $y_1 = 2\sin\frac{2\pi}{20}(x - 1000t); y_2 = 2\sin\frac{2\pi}{20}(x + 1000t)$

8. Water filled to a height of $\frac{7}{8}, \frac{5}{8}, \frac{3}{8}, \frac{1}{8}$ meter

9. 48 cm

10. $\frac{xl_2}{l_2 - l_1}, \frac{xl_1}{l_2 - l_1}, \frac{4xl_1l_2}{l_2 - l_1}$

11. 2.6 cm; 162.47 Hz

12. 220 Hz

13. 7.35 m

15. 0.229 m s^{-1}

18. $n < 0$.

19. 336 m/s

20. *F*

21. (*a*), (*b*), (*c*)

CHAPTER 7

1. $2\left(\sin x - \frac{\sin 2x}{2} + \frac{\sin 3x}{3} - \cdots\right)$

2. (*a*) $\frac{1}{2} + \frac{2}{\pi}\left[\sin x + \frac{\sin 3x}{3} + \frac{\sin 5x}{5} + \cdots\right]$; (*b*) $\frac{\pi}{4}$

3. $\frac{8}{\pi^2}\sum_{n=1}^{\infty}\frac{1}{n^2}\sin\frac{n\pi}{2}\sin nx$

5. $\frac{3}{2} + \frac{6}{\pi}\left[\sin\frac{\pi x}{5} + \frac{1}{3}\sin\frac{3\pi x}{5} + \frac{1}{5}\sin\frac{5\pi x}{5} + \cdots\right]$

6. $20 - \frac{40}{\pi}\sum_{n=1}^{\infty}\frac{1}{n}\sin\frac{n\pi x}{5}$

8. $\frac{A}{2} + \frac{A}{\pi}\left[\sin\omega t + \frac{1}{2}\sin 2\omega t + \frac{1}{3}\sin 3\omega t + \cdots\right]$

9. $\frac{2}{\pi} - \frac{4}{\pi}\left[\frac{\cos 2\omega t}{2^2 - 1} + \frac{\cos 4\omega t}{4^2 - 1} + \frac{\cos 6\omega t}{6^2 - 1} + \cdots\right]$

10. $\alpha B + \sum_{n=1}^{\infty}\frac{2B}{\pi n}\sin(\pi n\,\alpha)\cos n\omega t$

13. $\frac{8}{\pi}\sum_{n=1}^{\infty}\frac{n\sin 2nx}{4n^2 - 1}$

14. (*a*) $\frac{4}{\pi}\left(\sin\frac{\pi x}{2}-\frac{1}{2}\sin\frac{2\pi x}{2}+\frac{1}{3}\sin\frac{3\pi x}{2}-\cdots\right)$

(*b*) $1-\frac{8}{\pi^2}\left(\cos\frac{\pi x}{2}+\frac{1}{3^2}\cos\frac{3\pi x}{2}+\frac{1}{5^2}\cos\frac{5\pi x}{2}+\cdots\right)$

17. $g_s(\omega) = \sqrt{\frac{2}{\pi}}\frac{a}{\omega^2+a^2}$; $g_s(\omega) = \sqrt{\frac{2}{\pi}}\frac{\omega}{\omega^2+a^2}$

18. (*a*) $g(\omega) = \frac{1}{\sqrt{2\pi}}\left(\frac{\sin(\omega x)}{\omega x}\right)$

19. $-\omega^2 g(\omega) - i\frac{dg(\omega)}{d\omega} = 0$; $g(\omega) = A\exp\left(i\frac{\omega^3}{3}\right)$ (A = arbitrary constant)

20. $\frac{p^2}{2m}g(p)-\frac{1}{2}\hbar^2 k\frac{d^2g(p)}{dp^2} = Eg(p)$

22. $\hbar^2/2a^2$.

CHAPTER 8

1. 256 Hz

2. 219.6 Hz

3. 48 N

4. 214.29 cm

5. 35.36 Hz

6. 1 : 2 : 3

8. (*a*) Eqn. (8.13) with $E_n = 0$

(*b*) $y(x, t) = f(x - vt) + f(x + vt)$

where, $f(x - vt) = \frac{1}{2}\sum_{n=1}^{\infty} D_n \sin\frac{n\pi}{l}(x - vt)$

9. $0.01\cos\left(\sqrt{\frac{T}{\mu}}t\right)\sin x$

10. 5 kg s^{-1}, 10 kg s^{-1}; $\frac{8}{9}$

13. (*a*) As T increases v increases, so does the frequency.

(*b*) $v\pi\sqrt{260}$ (corresponding eigenfunctions are $F_{4\ 16}$, $F_{16\ 14}$)

16. 0.4 Hz

17. $\omega \propto \sqrt{T}$

18. 0.20, 0.47, 0.73

19. $0.1\cos\alpha_2 t\, J_0(\alpha_2 r)$

22. 1.91 Hz, 4.39 Hz, 6.89 Hz, 9.38 Hz.

23. (*d*)

24. 1.22 v

25. (*b*), (*c*)

CHAPTER 9

2. 312.4 Hz

3. 1098 Hz

4. 10 Hz

5. (*a*) 15 Hz (*b*) zero

6. 312.6 Hz

7. Before passing 544 Hz, after passing 423.5 Hz

8. 0.88 s

9. 8.53%

10. 16.4 kHz
11. 17.58 ft/s
12. 10 ft/s
13. 1013.86 Hz
14. 106 Hz
15. (*a*) 573.66 Hz (*b*) 583.78 Hz (*c*) 565.33 Hz
16. Zero (when the observer is between the wall and the source); 7.76 Hz when the source is between the wall and observer.
17. 0.073
18. $v = \frac{1}{7}c$ and hence it is not possible
19. $403.33 \leq f' \leq 484$ Hz
20. 1.2×10^6 m/s; receding.
21. (*d*)
22. 5.93
23. 30 m/s

CHAPTER 10

1. $\bar{a} = 0.069$, $T = 2.0$ s
2. 0.2
3. 4.58 s, 2.0 s
4. 240, 3.96 s
5. 1 W/m^2
6. 10 log (I_2/I_1) dB
7. 10^6
9. 60 dB
10. (*a*) 0.04 μW/m^2 (*b*) 46 dB
11. (*b*) 5.3×10^{-17} W/m^3
12. 17.07 m
13. Above 10 km from the ground.

CHAPTER 11

2. $n = Ke^{1/2}$
5. $-x$
6. (*a*) 150 MHz (*b*) *z*-axis, $B = 1\ \mu T$ (*c*) 3.14 m^{-1}, 9.42×10^8 rad/s (*d*) 120 W/m^2 (*e*) 1.2×10^{-6} N, 4×10^{-7} N/m^2
7. $E_{rms} = 1.55 \times 10^5$ V/m, $p = 0.21$ N/m^2
8. $H_x = 0$, $H_z = 0$, $H_y = 0.004 \cos [10^{15}\pi (t - z/c)]$
9. 0.4°C/s
10. (*a*) 6×10^8 N (*b*) Gravitational force = 3.6×10^{22} N
11. 4.51×10^{-10}
12. 1.0 m
13. 1.03 kV/m, 3.43 μ*T*
14. (*a*) $E_0 = 0.123$ V/m (*b*) $B_0 = 4.0 \times 10^{-10}$ T (*c*) 2.51×10^4 W
16. 1.19×10^6 W/m^2
17. 341.42 m
18. 1.1×10^7 N/m^2
19. (*a*) left circularly polarized
(*b*) linearly polarized wave with its polarization vector making an angle 135° with the *y*-direction
(*c*) right (clockwise)-elliptically polarized
20. (*a*) $\vec{E} = cB_0\left[\sin(kx - \omega t)\,\hat{j} + \cos(kx - \omega t)\hat{k}\right]$.

CHAPTER 12

1. 16.42 sin (ωt + 14.1°)
2. 26.82 sin (ωt + 8.5°)
3. 2.25 mm
4. The distance D must be doubled
5. 20 π
6. 0.01 rad
7. 5.38 × 10^{-5} cm
8. 2nd order
9. 0.1 mm
10. 666.7 nm
11. $I = I_0 [1 + 8 \cos^2 (\delta/2)]$,

    I_0 = Intensity due to light from the narrow slit and $\delta = \frac{2\pi}{\lambda} d \sin \theta$.
12. 0.54 m, 2.06 m, 5.63 m
13. 25 Hz
14. 330 Hz
15. (*b*) and (*d*)
16. (*d*)
17. (*a*) and (*c*)
18. (*c*)
19. (*i*) 0.117 cm (*ii*) 0.156 cm
20. 589.2 nm
21. 7 × 10^{-6} W
22. (*i*) 630 μm (*ii*) 1.575 μm
23. 0.08 mm
24. 589.7 nm
25. 27 μm
26. 1.76
27. 275 μm
28. 10
29. 516 nm
30. 0.195 λ
31. (*a*) 166.67 nm (*b*) No
32. 846.77 nm
33. 1.01′
34. 2.65 × 10^{-4} rad
35. 2 mm
36. 20.27″
37. (*a*) 1800 nm (*b*) 8
42. 0.83 cm
43. 1.36
45. 0.05 cm
46. 582.28 nm
47. 5.91 Å
48. 0.036 mm
49. 6 μm
50. (*a*)
51. (*a*)
52. 3.5 mm

CHAPTER 13

1. 3.74 μm
2. 454.5 nm
3. 4.5%, 1.62%, 0.83%
4. (*a*) 75 cm (*b*) $\frac{\lambda f}{a}$ = 1.13 mm
5. 0.2 cm
6. 0.030 seconds of arc
7. 33.55 × 10^{-8} radian
8. 3.12 × 10^{8} m
9. 10.17 cm
10. (*a*) 1.34 × 10^{-4} rad (*b*) 21.47 m
11. 3.04 m
12. 10.57 km
13. 52.62 m
14. 1.22 × 10^{-3} cm
16. (*a*) $\frac{d}{a} = 4$ (*b*) m = 8, 12, ...
17. (*a*) $\frac{d}{a} = \frac{11}{2}$ (*b*) 5
18. (*a*) 9 (*b*) 0.25
22. (*a*) 17.1276°, 17.1452° (*b*) 1.06 arc min

23. ± (10.20°, 20.73°, 32.07°, 45.07°, 62.25°)

24. 625 nm, 500 nm, 416.7 nm

25. 12500

26. (*a*) 5 μm (*b*) 1.25 μm (*c*) m = 0, 1, 2, 3, 5, 6, 7, 9

27. 972

29. 2.06°, m = 2, 4, 6,.....82

31. 2976

33. 983

34. 1.638 cm

35. (*a*) 2400 nm (*b*) 800 nm (*c*) m = 0, 1, 2

36. 484.93 nm, 484.97 nm, 3.386 cm

38. 0.236 nm

39. 2.69°

40. 7.0°

42. (*a*) $\frac{a_0}{\sqrt{2}}, \frac{a_0}{\sqrt{5}}, \frac{a_0}{\sqrt{10}}, \frac{a_0}{\sqrt{13}}, \frac{a_0}{\sqrt{17}}$

43. 4200 Å, 1.43

44. 2.945×10^{-4}

45. (*d*)

Index